Taxonomic Nomenclature

Taxonomic Nomenclature
What's in a Name – Theory and History

Igor Ya. Pavlinov

CRC Press is an imprint of the
Taylor & Francis Group, an **informa** business

First edition published 2022
by CRC Press
6000 Broken Sound Parkway NW, Suite 300, Boca Raton, FL 33487-2742

and by CRC Press
2 Park Square, Milton Park, Abingdon, Oxon OX14 4RN

© 2022 Igor Ya. Pavlinov

CRC Press is an imprint of Taylor & Francis Group, LLC

Reasonable efforts have been made to publish reliable data and information, but the author and publisher cannot assume responsibility for the validity of all materials or the consequences of their use. The authors and publishers have attempted to trace the copyright holders of all material reproduced in this publication and apologize to copyright holders if permission to publish in this form has not been obtained. If any copyright material has not been acknowledged please write and let us know so we may rectify in any future reprint.

Except as permitted under U.S. Copyright Law, no part of this book may be reprinted, reproduced, transmitted, or utilized in any form by any electronic, mechanical, or other means, now known or hereafter invented, including photocopying, microfilming, and recording, or in any information storage or retrieval system, without written permission from the publishers.

For permission to photocopy or use material electronically from this work, access www.copyright.com or contact the Copyright Clearance Center, Inc. (CCC), 222 Rosewood Drive, Danvers, MA 01923, 978-750-8400. For works that are not available on CCC please contact mpkbookspermissions@tandf.co.uk

Trademark notice: Product or corporate names may be trademarks or registered trademarks and are used only for identification and explanation without intent to infringe.

Library of Congress Cataloging-in-Publication Data
Names: Pavlinov, I. IA. (Igor' IAkovlevich), author.
Title: Taxonomic nomenclature: what's in a name: theory and history / Igor Ya Pavlinov.
Description: First edition. | Boca Raton: CRC Press, 2021. |
Includes bibliographical references and index. | Summary: "This book a conceptual examination of the role of nomenclature in systematics. It is not just another "instruction for use" of the nomenclature codes. The goal is to review theoretical foundations of taxonomic nomenclature and historical development of its fundamental regulating features"– Provided by publisher.
Identifiers: LCCN 2021023416 | ISBN 9781032015071 (hardback) | ISBN 9781032022437 (paperback) | ISBN 9781003182535 (ebook)
Subjects: LCSH: Biology–Nomenclature. | Plants–Nomenclature. | Animals–Nomenclature.
Classification: LCC QH83 .P395 2021 | DDC 578.01/2–dc23
LC record available at https://lccn.loc.gov/2021023416

ISBN: 978-1-032-01507-1 (hbk)
ISBN: 978-1-032-02243-7 (pbk)
ISBN: 978-1-003-18253-5 (ebk)

DOI: 10.1201/9781003182535

Typeset in Times
by Newgen Publishing UK

Things are so because they are called so.

　　　　　　　　　　　　Chuang-tzu

Contents

Preface ix

Introduction 1

Theoretical Part

Chapter 1 Taxonomic Theory and Nomenclature 7
          1.1   The Basics of Taxonomic Theory 7
          1.2   How Does Taxonomic Theory Matter? 11
          1.3   Taxonomic Hierarchy: A Case Study 15

Chapter 2 Basic Elements of a Theory of Nomenclature 21
          2.1   Basic Thesaurus 22
                2.1.1   Nomenclature Concepts, Systems, Codes 22
                2.1.2   Nomenclatural Objects 24
                2.1.3   Nomenclature Regulators 29
                2.1.4   Nomenclatural Activity and Tasks 33
          2.2   To the Philosophy of *T*-Designators 35
          2.3   Basic Nomenclature Concepts 42

Chapter 3 Basic Principles of Nomenclature 47
          3.1   General Regulatory Principles 49
          3.2   Cognitive Principles 52
          3.3   Linguistic Principles 57
          3.4   Juridical Principles 63
          3.5   Taxonomic Principles 76
          3.6   Pragmatic and Other Principles 93

Historical Part

Chapter 4 An Overview and the Beginning 97
          4.1   Main Historical Trends and Stages 98
          4.2   The Empirical Route 100
                4.2.1   Folk Nomenclature 101
                4.2.2   Language of Proto-Systematics 103

Chapter 5	The Essentialist Route	115
	5.1 Major Features	115
	5.2 Pre-Linnaean Stage	118
	5.3 Linnaean Reform	125

Chapter 6	The Nominalist Route	135
	6.1 Dawn of Nominalism	137
	6.1.1 Adansonean Reform: A Failed Attempt	138
	6.1.2 Affirmation of Binomial Nomenclature	139
	6.2 The 19th Century: Basic Issues	144
	6.2.1 Rank Fragmentation and Rank Dependence	144
	6.2.2 Priority *vs.* Usage	146
	6.2.3 Circumscription *vs.* Characters *vs.* Types	147
	6.2.4 Classicality *vs.* Arbitrariness of Names	149
	6.3 The 19th Century: Codification of Nomenclature	150
	6.3.1 First Codes	151
	6.3.1.1 Botany	151
	6.3.1.2 Zoology	156
	6.3.2 The "Great Schism": Multiplication of Codes	161
	6.3.2.1 Botany	161
	6.3.2.2 Zoology	171
	6.3.3 Prototypes of *BioCode*	178
	6.4 The 20th Century: Traditions and Innovations	182
	6.4.1 Major Trends	182
	6.4.2 Botany	187
	6.4.3 Zoology	190
	6.4.4 Microbiology	193
	6.4.5 Cultivated Plants	194
	6.5 Draft *BioCode*	195

Chapter 7	The Conceptualist Route	197
	7.1 Rational-Logical Nomenclature	198
	7.2 Phylogenetic Nomenclature	202

Instead of Conclusion: A General Outlook ... 207

References ... 211

Index ... 255

Preface

Taxonomic nomenclature (often called biological, but this is not fully correct; see Introduction) constitutes an important part of the professional language of biological systematics. The latter describes the diversity of organisms by arranging them in taxa of various levels of generality and assigning them scientific names. These names serve as one of the important aids of communications both between taxonomists themselves and between them and various users of taxonomic knowledge. The nomenclature manages all these names in a certain way by regulating how they are established, allocated, corrected, rejected, etc., with the principles and rules that are summarized by the rulebooks, usually called Rules, Laws, or Codes.

This explains the great attention paid by systematists to various issues concerning taxonomic nomenclature. Its principles and rules have been being elaborated since the 16th century, when biological systematics began developing its scientific method together with its professional language. At present, it is still subject to the active public discussions aimed at its improvement, with participation of not only professional systematic biologists but also philosophers, and the new Codes still appear, at least as projects (such as *PhyloCode*, *BioCode*, *Linz ZooCode*). The textbooks and manuals on biological systematics usually contain sections on nomenclature, and several books were published especially devoted to it. However, they are of empirical nature and mostly comment on the provisions and working principles of the current Codes.

With this book, I would like to bring to your attention a substantively different look at taxonomic nomenclature. It is provided for by a general understanding that the latter, being a part of the professional language of systematics, cannot be considered outside the meaningful context imposed by the whole of systematics, not only empirical but also theoretical. My main objective is to consider in depth some ideas of what the nomenclature is as a specific linguistic system, how it is structured and functions in this capacity, how and why it develops, and how the nomenclature concepts and principles correlate with each other and with certain regulators "external" to them. In this regard, one of the important objectives of the exploration of taxonomic nomenclature becomes consideration of certain general mechanisms that determine its basic structure and development. In particular, special emphasis is given to the "hidden" impact of the theoretical foundations of systematics on taxonomic nomenclature to make this impact more evident and thus explicitly interpretable.

Such a wide-scope largely conceptual consideration of taxonomic nomenclature allows the highlighting of the following important points in it that usually elude the attention of its explorers and interpreters bounded by a commonly accepted view:

- The nomenclature principles and rules, as they are summarized by the Codes, were established on a precedence basis and therefore are purely empirical; so the whole of nomenclature needs thorough theoretical analysis.

- The taxonomic nomenclature is theory-laden in that some of its fundamental concepts and principles are based on certain onto-epistemic premises about the nature of the objects it is applied to.
- The most significant historical changes in nomenclature were driven by conceptual changes in both understanding of the theoretical foundations of systematics and its basic descriptive means.
- The taxonomic nomenclature began not with C. Linnaeus, as is usually considered, but several centuries before him, as part of the professional language of biological systematics.
- The contemporary nomenclature is not "Linnaean" in its foundations, as Linnaeus developed its essentialist concept, while its nowadays dominant version is the nominalist one first explicitly proclaimed by M. Adanson.
- The taxonomic freedom declared by the Codes is but a "good wish" rather than a working device, since certain taxonomic decisions are strongly regulated by the Codes.

This book is largely based on my reviews of the theory and history of taxonomic nomenclature published in Russian in the early 2010s. However, it does not simply reproduce their contents, but exposes a more advanced understanding of the structure and development of taxonomic nomenclature that was worked out by me during the last several years. So I am happy to familiarize English-speaking nomenclaturists with my ideas and thus contribute to the discussion of certain general issues concerning the foundations, structure, functions, and development of taxonomic nomenclature.

I address this book to those who would like to go beyond the routine boundaries of the Codes and look at the taxonomic nomenclature from more general theoretical and historical perspectives.

Igor Ya. Pavlinov
Zoological Museum at Lomonosov Moscow State University
Moscow, Russia

Introduction

The specific knowledge developed by scientific disciplines is fixed in a diversity of forms, which are admissible to consider in general as an array of specific texts written in the professional languages of these disciplines [Croft and Cruse 2004]. With this, for these texts to be adequate to the studied objects, the structure of the languages employed for their composing should be adequate to the structure of diversity of these objects [Chomsky 1987]. Therefore, the normal functioning and developing of scientific disciplines involve their own professional, more or less formalized languages with specific sets of regulators for handling them.

Biological systematics (frequently, but not correctly also called taxonomy) provides no exception in this respect. It describes the diversity of individual objects (organisms) to represent this diversity by *taxonomic systems* (classifications) comprising generalized objects (taxa, their ranks, etc.). These systems and their objects may be exhaustively characterized by their *content* and *form* understood in a philosophical manner (see [Harrison 1973] on this). Their content encompasses substantive propositions about the diversity of organisms studied by systematics: how it is structured, how its structure is reflected in the array of taxa of different levels of generalty, how organisms are distributed among them, by which diagnostic features they are characterized, etc. Their form is a kind of specific text written in the specific language elaborated by systematics in order to turn the respective substantive propositions into adequate taxonomic descriptions by certain linguistic means [Chebanov and Martynenko 1998; Pavlinov 2015a]. It is the latter that constitute the professional language of systematics, so from the cognitive linguistics perspective, the functioning and development of descriptive systematics can be thought of as the functioning and development of its inherent language.

One of the important parts of this language is *nomenclature* dealing with the designations of the generalized objects studied by systematics. As far as nomenclature is developed by biological systematics, it is usually also called *biological* by tautology; this notion appeared probably in the last third of the 19th century [Cope 1878]. Such an extended interpretation of the nomenclature developed and applied by systematics is partly justified by the fact that the scientific classifications of organisms, in which they appear under specific names regulated by nomenclature, are relevant for the whole of biology. However, systematics studies only one of many manifestations of biological diversity, which is usually called *taxonomic diversity* (e.g., as it is adopted in the Global Taxonomy Initiative project). Other aspects of biodiversity are studied by other classifying biological disciplines with quite a wide scope (biogeography, phytocenology, sociobiology, genetics, etc.), which develop their own special languages (e.g., [Mirkin 1985; Weber et al. 2000; Ebach et al. 2008; Herrando-Pérez et al. 2014; Bruford et al. 2020]). Accordingly, the nomenclature systems developed in them are no less "biological" than in systematics.

Taking this consideration into account, the whole of the nomenclature developed by biological systematics should be appropriately called *taxonomic*. In fact, this term ascends to the notion of *systematic* nomenclature used by the authors of one of the first nomenclature Codes [Strickland et al. 1843a, 1843b], who echoed the ideas of the English naturalist and mathematician John Herschel (1792–1871) [Herschel 1830]. The term "biological" (= bionomenclature) is reserved here for the nomenclature system formalized by the recent *BioCode*. The traditional terms are preserved for the main subject area nomenclature Codes, viz., botanical, zoological, bacteriological, virological.

It is proposed that the discipline that studies the principles of organization and functioning of taxonomic nomenclature in both theoretical and historical aspects is called *onymology* or *taxonymy* [Dubois 2000, 2005, 2010a; Pavlinov 2014, 2015a; Aescht 2018].

Taxonomic nomenclature has two general meanings, or two basic parts. One of them encompasses an array of the taxonomic designations (names, symbols, etc.) themselves: this is the *nominative* nomenclature. Another part encompasses an array of the regulators (norms, principles, rules, etc.) controlling various manipulations with the designations: this is the *regulative* nomenclature. It is the nomenclature in its regulatory meaning that constitutes the most essential part of the nomenclature theory and, accordingly, the main subject of this book.

The regulative nomenclature is not so much a static as a developing system with its content and structure changing with time. Its historical development obeys certain regularities that are shaped by "external" and "internal" causes: the former belong to the philosophical and theoretical contexts of biological systematics, the latter are determined by structure of any linguistic system and by an aspiration of the taxonomic community toward elaboration of a properly organized device for taxonomic descriptions. Different theory-laden interpretations of the meaning of "being properly organized" were and still are among the main driving forces that promote and direct the historical development of taxonomic nomenclature along one or another route. Therefore, exploration of the history of nomenclature in a certain conceptualist framework is an important prerequisite of the understanding of both its historical dynamics and content at each of the stages of its development, including contemporary.

The structure of the book is determined by its key objective, specified in the Preface: to promote understanding of the structure and functions of the regulative taxonomic nomenclature as part of the professional language of systematics, including its conditioning by the history and theory of taxonomic research.

Chapter 1 considers the link between certain theoretical ideas of systematics and taxonomic nomenclature. For this, it explores how these ideas, as summarized by taxonomic theory, affect the functioning, structuring, and development of nomenclature.

Chapters 2 and 3 present the author's vision of the theoretical foundations of taxonomic nomenclature which may be considered as a "beginning" of the construing onymology/taxonymy. Chapter 2 provides an overview of basic notions of nomenclature, analyzes the content of nomenclatural activity, considers some philosophical issues concerning taxonomic designators, and outlines the main nomenclature concepts. Chapter 3 provides an analysis of basic principles of nomenclature designed to resolve certain nomenclatural tasks. These principles are grouped in five main blocks, each with its own regulatory framework, which correspond to the basic aspects of consideration of taxonomic nomenclature indicated above.

Chapters 4–7 present an overview of the history of taxonomic nomenclature that can be called, with some reservations, "conceptual." Chapter 4 briefly describes the main trends in the development of the basic concepts of nomenclature and characterizes the very first (pre-scientific) stages of its formation. Chapter 5 considers the essentialist stage in the development of nomenclature culminating in the Linnaean reform. Chapter 6 discusses in more detail the history of the development of the nominalist version of nomenclature, which currently dominates. Attention is paid to how the codified nomenclature was formed in the second half of the 18th and throughout the 19th centuries, when the foundations of the contemporary language of systematics were laid. The main trends in the development of nominalist nomenclature in the 20th and early 21st centuries are also characterized. Chapter 7 overviews two strongly conceptualized versions of nomenclature, rational-logical and phylogenetic ones. These chapters consider all more or less noticeable nomenclature systems and rulebooks that appeared from the 17th century up to current times to show how taxonomic nomenclature was gradually developing during this long period. This seems important for the sake of historical fairness, since excessive praise for C. Linnaeus in recent textbooks creates the impression that he was almost the only significant figure in the development of taxonomic nomenclature, which is certainly not the case.

The References section with about 840 entries can serve as a rather complete catalogue of bibliography on taxonomic nomenclature. It is perhaps to be noted that I have read nearly all of them (this striking opportunity appeared possible thanks to the Internet), so all my citations of them are documented.

Theoretical Part

1 Taxonomic Theory and Nomenclature

One of the main objectives of (almost non-existent) onymology/taxonymy is to explore the causes of functioning, structuring, and development of taxonomic nomenclature. This fundamental and most problematic issue involves uncovering these causes, including concrete acting forces and mechanisms of their direct or indirect effects on the descriptive language of systematics. An evident prerequisite of such an exploration is the consideration of nomenclature in as extended a causal context as possible in order to reveal and analyze all possible causes in question.

A significant portion of this overall context is shaped by the taxonomic theory in its general consideration [Pavlinov 2018, 2021] that defines both the objects studied by systematics and the ways of studying them. Evidently, as far as studying involves description, and the latter involves naming, a close consideration of the links between the principles of defining and naming the objects studied and described by systematics seems to be one of the important tasks of the above-mentioned onymology/taxonymy.

This chapter considers these links. It first outlines very briefly the basics of taxonomic theory necessary to understand the relationship between the latter and nomenclature theory. The next sections explore in more detail the ways in which they may be (and actually are) linked.

1.1 THE BASICS OF TAXONOMIC THEORY[1]

The *fundamentum* of any scientific discipline is its theory; in the case of biological systematics, this is *taxonomic theory* (TT). Its purpose is to properly rationalize: (a) the cognizable objects shaping the subject area of systematics; (b) the principles of their research; and (c) the ways of describing them. The TT is developed by the theoretical section of systematics, namely, *taxonomy* [Simpson 1961; Sneath and Sokal 1973; Pavlinov 2018, 2021], though this notion often denotes the whole of systematics [Mayr 1942, 1969] or just its "day-to-day" practice [Blackwelder 1967].

---

[1] The entire content of this section is borrowed from previous publications by the author [Pavlinov 2011, 2018, 2021], so neither the references to other sources nor consideration of other points of view are presumed for the sake of clarity and decisiveness.

Each TT, however formulated, is implemented and developed by the respective *research program*.

Natural science theories do not exist by themselves: their basis is shaped by the philosophy of science. Different versions of the latter provide general substantiation of particular theories by developing the principle of accounting their cognizable objects, their methodological consistency, etc. Therefore, in order to navigate the theoretical foundations of systematics and understand how they can affect the formation of its descriptive language, it is important to be aware of its general philosophical-scientific background.

According to one of the modern versions of the philosophy of science, based on the ideas of conceptualism, any cognitive activity is carried out within the framework of a three-component *cognitive situation*. Its *ontic* component corresponds to the studied object, i.e., what exactly is investigated by a given scientific discipline. Its *epistemic* component outlines how this object is to be explored and described; descriptive language is a part of this component. Finally, its *subjective* component refers to the subjects conducting research (scientists, scientific communities).

One of the fundamental properties of a cognitive situation is a certain interconnection of its components, which allows it to be represented metaphorically as a *cognitive triangle*. The latter illustrates how ontic and epistemic components are interrelated indirectly by means of the certain cognitive activity of a knowing subject. Their interrelationship is fixed by the *principle of onto-epistemic correspondence*. It is clear from this that, whatever ontologies of the studied objects might be, they are not "given," but result from certain particular interpretations imposed by a subject's cognitive activity.

The latter consideration provides one of the key ideas of contemporary conceptualism, according to which an object actually studied within a certain cognitive situation refers not to the *objective reality* as such, i.e., Nature in its wholeness, but to its certain particular manifestation (aspect, fragment, etc.). The latter is outlined by means of a certain basic substantive concept or theory outlining the ontic component of a cognitive situation. In particular, as stated in the Introduction, contrary to a commonly accepted view, biological systematics studies not the biological diversity (BD) in its entirety, but only one of its manifestations, namely *taxonomic diversity* (TD), with its other manifestations being studied by other biological disciplines (ecology, biogeography, sociobiology, genetics, morphology, etc.). If BD is admissible to consider a kind of "objective reality," then TD, as it is delineated by means of general TT, represents a *conceptual reality*, and may be conveniently called *taxonomic reality* (TR).

> There two basic aspects of consideration of the latter are recognized, *taxonomic* and *partonomic* ones. The former refers to the diversity of the organisms to be investigated, while the latter involves an analysis of the diversity of their properties.

The whole of TR, in turn, is "fractioned" by different particular TTs (typological, phylogenetic, phenetic, etc.) into certain particular manifestations. Each of them is delineated by a certain substantive conception, which specifies in a characteristic manner a more general metaphysical concept, with the latter resting on some fundamental philosophical understanding of both Nature and the ways of its cognition and description. For example, the realistic world picture assumes the objectivity (classical "reality") of the complexly structured TR, and the latter's structuredness is presumed to be conditioned by some processes occurring in Nature. This fundamental world picture is concretized in different ways by certain metaphysical concepts, be it, for example, an idea of the stationary Ladder of Nature or an idea of transformism. The former may be specified by supposing either an ascending or descending order of perfection, while the latter is concretized by particular evolutionary theories. The nominalist world picture presumes a much more simplified metaphysics, according to which there is nothing but physically perceived objects (including organisms) in Nature that deserve to be studied and described.

Thus, we have a hierarchically structured substantive background knowledge, which is based on the natural philosophical world picture and finalized by particular TTs with their concrete conceptions delineating particular manifestations of TR. This specific *conceptual pyramid* shapes the whole of the ontic component of the cognitive situation, and the latter's epistemic component is shaped so as to provide the most effective means of exploring and describing the concrete manifestation of TR.

The main objective of taxonomic research carried out within the particular cognitive situation is to elaborate scientific hypotheses about the structure of the given manifestation of TR, as it is defined by a particular TT. Each such hypothesis is represented by a *taxonomic system* (TS) of organisms, ordinarily called *classification*. Its most fundamental understanding is provided by the concept of the *natural system* (NS) as the one most adequately representing the structure of the studied manifestation of TR and thus corresponding to the most plausible (true) hypothesis about it. Accordingly, this general meaning of NS is then conceptualized by particular TTs: for example, NS can reflect a certain natural-philosophically understood System of Nature or the phylogenetic pattern or a kind of typological universum.

The main element of TS is a classification unit, generally referred to as a *taxon*. TS can be *hierarchical* or *parametric*; in the first more general case, its structure is determined by the inclusive (encaptic) hierarchy of the taxa of different levels of generality. This inclusive *taxonomic hierarchy* can be either *rankless*, if the levels of generality are not fixed, or *ranked*, if they are fixed some way (including terminologically). Each fixed level of generality is denoted as a *taxonomic rank*, and corresponding to it is a certain *taxonomic category* defined as a set of taxa of the given rank. Ideally, there is a one-to-one correspondence between the hierarchies of ranks and corresponding categories, which is formalized by the *principle of rank–category correspondence*.[2] Each taxon is described by specific *taxonomic characters* (traits, features, etc.) distinguishing it from others in the inclusive taxon of a certain higher rank. The most significant characters compose *taxonomic diagnosis*, and the latter functions as a *differential diagnosis* to the degree it is designed to differentiate (distinguish) most close taxa.

---

[2] However, in practice not all ranks established by a certain universal ranking scale are obligatorily implemented in the hierarchical structure of particular classifications.

The taxa, taxonomic categories, and characters are denoted in general as *taxonomic objects* (TOs): they have to be properly defined, first conceptually (theoretically) and then operationally (empirically), to become elements of a cognitive situation outlined by particular TT. The conceptual definition of TO provides its substantive individuation as an element of TR, including suggestion of its ontic status (real or nominal, phylogenetic or phenetic, etc.), and the development of such definition is one of the basic tasks of TT. The methods of operational definition of TO, as they are substantiated by cognitive philosophy, are of three main types, viz., *intensional*, *extensional*, and *ostensive* (ostensional). In the case of taxon, its intensional definition refers to the characters of organisms it encompasses; its extensional definition is based on its circumscription by listing the subtaxa or organisms it encompasses; and its ostensive definition presumes direct indication of these concrete organisms (and eventually subtaxa). In the case of taxonomic category, its position in the ranked hierarchy can be considered its intensional definition, and the list of particular taxa of the given category provides its extensional definition (the question of ostensive definition of the category is not yet clear).

The most fundamental characteristic of TOs as the elements of TR is their ontological status. According to classical dichotomy, they may be either *real* or *nominal*: the former are presumed to exist objectively in Nature and recognized by certain means, whereas the latter are "inventions" of the cognizing mind. Contemporary conceptualist treatment adds to this dichotomy that recognition of particular TOs is conducted within the context provided by particular TT: "ladderists," typologists, phylogenetists, phenetists, etc., ontologize them differently.

The historical development of the ways of comprehension of organismal diversity, from the point of view of contemporary conceptualism, can be thought of as a succession of the following main stages.

Pre-systematics began with the rise of human reasonable cognitive activities involving empirical descriptive languages. It is known as *folk systematics* characteristic of indigenous communities (as well as of all laymen).

Proto-systematics was immersed in a broader context of ideas about both Cosmos and its comprehending, which are ordered in a similar manner. The first rational onto-epistemic principles and aids of cognitive (including classificatory) activity were developed within this context, including the scholastic genus–species classification scheme. It began with antique natural philosophy and lasted up to the renaissance Herbal epoch of the 14th–16th centuries.

Scientific systematics began with the development of a new cognitive situation in which the main goal became exploration of the System of Nature with the rational means that were partly inherited from proto-systematics and partly developed by scientific systematics itself. The latter's scientification gave rise to an array of TTs and research programs implementing them. Among them, the following TTs appeared to be most involved in the conceptual development of nomenclature.

The first was *scholastic systematics* of the 16th–18th centuries presuming the unity of both classification and nomenclature embraced by a single essentialist interpretation of the whole of the System of Nature. It was replaced by *post-scholastic systematics*, which disjointed classificatory and nomenclatural tasks and developed a nominalist concept of nomenclature. *Natural systematics* of the second half of the 18th and first half of the 19th centuries was the first and most noticeable in this respect; it was aimed at knowing the System of Nature of living organisms based on the latter's non-essentialist treatment. *Classification Darwinism* of the second half of the 19th century, and *biosystematics*, stemming from it in the first half of the 20th century, concentrated on the lowest-rank units of taxonomic diversity and developed the language describing them in most detail. *Classification phenetics* of the first half to the mid 20th century denied the reality of any taxonomic system with reference to a positivist philosophy of science, and it contributed to the development of rational-logical nomenclature. *Cladistics* suggested a phylogenetic concept of nomenclature substantiated by the concept of phylogenetic pattern.

1.2 HOW DOES TAXONOMIC THEORY MATTER?

The contemporary Codes expressly emphasize their dealing with the names of taxa and ranks and not with the taxa (ranks) proper. In this regard, they declaratively refuse to regulate particular taxonomic decisions concerning the recognition and ranking of particular taxa, and this is formalized by the *principle of taxonomic freedom*. This is considered as a testimony of the theory-neutral nature of taxonomic nomenclature and is presented as one of the most fundamental principles of the whole of nomenclatural activity. This is partly—although only partly—true, as far as certain individual opinions concerning particular taxa are involved. However, when one considers the foundations of nomenclature from a more extended historical-philosophical perspective and in its whole capacity, it turns out that fairly complex, albeit not always obvious, links exist between taxonomic and nomenclature theories [Pavlinov 2014, 2015a, 2019].

Generally speaking, consideration of these links is part of the fundamental problem of the relationship between perceived reality, the character of its perception, and the language describing the results of perception. It is actively studied by contemporary philosophy of science and cognitive science [Chomsky 1987; Devitt and Sterelny 1999; Croft and Cruse 2004; Kubryakova 2004; Barrett and LaCroix 2020]. The following propositions in this issue should be highlighted as the most significant for the main topic of this section.

One of the key ideas of cognitive linguistics is that the relation between the cognized world and the language of its description, in general, cannot and must not be accidental. In fact, the structure of the language designed to describe the cognized world is to be adequate, as much as possible, to the latter's structure [Chomsky 1987; Hill and Mannheim 1992; Devitt and Sterelny 1999; Allard-Kropp 2020]. With the absence of at least partial adequacy between the described reality and the language with which it is described, how could we ever assume that our descriptions correspond to what we intend to describe? This adequacy provides a certain unity between content and form of the knowledge mentioned in the Introduction.

From this viewpoint, the relation in question provides a specific *linguistic world picture* that reflects the cognized world by specific linguistic means in a conceptual space shaping the ontic component of a cognitive situation [Gumperz and Levinson 1996; Devitt and Sterelny 1999; Talmy 2000; Uryson 2003; Kubryakova 2004; Yoon 2010]. Accordingly, as far as biological systematics is concerned, the classifications elaborated by it can be treated as specific texts written with the specific linguistic means elaborated by systematics just for this purpose. Therefore, it would not be a great exaggeration to say that these classifications-as-texts are specific linguistic pictures of "taxonomic reality" studied by systematics [Slaughter 1982; Pavlinov 2015a, 2019]. For example, for the hierarchically organized diversity to be properly described by classifications, it is necessary to operate with a hierarchically organized language with the propositions and notions of various levels of generality.

As emphasized in Section 1.1, natural science studies not Nature in general (the universe as such), but the concrete conceptual realities outlined by particular theories. So, they are these particular realities that are reflected in the particular linguistic world pictures. In biological systematics, these are particular "taxonomic realities" that are outlined by means of particular taxonomic theories as specific manifestations of biological diversity. Essentialism, typology, phenetics, phylogenetics, etc., are examples of such theories, on which basis particular "taxonomic realities" are first defined and then described by means of the specific cognitive means, including linguistic ones.

It is obvious from both philosophical and cognitive science perspectives that each particular "taxonomic reality," conceptualized in one way or another, requires a specific descriptive language most adequate to its structure. This implies that the diversity of taxonomic theories entails a variety of linguistic means of describing these specifically defined particular "realities." As a result, the unified language of taxonomic descriptions becomes fragmented into "dialects," i.e., particular nomenclature concepts and systems. The content of each of them, according to the initial provisions, matches as closely as possible the structure of each of the particular "taxonomic realities."[3] And since the latter are defined within the framework of particular taxonomic theories, the respective nomenclature concepts and systems appear to be indirectly dependent on the contents of these theories.

It follows from this general consideration that a certain irremovable, although not too rigid and not too obvious, connection exists between the theories defining particular "taxonomic realities" and the languages designed to describe them [Pavlinov 2015a, 2019]. In other words, this means that taxonomic nomenclature, being a part of the descriptive language of systematics, depends to a certain degree on taxonomic theory: they both function and develop more or less interconnectedly. Therefore, the commonly accepted view that taxonomic nomenclature is or should be "neutral" with respect to any taxonomic theory, as part of its general *deontology* [Dubois 2005; Aescht 2018; Dubois et al. 2019], is nothing but a kind of wishful thinking based on a "local" consideration of particular nomenclature systems rather than on a "global" view of taxonomic nomenclature in general.

---

[3] The most pessimistic viewpoint presumes "inadequacy of *any* nomenclatorial system for the task of describing organic diversity" ([Ehrlich 1961: 157–158; italics in the original]). Obviously, this radical idea is completely baseless from a cognitive perspective, since it deprives systematics, and with it also the whole of biology, of a unified properly organized descriptive language.

The connection in question is manifested in that the basic theoretical ideas developed by particular taxonomic theories serve not only to elaborate the principles of classification, but also affect at least some significant principles of nomenclature. Actually, both arrays of principles are considered on a unified basis of the specific understanding of what the structure of "taxonomic reality" is, what the objectives and methods of taxonomic research are, and how their results should be expressed by specific linguistic means. It is this connection that forces the development of taxonomic nomenclature to make it more appropriate for the needs of taxonomic theory; here it is worth recalling an idea of the American zoologist Ernst Mayr (1904–2005) that "the rules [of nomenclature] will have to be adjusted to the conceptual development of taxonomy" [Mayr 1969: 300].

The link between taxonomic theory and nomenclature manifests itself most clearly in the macro-events of the conceptual history of both systematics and nomenclature [Pavlinov 2015a, 2019]. Scholastic systematics of the 16th–18th centuries had begun to develop with the shift from mostly empirical to essentialist nomenclature, and this had been basically triggered by its mastering the essentialist world view of neo-Platonic kind. Accordingly, the names denoting groups of organisms should reflect their essences. This key feature was directly indicated by J. Pitton de Tournefort, who argued that "an idea of character, essentially distinguishing some plants from others, should be invariably associated with the name of each plant" [Tournefort 1694: 2]. C. Linnaeus echoed him by stating that "systematic names ought to be essential ones" and, in particular, "the specific name is the essential definition" [Linnaeus 1751: § 252, 257]. The transition from scholastic to post-scholastic schools of thought at the end of the 18th century has involved a significant change in the onto-epistemic foundations of all the systematics, and this led to the substitution of essentialist with basically nominalist nomenclature: the names ceased to express the essences of organisms, and they simply became their "labels." This fundamentally new interpretation of taxonomic nomenclature, with its key idea of "a name is just a name," inevitably led to the simplification of taxonyms: essential verbose descriptors became formal one-word signifiers [Adanson 1763; Murray 1782; Fabricius 1801]. Later, an inclusion of elements of a Darwinian model of evolution in the theoretical basis of systematics in the second half of the 19th century led to the recognition of the fundamental importance of the infraspecific units, which gave way to suggestions to replace the then already recognized binomial nomenclature with tri- and quadrinomial ones [de Candolle 1883; Coues et al. 1886; Banks and Caudell 1912]. Positivist revolution in systematics of the first half of the 20th century activated new ideas of logic-rational nomenclature, which resulted in the appearance of numericlature [Michener 1963; Hull 1966, 1968]. Contemporary phylogenetics (cladistics), which declared its theory the most recent revolution in systematics, found it necessary to replace the whole of traditional nomenclature with the phylogenetic one [de Queiroz and Gauthier 1990, 1994; de Queiroz and Cantino 2001; de Queiroz 2005, 2012].

However, the relation between taxonomic and nomenclature theories is not that strict and straight-forward, so there is no one-to-one correspondence between their particular versions. This means that different taxonomic theories can develop the same or similar nomenclature systems; and conversely, different nomenclature concepts/systems can be defended by adherents of the same taxonomic theory. Both variants are illustrated by the development of nomenclature systems associated with the taxonomic hierarchy. The rankless hierarchy with its corresponding rankless nomenclature was developed by theoreticians of scholastic systematics of the 16th–17th centuries [Cesalpino 1583; Morison 1672; Ray 1682, 1696], whereas ideologists of phenetic and cladistic systematics of the 20th century also voted for it [Sokal and Sneath 1963; Sneath and Sokal 1973; de Queiroz and Gauthier 1990, 1994; de Queiroz 2005, 2012]. On the other hand, among the adherents of cladistics, there are supporters of not only the phylogenetic nomenclature, but also the traditional one [Benton 2000; Nixon and Carpenter 2000; Keller et al. 2003].

A "substantive burden" of some issues in nomenclature may not be explicit and strictly theoretical: sometimes they just presume specific biological generalizations about the named organisms. In its most common case, this includes consideration of the adequacy of the descriptive language of systematics to the real biological pattern of taxonomic diversity (i.e., [Lane and Marshall 1981; Pavlinov 2015a, 2019]). In more particular cases, it may involve substantive justification for a certain nomenclatorial act, which is exemplified by the very strange decision by the International Commission on Zoological Nomenclature to denote domestic animals by the names of their wild ancestors, contrary to the provisions of the principle of priority [Opinion 2027 2003]. The reason for this was that it seemed "inappropriate" for applicants and decision-makers to denote the wild forms by the names of their domestic descendants [Gentry et al. 2004]. It is evident that this decision is very far from meeting the basic conditions of contemporary nominalist nomenclature according to which "a name is just a name."

The relation in question is provided for by dependence of certain nomenclature principles on classificatory ones, according to which some important properties of classifications are defined, such as their ranked hierarchy, substantive status of taxa, etc. This relation may be *complete* or *partial*, i.e., covering either the nomenclature system as a whole or only some of its parts. In general, the essentialist concept is based on the essentialist vision of Nature rooted in Platonic natural philosophy. The early rational-logical nomenclature is based on the concept of the universal "philosophical" language of science (see Section 7.1). Among the recent nomenclature concepts, such is the phylonomenclature: its developers expressly state that they suggest "a system of nomenclature [...] that is more concordant with evolutionary concepts of taxa [and] more closely conform to the manner in which they are conceptualized" [de Queiroz and Cantino 2001: 269]. Among particular theory-dependent principles of fairly general order, the rank dependence of the traditional nomenclature should be mentioned: it was originally substantiated by recognition of different ontologies of groups of different levels of generality (see below).

The connection between particular taxonomic and nomenclature theories may be either *explicit* or *hidden*. In the first case, its clear declaration occurs; this is exemplified by the introduction of trinomial nomenclature with reference to the Darwinian evolutionary concept [Coues et al. 1886] or the substantiation of phylonomenclature with reference to phylogenetic theory [de Queiroz and Gauthier 1990, 1994; Queiroz and Cantino 2001]. In the second case, the relationship has a deeper and therefore not as obvious character: this is instantiated by the above-mentioned rank-dependent ways of denotation of taxa of different levels of generality in essentialist nomenclature.

Finally, this connection may be either "*actual*" or "*relict*." The point is that the particular principles and rules were initially formed under the influence of certain theoretical considerations, which may later lose their contextual meaning. An example is again the general principle of rank dependence of nomenclature: it was initially based on essentialist ontology, which is irrelevant now. As already said, the respective method of naming taxa of different ranks was initially associated with the essentialist ontology attributed to them by scholastic taxonomists. Subsequently, this ontological background disappeared, but the rank-dependent method of naming remained an important part of the traditional nominalist nomenclature. Due to this "atavistic" element, a theoretical burden appeared to be replaced by the historical one in the professional language of biological systematics.

It should be noted that a certain theoretical burden of nomenclature principles is not their immanent and constant characteristic. On the one hand, the same principle may have a different theoretical sound in different nomenclature systems. For example, the ontology dependence of the formation of taxonyms has a significant metaphysical background in essentialist nomenclature, but is largely free from it in nominalist nomenclature. On the other hand, different authors may give significantly different evaluations of the theoretical burden of the principles. Thus, among the versions of the general principle of taxonomic certainty (see Section 3.5), some nomenclaturists consider the principle of typification theory-dependent, being inherently associated with the metaphysical concept of type [Ogilby 1838; Simpson 1961; Mayr 1969; Rasnitsyn 2002], while for others, it is the alternative principle of diagnosing that is associated with the metaphysical (essentialist) interpretation of taxa [Cook 1898, 1900; Dubois 2008a, 2011a]. Even the seemingly theory-neutral principle of priority can be, with a certain intent, considered "theoretical" as opposed to the "practical" principle of usage [Lewis 1871, 1875; Robinson 1895].

1.3 TAXONOMIC HIERARCHY: A CASE STUDY

The relationship between taxonomic and nomenclature theories is most clearly manifested in the hierarchy-dependent character of the formation of taxonomic names. In this regard, an analysis of the significance ascribed to taxonomic N-objects (see Section 2.1.2 on their definition) and their names depending on their position in the taxonomic hierarchy by different metaphysically founded nomenclature concepts is of special importance.

To begin with, it is to be emphasized that the very hierarchical principle of organization of classifications-as-texts is a consequence of a certain metaphysical world picture, in which phenomena of various levels of generality occur; in our case, this is the hierarchy of "taxonomic reality." An adequate description of such a hierarchical pattern requests specific organization of the descriptive language of systematics with hierarchically arranged notions and terms denoting them. Such a lexical structure is evidently non-accidental with respect to the hierarchy of "taxonomic reality." On the contrary, a nominalist world picture denies the hierarchy of natural phenomena, which is exemplified by the concept of the Ladder of Nature; accordingly, the lexical structure of taxonomic descriptions might be of any kind, and its hierarchical structure is but one possible option. So, it is evident that both acknowledging and denying the hierarchical pattern of Nature and the respective descriptive language most adequate to it is theory-dependent.

An initial elaboration of the natural method of systematics, of which an important part became essentialist nomenclature, was initially charged with a profound natural philosophy (see Section 4.2.2). It was based on the Platonic world picture as a sequential emanation of the *eidê* of different levels of generality. A postulated isomorphism between the principles of the organization of Nature and the principles of its cognition (antique rationalism) was a part of this overwhelming world picture. The latter, at the level of epistemology, gave rise to a hierarchically arranged deductive description of the results of the emanation shaped as a hierarchy of notions designating respective *eidê*. At the same time, these notions referred to particular phenomena of various levels of generality, which were groups of organisms implementing these *eidê* and endowed with the respective essences. Subsequently, neo-Platonists and scholastics embodied this hierarchy in the logical genus–species scheme with rankless hierarchy arranging concurrently and in a similar manner both the notions and the groups denoted by them.

One of the most significant consequences of this natural-philosophically justified natural method appeared to be the recognition of different ontologies of groups of organisms of various levels of generality, with reference to which different nomenclatural significance of respective taxonomic objects was substantiated. Within the scholastic cognitive tradition, the basic elements of the logical genus–species scheme were genera, whereas species appeared as detailing of generic essences. In discriminating genera and species according to their ontological status, scholastic systematists paid most attention just to the former, while the latter seemed to be of subsidiary meaning [Cesalpino 1583; Tournefort 1694; Ray 1696; Linnaeus 1751]. For Linnaeus, genera and species were natural as "works of Nature," while classes and orders were mostly artificial as "works" not only of Nature but also of systematists' Art. Therefore, he addressed his nomenclature canons mainly to the genera and species, while suprageneric categories were only briefly commented on by him. However, P. Magnol, by introducing the category of family in taxonomic hierarchy, emphasized the natural status of the groups of organisms recognized at this level [Magnol 1689; Adanson 1763].

More recent Codes conceptualize the significance of focal taxonomic *N*-objects they deal with in quite a similar metaphysical manner. They are defined as "natural groups" in *Paris Laws* of botanists and in the *American Association Code* [de Candolle 1867, 1868; Dall 1877], whereas clades are recognized by *PhyloCode* as the only "natural" entities to be individualized and named [Cantino and de Queiroz 2010, 2020]. As was said, it is clear that these judgments about the "reality" of taxa are theory-dependent: the latter implies that the taxa are elements of respective "taxonomic realities" specified by particular taxonomic theories. The suggestion to describe and name only "real" and not "hypothetical" taxa in the Zoological Code [Anonymous 1999] seems to fall into this kind of argument, as far as a certain ontological status of taxa is also indirectly presumed.

> The last distinction is amusing in that it does not take into account the hypothetical nature of scientific knowledge: from this philosophical standpoint, any judgments about taxa are elements of the taxonomic hypotheses about the structure of "taxonomic reality" (see Section 1.1). The most amusing thing is that the fossil traces presumed to have been left by hypothetical extinct animals are considered "real" while those of no more hypothetical Bigfoot are not.

In the mid-20th century, another partial modification of the ranking concept happened: species became considered a fundamental unit of both evolution and evolutionary systematics [Stebbins 1950; Mayr 1963, 1969]. This idea seems to occur in that the species rank is explicitly proclaimed fundamental in the botanical nomenclature and is presumed in the zoological nomenclature [Anonymous 1999, 2012]. This means, for instance, that a genus cannot be considered properly established if no one species in it is indicated in its original description.

The above-considered natural-philosophical background in distinguishing taxa of different ranks by their ontology can also be seen in certain lexical rules for the formation of their names. These rules are rather diverse and concern different aspects of the taxonomic nomenclature.

First, in line with the preconditions of the natural-philosophically deduced hierarchical genus–species scheme, each group of organisms was denoted by early systematists based on the scholastic formula "generic common and species particular." This standard was formalized by the *principle of binarity*, according to which the complete name of a group should consist of two parts, each describing distinctive "generic" and "specific" features of the group and thus being usually multi-word. The contemporary binomial nomenclature is a simplified version of this binary system retained only for species: it differs from the original multi-word binarity in that each of the two parts of a complete species name must be strictly one-word. Thus, the *principle of binomiality*, considered fundamental in contemporary traditional nomenclature, is a "relict" that preserved certain traces of the initial formation of the essentialist language of systematics.

Second, it was important for Linnaeus and his devoted disciples that the complete designator of every species would include the name of the genus to which it belonged [Linnaeus 1751]. On the contrary, there was no particular need for him and his followers to indicate the belonging of species to classes and orders by corresponding lexical means, although this was suggested several times (e.g., [Lang 1722; Petit-Thouars 1822], etc.). Moreover, within the framework of this natural philosophy, the specific epithet as such did not matter much unless it was not combined with the name of the "true" genus [Tournefort 1694; Linnaeus 1736, 1751; Bentham 1858, 1879; Van der Hoeven 1864; Dunning 1872].[4] However, another natural philosophy presumed different treatment of the genus–species designation, in which most attention was paid to the species category. For instance, the English botanist David Sharp (1840–1922) stated that "while maintaining constant species names, the absence of fixed genus names at present will not bring any harm to science [...] generic names are of secondary importance" [Sharp 1873: 24].

Finally, the generic name is a noun in the singular, which reflects an idea of the genus as a fundamental unit of the System of Nature. The descriptive specific epithet (the second part of the complete binomen) is an adjective: as was noted above, this reflects an idea of species as a kind of detail of the generic essence. The descriptive name of a suprageneric taxon is usually a plural noun: this reflects an idea of an order or a class being not so much a real natural unit of its own as an aggregate of genera. All these "atavistic" lexical features are preserved in contemporary taxonomic nomenclature.

The post-Linnaean development of taxonomic nomenclature was characterized by an increase in its rank-dependent character at both higher and lower levels of taxonomic hierarchy (see Section 6.2.1). This was caused by various factors and manifested in different ways, but all these reasons were of a metaphysical kind assuming certain substantive considerations.

At higher levels, this increase was caused by strengthening of the non-essentialist treatment of taxa, in which context the nominalist treatment of taxonomic names was developed. The descriptive function of the latter was significantly reduced, and they became derived from the names of the nomynotype genera instead [Adanson 1763; de Candolle 1813, 1819]. This fundamentally new feature of nominalist nomenclature subsequently became the general standard in all nomenclature Codes.

At lower levels, the formation of rank-dependent nomenclature was due to increased attention to intraspecific units due to inclusion of the Darwinian evolutionary concept in taxonomic theory in the second half of the 19th century and later. This was the reason for the proposal to legitimize trinomial and quadrinomial nomenclature systems [de Candolle 1883; Coues et al. 1886; Banks and Caudell 1912]. The realization of this general trend in the first half of the 20th century led to the development of a very detailed hierarchy of intraspecific categories by biosystematics, including up to ten fixed and termed ranks [Du Rietz 1930; Camp and Gilly 1943; Sylvester-Bradley 1952; Valentine and Löve 1958]. Trinomiality is preserved in most of the contemporary traditional Codes, and the fractional intraspecific hierarchy is preserved in contemporary botanical nomenclature.

In the contemporary Codes, the rank-dependent nature of nomenclature does not imply reference to any natural-philosophical (metaphysical) background, whether essentialist or evolutionary or something else. But this does not mean they lack the influence of certain substantive taxonomic theories. Just a connection with them now changed from the "actual" to "relict," and from the "explicit" to "hidden," so their initial theoretical burden was largely transformed into a historical one.

---

[4] This understanding is retained by contemporary traditional nomenclature, in which the generic designator is a "name," while the species designator is simply an "epithet" [Anonymous 2012].

In this regard, it should be emphasized that the entire criticism of the ranked hierarchy and the rank-dependent nomenclature is also metaphysically burdened. Thus, adherents of the idea of a continuous Ladder of Nature put forward their arguments against meaningfulness of ranked hierarchy at the turn of the 18th to 19th centuries (e.g., [Lamarck 1809]). This position is now defended by the pheneticists referring to the nominal status of the entire taxonomic hierarchy [Sokal and Sneath 1963; Sneath and Sokal 1973], while the phylogeneticists refer to the rankless hierarchy of phylogenetic pattern [de Queiroz and Gauthier 1994; de Queiroz 1997, 2005, 2012].

2 Basic Elements of a Theory of Nomenclature

The theory of taxonomic nomenclature, or *nomenclature theory*, of whatever particular content, by an initial condition, is elaborated to uncover and explain the general structure and mechanisms of generation and functioning of this part of the professional language of systematics. This theory does not exist yet, although the need for it is obvious [Ride 1988; Chebanov and Martynenko 1998; Dubois 2008a, 2011a; Pavlinov 2015a, 2019; Aescht 2018]. At present, it is especially required because of the contemporary diversification of nomenclature systems caused by the need to solve specific nomenclatural problems and tasks. With the absence of such a theory, there might only be isolated empirical solutions not integrated in the regulative taxonomic nomenclature as a whole.

The main reason for the absence of an even weakly developed nomenclature theory seems to be the domination of an empirical bias in its traditional consideration. In fact, its recent reviews involve an analysis and commentaries on the current or projected Codes [Schenk and McMasters 1956; Blackwelder 1967; Mayr 1969; Alekseev et al. 1989; Hawksworth 1991; de Queiroz and Gauthier 1992; Jeffrey 1992; Shipunov 1999; Spencer et al. 2007; Dubois 2008a, 2011a, 2015; Aescht 2018; Turland 2019; Sosef et al. 2020]. In them, discussions and comparisons of the particular nomenclature systems and Codes involve, above all, pragmatic issues aimed at ensuring the stability and universality of both the names and their regulators. Accordingly, the whole consideration of taxonomic nomenclature in these reviews is based more on historically acquired experience than on a certain consistent conceptual background, i.e., on the precedents rather than on the statute (in their juridical sense). Due to this, the principles and rules aimed at resolving this basic task are rated according to their contribution to this demand. As a consequence, purely conventional rules, such as independence of Codes, priority of names, etc., are introduced into the preambles of Codes as the fundamental "laws of nomenclature." However, an attempt to check these "laws" by means of a kind of "computational logic" [Franz et al. 2017] reveals that such an empirical consideration of nomenclature is insufficient for its consistent comprehension and, as a result, for its consistent development.

An interest in the theoretical aspect of taxonomic nomenclature became evident in recent years; this was largely stimulated by discussion of the new nomenclature systems, especially numericlature and phylonomenclature (see Chapter 7).

Questions are now being raised about concepts of nomenclature and the relative significance of its key principles and rules, about semantics of taxonomic names, about the need to develop in detail special terminology, etc. [de Queiroz and Gauthier 1990, 1994; Rasnitsyn 1992, 2002; Chebanov and Martynenko 1998; Härlin and Sundberg 1998; Kluge 1999a,b, 2020; Dubois 2000, 2005, 2008b, 2011a,b, 2015; Ereshefsky 2001a; de Queiroz 2005; Härlin 2005; Rieppel 2008; Pavlinov 2014, 2015a; Aescht 2018]. As indicated in the Introduction, the ideas developed by these considerations gave rise to an initial shaping of a discipline, called *onymology* or *taxonymy*, that explores organization, operation, and eventually the historical development of regulative nomenclature [Dubois 2000, 2005, 2010a; Pavlinov 2014, 2015a; Aescht 2018].

This chapter presents something like an introduction to the basic issues of what may be called a prototype of nomenclature theory, as it is understood by the author. Its main purpose is to introduce the basic thesaurus that shapes the conceptual space of this (as yet non-existent) theory, and no detailed discussion of other treatments is supposed. Another important part of this theory dealing with the development of the basic principles of taxonomic nomenclature is considered in Chapter 3.

2.1 BASIC THESAURUS

The starting point of the theoretical analysis of regulative nomenclature involves consideration of its basic thesaurus, which is a set of sufficiently clearly defined concepts and notions specific to it. This is an indispensable condition for the rationalization of the whole of nomenclature, therefore the Glossary/Vocabulary constitutes an important part of every contemporary Code.

This section reviews the basic concepts and notions of taxonomic nomenclature that shape the conceptual core of its thesaurus [Pavlinov 2014, 2015a]. It should be pointed out that only a basic set of the most important notions is considered here, comprising a fundamentum of the notional apparatus (basic thesaurus) of the forthcoming theory of taxonomic nomenclature. It is not my goal to consider all terms used in the taxonomic nomenclature in the manner in which they are comprehensively reviewed in the recent *Terms Used in Bionomenclature* [Hawksworth 2010]. With this, I tried not to overload this consideration by introducing new terms; those not in active use by systematic biologists are borrowed mainly from cognitive science and linguistics. As the recent cycle of numerous works by the French zoologist Alain Dubois shows (i.e., [Dubois 2000, 2005, 2006a,b, 2011a, 2013, 2015; Dubois et al. 2019]), the scope for inventing new words in the nomenclature is really enormous.

2.1.1 Nomenclature Concepts, Systems, Codes

The core of nomenclature theory is made up of the *nomenclature concepts*, which are theoretical constructs developed on the basis of some general ideas about the structure of "taxonomic reality" and the ways of describing it. Accordingly, the differences between concepts are mostly theoretical, even if a certain basic theory is not explicitly stated. Examples are essentialist and nominalist nomenclature concepts based on different natural philosophies: the first connects taxonomic designators with the essences of organisms, whereas the second denies an essentialist vision of Nature. These concepts are considered in a more detail in Section 2.3.

The practical basis of nomenclature is shaped by the *nomenclature systems* developed within particular concepts. Every system (ideally) is an integrated set of mutually coordinated nomenclature regulators, with the basic requirements being completeness, consistency, universality, and operationality presuming its unambiguity, automaticity, and repeatability [Pavlinov 2014, 2015a; Dubois 2015]. Their development begins with *incomplete* nomenclature systems (pro-system), which do not cover all nomenclatural tasks (mero-systems) or leave too much room for personal interpretations and subjective decisions (pseudo-systems) [Dubois 2015]. This development is finalized by a *comprehensive* (holo-system) nomenclature system meeting the just-listed criteria; it is presumed that different authors resolving the same nomenclatural task based on such a system should achieve the same result [Dubois 2015].

Every comprehensive nomenclature system is at least two-leveled. Its basic level comprises the general principles (norms), which are declarative to a reasonable degree and therefore do not imply direct application for the decisions of the particular nomenclatural tasks. Its applied level comprises the operating principles and rules, as well as other more specific regulators that interpret the more general principles in such a way as to accomplish certain working instruments for resolving concrete nomenclatural tasks. Every general "theoretical" principle potentially allows for multiple practical interpretations, which inevitably results in the multiplicity of particular nomenclature systems. Thus, within the framework of a nominalist concept, several nomenclature systems were developed that differ in the ways of nomenclatural definition of taxa or in setting the rules for the validity (correctness) of taxonomic names.

The multiplicity of both nomenclature concepts and systems allows the whole situation in taxonomic nomenclature to be designated as *nomenclatural pluralism* yielding *nomenclatural uncertainty*. Taking into account its basic causes (see Section 2.2), one may suppose that elaboration of an overwhelming and, therefore, single nomenclature holo-system is an unattainable ideal, so attempts to achieve it can be considered as a kind of pseudo-task which in principle has no rational solution (in the mathematical sense, see [Perminov 2001]).

The differences between nomenclature systems may be of two kinds. In some cases, they may have serious theoretical connotations: examples are the systems based on the above-mentioned methods of nomenclatural definition of taxa, ranked (ranked-based) *vs.* rankless systems, etc. In other cases, they are predominantly empirical: this is instantiated by difference between the particular *subject area* systems developed in botany, zoology, bacteriology, etc.

A peculiar example of the empirically distinguished nomenclature systems is provided by *orthonomenclature, paranomenclature*, and *ichnonomenclature*. The former refers to the "traditional" taxa (orthotaxa) that encompass, on a presumptive basis, the whole organisms and are recognized and named in a traditionally established manner. Paranomenclature deals with naming para- and morphotaxa that encompass the isolated parts of fossil organisms (such as plant pollen and leaves, animal egg shells, etc.). Ichnonomenclature deals with the naming of ichnotaxa that encompass the isolated fossil traces of animals.

Within the framework of nomenclature systems, the *nomenclature Codes* are developed as the standard working documents regulating practical nomenclatural activity of the members of taxonomic communities that adopted them. For this, they contain more or less expressly stated provisions (conditions, principles, rules, etc.) that allow the proper determination of the nomenclatural status of the *N*-objects considered by respective Codes, as well as vocabularies with explicit definitions of all relevant terms. Unlike nomenclature systems, the Codes should be as explicit as possible to make them more usable. These are the Codes that are applied, discussed, and improved in practical systematics.

2.1.2 Nomenclatural Objects

Biological systematics studies specific objects as the particular elements of conceptually defined "taxonomic reality" with specific epistemic means, including its descriptive language (see Section 1.1). When studying this "reality," systematics individualizes, characterizes, and denotes respective objects to arrange them in a classification (taxonomic system). Some of these objects and operations with them fall within the scope of regulative taxonomic nomenclature. To emphasize the specific aspect of their consideration from the latter's standpoint, they may aptly be denoted as *nomenclatural objects* (aka *N*-objects) [Pavlinov 2014, 2015a].

The set of concepts and notions associated with them have been being developed from the very first stages of the development of scientific systematics and its professional language, but it is not yet in a very satisfactory state [Chebanov and Martynenko 1998; Dubois 2005, 2011a, 2016; Pavlinov 2014, 2015a; Aescht 2018]. Here I mean not so much the specific terms denoting these objects (there are plenty of them), but the general notions that allow particular terms to be combined within a single basic thesaurus. In this section, a certain step towards development of the respective part of such a thesaurus is undertaken.

There are three main groups of *N*-objects that are theoretically defined with reference to the taxonomic ones, as set out in Section 1.1. The *taxonomic N*-objects belong to the "taxonomic reality," they are certain realistically or nominalistically treated "bodies" (real or nominal), to which the whole of the cognitive activity in systematics is basically directed; these are taxa, ranks/categories, taxonomic characters, and also particular kinds of specimens. The *linguistic N*-objects represent certain designations of these "bodies," so they are called *taxonomic designators* (aka *T*-designators); these are: (a) general notions/terms (taxon, rank, type, etc.) that constitute a significant portion of the conceptual space of systematics; and (b) individual designations of particular *N*-objects (particular taxa, etc.). The *N*-objects of these two basic kinds are interrelated in a specific manner made up by a semantic triangle (in the sense of [Carnap 1969]), and their interrelation is relevant to the discussion of what exactly,

Basic Elements of a Theory of Nomenclature

taxa or their designators, is defined by the nomenclatural means (see Section 2.2). These two groups are complemented by that of *acting N-objects* recognized within the nomenclatural activity proper. It comprises, first of all, the *nomenclatural acts* that shape some formal procedures presumed by specific nomenclature regulators. Besides, it is reasonable to distinguish specific *nomenclature documents* in this group; these are, e.g., the Official Lists and/or Registrars of names and works that are ascribed particular nomenclatural status (approved, conserved, protected, rejected, etc.).

The most important characteristic of any *N*-object, from a nomenclature perspective, is its *nomenclatural status* (*N*-status). According to the latter, a *N*-object can be *nomenclaturally consistent* if it complies with the conditions of a nomenclature system in which it is considered; otherwise it is *nomenclaturally inconsistent* (the same as "juridically void") [Pavlinov 2014, 2015a]. For the linguistic and acting *N*-objects, this pair of notions is specified as their nomenclatural *legitimacy vs. illegitimacy* [Parkinson 1984; Tindall 2008; Young 2009]. The nomenclatural consistency of a *N*-object is provided by its proper *nomenclatural definition* (*N*-definition), which constitutes the essential part of its operational definition.[1] Although taxonomic and linguistic *N*-objects are substantively different and are defined using different operations and notions, their complete *N*-definitions are nevertheless interconnected. In fact, the respective definition of a taxonomic *N*-object includes its individuation (fixation) by both reference to other taxonomic *N*-objects (characters, types, etc.) and denotation by a specific *T*-designator, while such definitions of a linguistic *N*-object include both its shaping as a linguistic unit (a word or else) and reference to the corresponding taxonomic *N*-object. In theory, each of the particular taxonomic and linguistic *N*-objects may be thought of and their specific properties may be analyzed as independent entities. But in nomenclatural practice, they make sense by their interrelation only; this is because neither non-named taxonomic *N*-object nor linguistic *N*-object unassigned to the latter is considered nomenclaturally consistent.

Taxonomically understood *N*-objects ("bodies") are of the following main groups: (a) nomenclaturally important specimens; (b) particular taxa; (c) taxonomic characters; (d) taxonomic ranks; and (e) taxonomic categories. They become *established* (nomenclaturally consistent) when properly *N*-defined.

The particular *specimens* considered by the Codes are of two main kinds. First, these are the type specimens (type series) of the species-group taxa; they are *N*-defined by indicating their particular status (holotype, syntype, etc.) together with the indication of taxa being typified. Such types are not defined directly in phylonomenclature, but are presumed indirectly by mentioning typified species. In a broader treatment, *voucher* or *reference* specimens may also be considered together with the types. Another kind of specimen refers to *hybrid* ones recognized in some Codes as particular *N*-objects; they are *N*-defined taxonomically by indicating species allocation of their parental organisms.

As far as *taxon* is concerned, it is suggested that there should be a distinction between two interpretations of them: *systematic* taxa are considered in the context of

---

[1] It is suggested that taxon *N*-definition should be called *taxognosis* [Dubois 2017a].

classificatory activity, whereas *nomenclatural* taxa are considered in the context of nomenclatural activity [Kluge 2020]; they are evidently interdependent. The nomenclatural taxon (as *N*-object) is *N*-defined by indicating its *specifiers* which are different in particular nomenclature systems. In traditional ones, these specifiers are circumscription, characters, type, rank, and unique taxonym; in phylonomenclature, they do not include type and rank, but do include phylogenetic status and ancestor (reference point). Some of these specifiers are obligatory and others are facultative: the former are included in *N*-definition on a mandatory basis (taxonym is among them), and the latter's inclusion depends on a particular nomenclature system. A properly *N*-defined taxon is called *nominal* (named); it should be distinguished from a taxon serving as a nominotype of the inclusive taxon called *nominotypical* (*nominotype*).

The importance of *taxonomic character* (trait, etc.) in nomenclature is determined in that it shapes taxonomic diagnosis as one of the possible specifiers constituting *N*-definitions of respective taxon. However, it is not usually considered as a particular *N*-object, largely due to the fact that the "philosophy of character" remains very poorly studied (e.g., [Sereno 2007; Winther 2009; Pavlinov 2018]).

For *taxonomic rank*, its *N*-definition is provided by an indication of its position in the respective ranking scale, which defines its position in the inclusive (encaptic) hierarchy, both together with its terminological denotation. For the *taxonomic category*, its *N*-definition comprises an indication of its correspondence to the respective rank in inclusive hierarchy and the taxa allocated to it. The consecutively subordinate *taxonomic* ranks/categories of the hierarchy, defined in the context of a taxonomic theory, are reflected on to the *nomenclatural* ones considered in the context of nomenclature. A one-to-one correspondence is presumed between taxonomic and nomenclatural subordinations. It is to be noted that different Codes may acknowledge different ranking systems, while some of them acknowledge rankles hierarchy.

The main purpose of *T-designators*, as said above, is to designate (denote) particular taxonomic *N*-objects. Depending on the particular kinds of the latter indicated above, the following basic kinds of *T*-designators should be distinguished [Pavlinov 2014, 2015a]: *taxonyms* designate taxa, *ranknyms* designate ranks and categories, *typonyms* (not in the sense of [Cook 1902a; Hawksworth 2010]) designate nominotypes; no specific *T*-designator for denoting taxonomic characters has yet been invented. From a nomenclature point of view, it is important to recognize *coordinate T*-designators; this status is given to designators that are controlled by the same nomenclature regulators; an example is coordinate denotation of taxa of the same rank-group (such as species-group, etc.).

Taxonyms used in the professional language of systematics are considered *scientific*; they are formed on a regular basis according to certain rules constituting the core of every nomenclature system. They are contrasted with *vernacular* names emerging spontaneously within the framework of folk nomenclature.[2] In traditional verbal nomenclature, a taxonym is a *name* in its quite trivial understanding; it is sometimes called a *nomenclatural name* [Hawksworth 2010] or *nomen* [Dubois 2005,

---

[2] Most Codes do not consider vernacular names; an exception is nomenclature of cultivated plants, which allows the use of such names [Brickell et al. 2009].

2006b, 2012, 2015; Dubois et al. 2019]; it is called *phylonym* or *phyloreference* in phylonomenclature [Cantino and de Queiroz 2010, 2020; Cellinese et al. 2021]. It may consist of one word or a certain combination thereof borrowed from natural languages; in numericlature, it is shaped by a combination of numerals that can hardly be treated as a name (in its traditional understanding). In some nomenclature systems, two parts of the genus–species binomen are distinguished terminologically: its generic part is the *name* proper, whereas its specific part is an *epithet* [Hawksworth 2010]. In an extended interpretation, taxonym may include not only the name/epithet itself, but also related metadata on it: authorship, publication date, etc. [Lanham 1965; Dubois 2000; Gradstein et al. 2001; Dayrat et al. 2004]. Such *composite* taxonym is suggested to be called *nominal complex*, with the name proper (nomen) being its essential part [Dubois 2000, 2005, 2012; Dubois et al. 2019].

Ranknyms fix the names of both taxonomic ranks and categories corresponding to them. They are borrowed from various sources, most of which are compiled by traditional nomenclature systems historically (i.e., with minimal rational considerations), and they are rather strictly regulated by Codes. One of the most important is the *rule of rank–category homonymy* of T-designations of taxonomic ranks and categories: they are denoted by the same terms, which also denote the taxa allocated to respective categories (species, genus, family, etc.). Ranknyms are omitted in the rankless nomenclature systems such as phylonomenclature.

Typonyms denote basically the nomenclatural types of the species-group taxa. Diversity of their different kinds began with several terms, and currently there are about a hundred of them [Schuchert and Buckman 1905; Burling 1912; Frizzell 1933; Evenhuis 2008a; Hawksworth 2010], but only a few are acknowledged in the acting nomenclature Codes (see Section 3.5 for further detail).

The meaning of *nomenclatural act* as a specific N-object is determined by its being involved in the solutions of certain nomenclatural tasks, so it is of paramount importance to provide taxonomic and linguistic N-objects with their respective nomenclatural status. The designations of particular groups of acts reflect their content, e.g., publication, type fixation, etc. Some nomenclature regulators (such as the principles of certainty, priority, etc.) are applied to nomenclatural acts in the same manner as to other N-objects.

Taxonyms are considered in most detail with respect to both their N-status and N-definitions in all nomenclature systems. Accordingly, special terminology for them is very rich with diverse terms; some of them are uniform for all nomenclature systems, whereas others are specific to particular ones. The basic terms for them are considered briefly below.

The T-designators, considered from a purely practical standpoint shaped by the principle of monosemy, are divided into two main classes, *homonyms* and *synonyms*, established by the specific relationships of homonymy and synonymy, respectively. *Homonymy* means denoting different taxonomic N-objects by the same T-designator, and *synonymy* is designation of the same taxonomic N-object by different T-designators. Homo- and synonymy can "overlap"; examples are *isonymy* (the same name given to one taxon by different authors) and *zygography* (typographical error, different transliteration, etc.) [Dubois 2013, 2015; Dubois et al. 2019].[3] Several special cases are considered as versions of homonymy in its extended treatment in some nomenclature systems. *Pleonasms* are different names that are identical semantically by denoting the same object: for example, referring to the same organism in different languages (*Bos* and *Taurus* in zoology), different eponyms/patronyms derived from the parts of a compound name of the same person (*Pittonia* and *Tournefortia* in botany), etc. *Omophony* is the coincidence of the sound (pronunciation) of names of different spellings, which makes them indistinguishable in oral speech; it is sometimes called *parahomonymy*. These particular versions of homonymy are suppressed in some Codes. Generic names differing in gender ending only are considered homonyms or not in particular nomenclature systems.

The distinction between two special classes of homonymy is proposed in rank-dependent nomenclature [Kubanin 2001].

Horizontal homonymy corresponds to the identity of names of taxa of the same rank (rank-group). This kind of homonymy of the names of suprageneric taxa is considered with respect to their stem parts only, while their differences in group-specific prefixes and suffixes are not taken into account. Identical generic names of animals, plants (traditionally understood), and prokaryotes, covered by different Codes, are proposed to denote *hemihomonyms* [Starobogatov 1991; Shipunov 2011]. Two basic versions of horizontal homonymy are distinguished for each of the species-group and higher-rank taxonyms, viz., *primary* and *secondary* homonyms. In the case of species-group taxonyms, coincident specific epithets are primary homonyms if combined with the same generic name initially, and they are secondary homonyms if they are subsequently combined with the same generic name. For the family-group and higher-rank taxonyms, the status of homonymy depends on the homonymy of the generic names of their respective nomyntypes: their total identity corresponds to the primary homonymy and the identity of their stems only corresponds to the secondary homonymy.

Vertical (cross-rank) homonymy means identity of the names of taxa of different ranks or rank-groups. One of its variants is *autonymy*, i.e., coincidence of coordinate names designating subordinate taxa of the same rank-group based on the same type; for example, genus and its nominative subgenus. Another variant is *genus–species tautonymy*, i.e., coincidence of names of the genus and a species in it (e.g., *Pica pica*, *Gulo gulo* in zoology). Historically, such tautonymy usually appeared when a species was separated in a distinct genus and its specific epithet became the generic name. In considering homonymy of names of suprageneric taxa, their differences in rank-specific suffixes are not taken into account.

Two basic forms of synonymy are distinguished traditionally in taxonomic nomenclature. In one of them, taxonyms are established originally for the same taxon based on the same specifiers; in another, they are established originally for different taxa (based on different specifiers) subsequently united into one of the same rank. These variants of synonymy are respectively called: *homotypic* (*nomenclatural*)

---

[3] In this regard, it seems reasonable to distinguish between the *name* itself as such and its *spelling* [Dubois 2010b, 2013].

Basic Elements of a Theory of Nomenclature   29

and *hetetotypic* (*taxonomic*) in botany, *objective* and *subjective* in zoology; they are termed in *homodefinitional* and *heterodefinitional* in phylonomenclature.

Of the terms denoting more specific variants of N-status of taxonyms in contemporary nomenclature, the following are most significant [Hawksworth 2010]; they can be arranged according to the protocol of consideration of their nomenclatural consistency [Smith 1962; Pavlinov 2014, 2015a; Dubois et al. 2019]:

- *available* = *legitimate* = *established* name which is validly (effectively) published in accordance with the provisions of the relevant Code
- *unavailable* = *illegitimate* name which is opposed by its meaning to the previous one
- *valid* = *correct* = *accepted* name which is used to designate "officially" a taxon in a particular classification
- *approved* = *registered* name which is accepted by the relevant international registration authority
- *protected* = *conserved* name which is officially legitimated to be used as the valid (correct) one contrary to certain standard provisions of a Code; also *sanctioned* name
- *rejected* = *suppressed* name which use as the valid (correct) one is prohibited by certain official decisions contrary to standard provisions of a Code
- *converted* name (in phylonomenclature) which is "borrowed" from another nomenclature system
- *senior* or *junior* synonym/homonym which is a name competing as potentially valid (correct) for a given taxon; junior homonym is also called *preoccupied* name
- *replacement* name which is established expressly to replace an already established name
- *emended* name, where the spelling is expressly and justifiably emended (corrected); otherwise it is just an "incorrect spelling"
- *coordinate* name which is one of several considered within the same nomenclatorial rank-group; all coordinate names associated with a given taxon and with all its subtaxa of the same rank-group are treated as synonyms

2.1.3 NOMENCLATURE REGULATORS

The regulative nomenclature can be presented (by tautology) as an array of the *nomenclature regulators* which are norms, precepts, "laws," principles, rules, etc. [Pavlinov 2014, 2015a, 2019]. They regulate in a specific way N-definitions of N-objects by indicating their characteristic specifiers and assigning respective denotations (names, terms) to them. The basic purpose of such regulation is to make both the N-definitions and the results of their applications nomenclaturally consistent.

Nomenclature regulators are elaborated as an important constructive part of the nomenclature concepts, which, in turn, are formed based on theoretical (and eventually certain other) premises. Those elaborated by taxonomic theory define in a specific manner the taxonomic N-objects systematics deals with, while nomenclature theory specifies three basic functions assigned to *T*-designators, viz., designative, descriptive, and classificatory (see Section 2.2). As a result, each nomenclature concept is characterized by a specific formulation and interpretation of the main nomenclature regulators.

The latter serve (again by tautology) as the major means of regulating the methods of posing and solving nomenclatural tasks. Their development constitutes the principal objective of the regulative nomenclature, and their application constitutes the main content of the nomenclatural activity. From a cognitive (theoretical) viewpoint, they can be divided into two basic groups, namely *"internal"* and *"external"* regulators. The former includes general cognitive and linguistic principles of the organization of descriptive language of systematics; they are largely inherited from natural language. The latter includes taxonomic and partly juridical principles that are developed intentionally by rational (in a broad sense) nomenclature. From a pragmatic (practical) viewpoint, nomenclature regulators are divided into another pair of basic groups, viz., *primary* and *secondary* ones: the former are put into action as such, while the latter correct their action in one way or another. Internal and external regulators are considered in Chapter 3, while primary and secondary ones are characterized below.

The *primary regulators* comprise norms, principles, rules, and recommendations; their order in this list reflects their levels of generality. The *nomenclature norm* is a kind of rather informal precept defining the general content of posing and resolving possible nomenclatural tasks. In this understanding, norms can be seen as the "framework" principles that establish certain rational conditions for nomenclatural activity. The *nomenclature principle* is a narrower interpretation of a norm by explicit formulation of the ways of posing and solving standard nomenclatural tasks in certain typical situations. The *nomenclature rule* implements the principle for the practical solution of concrete nomenclatural tasks in the course of particular taxonomic research.[4] Finally, the *recommendation* indicates a preferable way to solve such task in not quite a typical situation taking into account previous practice; according to its status, it functions as a "precedent by-law." Each recommendation is usually accompanied by *examples* (cases) showing how the respective rule was correctly applied previously in a similar situation.

The "adjoining" levels of generality of primary regulators—norms and principles, principles and rules—partially overlap and therefore cannot be strictly demarcated. This reflects an incomplete rationalization of the regulative nomenclature, which is an inevitable and irreparable consequence of its being rooted in the natural language of folk systematics.

The *secondary regulators* of nomenclatural activity are introduced when the particular nomenclatural tasks cannot be resolved in a standard manner based on the direct application of the primary regulators. Such impossibility yields specific uncertainty expressed in different possible ways of resolving these tasks, which entails

---

[4] In some Codes, all such regulators are called rules. Previously, the most important were termed "laws" to emphasize their significance and inviolability; the working principle/rule of priority most often appeared in this capacity. Such high-rated principles, even if purely technical and conventional, are usually summarized in the introductory sections (preambles) of Codes, reflecting their basically empirical foundation and designation.

Basic Elements of a Theory of Nomenclature 31

a potential violation of universality and stability of the nominative nomenclature. Therefore, the secondary regulators are brought into action, and their main objective is to "adapt" applications of the relevant primary working principles/rules to such unclear cases; the most frequent and typical example is suspension of the direct action of the principle/rule of priority. With this, it is important to keep in mind that the enactment of such secondary regulations is not arbitrary: it is rationally regulated by Codes.

These secondary regulators comprise, first of all, various kinds of specially stipulated *exceptions* to the established primary principles and rules; they, like recommendations, are based on precedents and are also indicated directly in Codes. Unlike these, *interpretations* ("decisions," "opinions," etc.) of the applications of standard nomenclature regulators are proposed *ad hoc* for each individual case by nomenclature bodies (commissions, committees, etc.) especially authorized to do so (analogs of arbitration courts).

Unregulated interpretations of a special kind, based on *personal opinions* of individual systematists on how to apply particular principles and rules in ambiguous situations, should also be mentioned. They are not envisaged by the Codes that seek to suppress such kinds of personal initiatives. However, the latter are inevitable because: (a) many provisions of the Codes are not formulated rigidly enough to exclude different interpretations; and (b) nomenclatural activity is carried out by living people who usually interpret such fuzzily stated clauses guided by their personal preferences and intentions.

It is clear from the foregoing that nomenclature regulators are of different levels of generality, which allows them to be ordered hierarchically depending on the importance of the functions they carry out in the nomenclatural activity. This importance is determined by the extensiveness and significance of the nomenclatural tasks that are resolved by means of respective regulators. It should be noted that the hierarchical assessment of nomenclature regulators by their importance, presented here based on a theoretical consideration of regulative nomenclature, is far from their rating scale adopted in traditional nomenclature systems. As far as the latter's having been developed mostly on a pragmatic basis, quite technical rules are usually considered most important in them (such as the "law of priority").

The hierarchical ordering of regulators by their importance is twofold. On the one hand, regulators related to the resolving different nomenclatural tasks form a *linear hierarchy*, which is determined by the relative significance of the tasks themselves. For example, the task of the correct N-definition of taxa seems to be more significant than the task of selecting particular T-designators for them: without properly solving the first task, the second is simply meaningless. On the other hand, regulators related to the solution of the same nomenclatural tasks form an *inclusive* (*nested*) *hierarchy*, in which they are arranged as higher- and lower-order ones. The former (more general) define the ways of solving these tasks in a general case, while the latter (more particular) outline the procedures of practically solving them. An example is the hierarchy of regulators designed for the correct N-definition of taxa: they are arranged as declarative principles and working rules. Both linear and inclusive hierarchies may be construed in different ways in different nomenclature systems, depending on their specific assessments of the significance of both the tasks and the methods of their solution. For instance, nomenclature stability is considered most important in some systems, while its expediency is given more "weight" in others.

The array of interactions between nomenclature regulators of various levels of generality is shaped by ascending and descending flows of causalities. In the first case, the main role is played by the practical need for a stable universal nominative nomenclature that stimulates the development of no less stable and universal primary regulators. In the second case, the most general norms of rational organization of the professional language of systematics, as a natural science discipline, are decisive for the elaboration of specific norms and principles of taxonomic nomenclature that would be most adequate for the basic premises of respective cognitive situations. With this, as far as the most general and final goal of the whole of the taxonomic nomenclature is to provide stable and universal N-designators for N-objects, these ascending and descending hierarchies are largely crossed.

At each level of generality, every nomenclature regulator admits several particular interpretations which yield together sequential multiplication of these regulators passing from higher to lower levels of generality. This results in a specific "pyramidal" logical structure of the whole of the regulative nomenclature: in this "nomenclature pyramid," the most general norms take top position, whereas particular working rules and recommendations are placed at the base. By this, the entire construction of such a "pyramid" is quite similar to that produced by the hierarchy of taxonomic theories that substantiates taxonomic pluralism [Pavlinov 2011, 2018, 2021]. Respectively, it is possible to speculate about *nomenclature pluralism* as an immanent property of the professional language of systematics being a specific linguistic system shaped by a complex interaction of non-coinciding factors of various levels of generality.

This general consideration can be illustrated by the following example. One of the most general nomenclature norms is shaped by the coordinated requirements for the stability and universality of the nominative nomenclature that is ensured by denoting each taxon by a unique, universally acknowledged taxonym. It is concretized by the *principles of stability* and *certainty*, each being implemented by several alternative working principles. For instance, the main condition of the general principle of stability is ensured by certain working principles, which apply specific criteria for choosing one synonym from several possible ones to be used as a valid (correct) one for a particular taxon. The principles of *priority* or *usage* of taxonyms are most commonly adopted for this purpose. The principle of priority is additionally specified at the next lower level by more particular rules that regulate its application: it may be either fixed or absolute, and the starting date for the fixed priority was arbitrarily selected as 1758 in zoology, and 1753, 1789, and 1820 (for various taxonomic categories and major groups of organisms) in botany. Finally, certain recommendations provide practical solutions for the discrepancies involving, for example, simultaneously published taxonyms.

Such a consideration of how the importance of particular regulators may be assessed from a general perspective is opposed to the traditional way of highly rating particular lower-level (operational) rules to resolve particular nomenclatural tasks based on conventions between members of particular taxonomic communities. It should be emphasized that this traditional way has no solid theoretical foundation and is, therefore, vulnerable to criticism, which makes it potentially unstable.

2.1.4 Nomenclatural Activity and Tasks

The procedure of descriptive systematics, in a fairly simple understanding, can be thought of as an accomplishment of the particular *taxonomic descriptions* of particular *N*-objects [Chebanov and Martynenko 1998]. In a more general sense, it may be defined as the *nomenclatural activity* as an array of posing and resolving particular *nomenclatural tasks* that aimed to bring these descriptions in line with certain standards by proper application of nomenclature regulators [Pavlinov 2014, 2015a]. The sequence of operations implementing this activity is referred to as the *nomenclatural process* consisting of three main steps providing for availability, allocation, and validity of *N*-designators [Dubois 2011a; Aescht 2018; Dubois et al. 2019]. One may conclude that it is the provision of this activity with the necessary nomenclature regulators that is the main purpose of all of regulative nomenclature. The nomenclatural activity is applied to the respective *N*-objects and is conducted by the *nomenclature subjects* (scientists, their communities, etc.).

A particular fulfillment of a certain nomenclatural task by the application of relevant nomenclature regulators to particular *N*-objects makes up the *nomenclatural act*. These acts constitute the above-mentioned acting group of *N*-objects, and they are regulated in a standard manner along with other *N*-objects. They involve certain manipulations with *N*-objects (including acts themselves) that change either the very objects or their *N*-status or both, in some way or other. These manipulations may be nomenclaturally *significant* or *insignificant*; only the former are endowed with a specific *N*-status. A relative importance of nomenclatural acts may be estimated differently in different nomenclature systems: for instance, taxon allocation to a particular category is considered significant in traditional nomenclature but not so in phylonomenclature.

Nomenclatural activity is generally believed to be aimed almost exclusively at regulating *T*-designators denoting taxonomic *N*-objects and does not concern these objects themselves. However, such a simplified understanding of the activity in question is not correct: in fact, nomenclature systems regulate the tasks associated not only with *T*-designators (names etc.), but also with taxonomic *N*-objects (taxa, ranks, etc.) [Pavlinov 2014, 2015a]. In the first case, the main questions concern how denotations of taxa and their ranks should be fixed (established, applied, changed, rejected, etc.) on certain nomenclaturally consistent grounds. These are *nominative* tasks which are traditionally paid most attention. In the second case, the main questions concern how taxa and ranks should be fixed on the same grounds; in this case, the principles of theoretical (with reference to a presumed "reality") and operational (by application of particular specifiers) definitions of taxa and their ranks/categories are consistently regulated. These are undoubtedly the *classificatory* tasks, the formulation and solution of which constitute an important part of nomenclatural activity.

It is noteworthy that, in some cases, the particular tasks of these two main groups may be very closely linked to each other. This is most evident in the case of recognition and denotation of ortho-, para-, morpho-, and ichnotaxa, to which some important principles (for example, those regulating synonymy) are applied dependent on their taxonomic status. In phylonomenclature, the same holds true for the clades, which are only subject to application of the phylonyms. Thus, every sufficiently developed nomenclature system presumes explicit regulation of classificatory tasks—of course, to the extent that they are associated with the solution of the nominative tasks.

In this regard, it should be emphasized again that a certain conjugacy of regulators dealing with N-objects and their T-designators is fundamentally important for normal functioning of the professional language of biological systematics. It is this conjugacy that provides a closer interconnection between taxonomic N-objects and their denotations, which is one of the main provisions of every nomenclature system.

Thus, the practical issues that concern N-definitions of N-objects and ascribing them certain N-status are encompassed by the nomenclatural tasks as a special kind of research task. It is reasonable to argue that the nomenclatural activity begins with the grasping and posing of particular tasks and proceeds with their resolution by application of certain regulators especially elaborated for these purposes. Thus, the correct outlining and ordering of nomenclatural tasks is of prime importance for a proper organization of the whole of nomenclature activity.

Considered as the constituent elements of the above-mentioned ordered nomenclatural process, nomenclatural tasks may be classified into two main kinds. One of them are *primary*; they are posed by the very practice of taxonomic research that requires application of certain nomenclatural regulations. Others are *secondary*; they involve application of these regulators to resolve primary tasks.

The overall variety of nomenclatural tasks that are posed and resolved mostly within traditional nomenclature systems, can be generalized by dividing them into three major groups: classificatory, nominative, and procedural [Pavlinov 2015a, 2019]. Each deals with manipulations of a particular kind of N-object, characterized above. The taxa and ranks/categories are first distinguished and defined operationally, including ranking of the former. After that, their T-designators are considered with regard to their nomenclature consistency, allocated to them, and changed when needed. Last, the results of solving these tasks are properly formatted and published, and additional clarifications are made if necessary.

The **classificatory group** comprises the tasks that deal with taxonomic N-objects and are associated with their recognition, N-definition by means of respective specifiers, and change in their treatment (position in the classification, circumscription, diagnosis). The main purpose of these tasks is to make taxonomic N-objects nomenclaturally available (legitimate), which is a prerequisite to recognizing their T-designators also available (legitimate). It should be emphasized again that these classificatory tasks are nomenclaturally significant and precede the nominative ones: for a taxonym to be assigned properly to a taxon, the latter must be first fixed by respective specifiers, otherwise the former appears "naked" (*nudum*). Deposition of the type material seems to be part of this task group [Pavlinov 2015a].

The **nominative group** comprises the tasks affecting T-designators; they involve the latter establishing, allocating, and occasionally changing (when needed) to make them nomenclaturally consistent, unambiguous, grammatically correct, stable, etc. The tasks of this group, especially those dealing with taxonyms, are traditionally considered the main ones in taxonomic nomenclature. However, they largely depend on solution of the classificatory tasks and therefore, as noted above, are secondary to them. This means that the need for nominative tasks arises in most cases in connection with certain prior classificatory decisions and has no justification without them.

The **procedural group** comprises largely the nomenclatural tasks providing for the correct execution of nomenclatural acts (publications, registrations, decisions by authorized collegial bodies, etc.) and ensuring their nomenclatural consistency, i.e., compliance with the conditions of respective Codes. This group has a purely "service" function and has no special standing outside of solving the tasks of the above-considered basic groups.

2.2 TO THE PHILOSOPHY OF *T*-DESIGNATORS

The taxonomic and linguistic N-objects can reasonably be considered as elements of the ontic and epistemic components of cognitive situation, respectively (see Section 1.1). One of the latter's most fundamental properties is its orderliness assured generally by the principle of onto-epistemic correspondence. In particular, with regard to these N-objects, it is provided by their interconnection established by the *reference* of T-designators (linguistic N-objects) to the respective denoted objects (taxonomic N-objects) [Hull 1983; Bolton 1996; Shatalkin 1999; Pavlinov 2015a].[5] The particular references link interrelated taxonomic and linguistic N-objects into ordered pairs called *concepts*, the totality of which constitutes the *semantic space* of practical systematics [Nikishina 2002; Popova and Sternin 2007]. The absence of references means that the T-designators are "empty" and the taxonomic N-objects remain non-individualized linguistically. Therefore, as one of the Linnaean canons says, "nomenclature, [...] after the arrangement has been done, should first of all apply the names" [Linnaeus 1751: § 210]; this is confirmed by all contemporary Codes. The nature of this relationship is defined specifically within the particular basic nomenclature concepts which differ in their inherent "philosophy of name" [Pavlinov 2014, 2015a].

Every taxonomic N-object, considered semantically, may be decomposed into two basic components, viz., its *intension* and *extension*, defined by two aspects of its consideration—with respect to its "quality" (what the object is) and "quantity" (what elements are contained in the object). This two components together with the object's designative notion constitute the so-called *semantic triangle* (in the sense of [Carnap 1969]), in which the object's intension, extension, and notion correspond to the triangle's vertices and interrelations among them correspond to its edges. If this triangle is treated linguistically, a notion may be substituted with a T-designator signifying this notion. For example, in the case of taxon, elements of the semantic triangle correspond to its

---

[5] In botanical nomenclature, it is *application*; and in the *Linz ZooCode Proposal*, it is *allocation* of names to respective taxa [Anonymous 2012; Dubois et al. 2019; Dubois 2020a]

diagnosis, circumscription, and taxonym, respectively [Meyen and Shreyder 1976; Pavlinov 2014, 2015a, 2018]. It is this "triadic" concept that explains metaphorically the above-mentioned intercorrelation of N-definitions of taxonomic and linguistic N-objects.

The relationships between taxa and their taxonyms established by respective references are subject to serious and manifold considerations. Some relate to the practical issues of the use of taxonomic descriptions, while others are burdened with a profound philosophical meaning.

From a practical perspective considered by nomenclature Codes in most detail, one of the key requirements for the reference is its unambiguity formalized by the *principle of monosemy*, which in the case of taxonomic N-objects is expressed by the formula "one object—one name." It serves as the main tool to ensure the universality and stability of taxonomic names and hence the entire nominative nomenclature. It was with the formulation of this principle in the 17th century that regulative taxonomic nomenclature began to develop purposefully [Kupriyanov 2005; Pavlinov 2013, 2014, 2015a] (see Section 5.2).

From a more philosophical consideration of the reference unambiguity, T-designators may be considered *rigid* or *non-rigid* (in the sense of [Kripke 1972; LaPorte 2018]). In the case of taxa, being rigid means one-to-one correspondence between a particular taxon and its taxonym: the respective reference uniting them in a concept remains constant under any circumstances, as far as the respective taxon remains the same. Being non-rigid means non-strict correspondence: the reference in question changes depending on the contexts of the taxon consideration set by different circumstances and expressed by its specific N-definition. These circumstances may be treated as different "*possible worlds*" (in the sense of [Kripke 1972]) into which a taxon may be "placed" by its different theoretical definitions; in general, particular "taxonomic realities" outlined by particular taxonomic theories customarily are considered as such "worlds" [Pavlinov 2011, 2018]. Different N-definitions of taxa in them may entail changes in the properties of both the references of their respective taxonyms and occasionally in the latter themselves.

This issue evidently depends on possible answers to the fundamental questions as to under which circumstances and to what extent taxa may be considered "the same" if their conceptual and/or operational definitions change. For instance, might different theoretical definitions of a taxon, say, typological or phylogenetical, be enough to consider it "different"? Or might its different operational definitions, say, by typification or ancestration, be enough to consider it "different" providing that it is the "same" with respect to its circumscription? The answers will depend evidently on the priorities given by different nomenclature systems to the particular specifiers employed for the N-definitions of taxa. The early nomenclature systems presumed that changes in circumscription or diagnosis of a taxon meant that the latter ceases to be "the same" and needs a new name. This provision is excluded from the contemporary traditional Codes that are type-based, but coincidence of circumscription of the taxa defined typologically or phylogenetically is considered critical for acknowledging them as the "same" deserving being named by the same T-designator. With this, however, a reservation may be made as regards indication of the secondary authorship as part of the compound nominal complex following change in treatment of the denoted taxon, while the name proper remains the same.

In the general problem of reference, one of the central issues is posed by the question of what exactly is defined (in the logical sense) by respective specifiers, whether taxonomic *N*-objects or *T*-designators, in particular, taxa or their taxonyms [de Queiroz and Gauthier 1990; Ghiselin 1995; Chebanov and Martynenko 1998; Stuessy 2000; de Queiroz and Cantino 2001; Moore 2003; Rieppel 2008; Béthoux 2010; Pavlinov 2015a; Aescht 2018]. This particular problem is caused by the abovementioned duality of *N*-object, which can be understood either taxonomically or linguistically (nomenclatural acts seem not to be involved in this issue). The theoretical positions can be summarized by three main points of view. According to the essentialist concept, the separation of a "thing" and its "name" contradicts their closely interrelated nature, therefore *N*-definition is coherently applied to both a taxon and its taxonym. In modern literature, far from this natural philosophy, two other possibilities are considered. Some authors believe that names (taxonyms) and not taxa are defined: on this basis, in particular, it is argued that nomenclatural type is the "type of name" and not the "type of taxon," so it is called a "name-bearer" or *nominotype* (see Section 3.5).[6] Others believe that they are the taxa that are defined as the objects, whereas their names are only "labels" that do not need any definitions; as Erna Aescht formulated it aphoristically, taxa are defined while names (*nomina*) are allocated to them [Aescht 2018].

The latter position seems to be more correct from the point of view of the aforementioned Carnap's concept of semantic triangle [Pavlinov 2015a]. In fact, both intensional and extensional definitions refer to the taxonomic *N*-object and not to its *T*-designator. That is why taxon can be denoted differently (hence the problem of synonymy) though still retains the same substantial interpretation set by whatever its specifiers are and, in this sense, remains the "same" (i.e., self-identical) as an element of "taxonomic reality."

Over the past few decades, the issue of ontological status of *T*-designators is discussed with regard to the ontological status of taxonomic *N*-objects they refer to, with the taxa being paid most attention [de Queiroz and Gauthier 1990, 1994; de Queiroz 1992; Shatalkin 1999; Härlin 2005; Brzozowski 2020]. If a taxon is interpreted as a class (in the logical sense), its designator is a notion (*attributive* denotator), which is defined logically. It cannot be rigid because of possible changes in taxon circumscription and/or diagnosis due to its different theoretical and/or operational definitions. Unlike this, if a taxon is interpreted as an individual-like entity, which is by initial assumption self-identical under any circumstances ("possible worlds"), it is not defined but just pointed to, so its taxonym is a rigid designator in the sense of a proper name (*referential* denotator). However, this "class *vs.* individual" dichotomy is a simplification: in fact, interpretations of the taxon's ontology are more diverse [Rasnitsyn 2002; Rieppel 2008; Pavlinov 2011, 2018, 2021], and this should be taken into account. Thus, if a taxon is regarded as a natural kind defined by its essential property (in the sense of [Quine 1969]), an essentialist understanding of the "possible worlds" presumes that a designator of such a taxon can be considered *conditionally rigid* [Kravetz 2001; LaPorte 2018]. On the other hand, it seems reasonable to

---

[6] The general limitation of this viewpoint is that only type is considered in its justification, while other specifiers are ignored.

consider *T*-designators of phylogenetically defined taxa (clades) and species, if they are treated as individual-like historical entities, also as conditionally rigid proper-like names [de Queiroz and Gauthier 1990, 1994; de Queiroz 1992; Härlin 2005; Ereshefsky 2007a, 2017; Pavlinov 2018, 2021].[7]

The situation with the ranks, taxonomic categories, and their ranknyms may be represented as follows. The ranks are ordered linearly along a certain fixed ranking scale, so each rank is unitary and unique by its position in that scale, and its ranknym seems to be its proper name (but see below about probabilistic definitions). However, taxonomic categories are always classes (in the logical sense) of taxa of certain hierarchical ranks; for instance, the "species category" encompasses all taxa that are ranked as species. Therefore, ranknym referring not to a rank, but to a particular category, is always a notion that, when ascribed to particular taxa, indicates their allocation to this particular category.

For typonyms, the following interpretation seems to be plausible [Pavlinov 2015a]. Every nomenclatural type, as a specifier of a particular taxon, is endowed with a single nomenclatural meaning and thus functions as a discrete and integrated nomenclatural unit. Accordingly, it can be treated as an individual-like *N*-object, so its complete typonym, such as "holotype of the species *A-us b-us*," is a kind of proper-like composite name. This fundamental property seems to be an attribute of any nominotype regardless of the particular circumscription and rank of taxa they typify. However, with regard to species-group taxa, some nuance may arise in the case of the type series including several specimens (e.g., syntypes). This is due to the fact that different elements of the original composite type series may subsequently be proven members of different particular taxa, so the whole-type series turns out to be a set. But as soon as the "alien" specimen(s) is/are excluded from it, the remaining part of the type series regains its individual-like status, with its composite typonym being still a proper-like name.

The foregoing consideration allows attention to be drawn to the following important points concerning the general character of both *T*-designators and their relationships (references) to taxonomic *N*-objects.

Firstly, the taxon as an *N*-object of "taxonomic reality" does not exist outside how it is considered within a certain cognitive situation imposed by the subject of taxonomic research [Pavlinov 2018, 2021]. Thus, the referential function of *T*-designators is unavoidably subject-dependent, so it cannot be treated out of context posed by these subjects as components of a cognitive situation. Therefore, this seems to turn a semantic triangle, in which it is conventionally considered, into a "semantic square," with its fourth irremovable vertex being a subject with its cognitive intention [Raclavský 2012].

Secondly, such contexts may be complicated by a probabilistic view of the whole cognitive situation. This presumes the probabilistic status of *N*-definitions of any *N*-objects: their contents depend on the particular probabilistically given contexts of consideration of the fragments of "taxonomic reality" and respective semantic spaces. In such a situation, a number of context-dependent classifications can be elaborated

---

[7] It is of interest that the author of one of the first Codes, H. Strickland (see Section 6.3.1.2), without explicitly addressing whatever ontology, wrote that "a complete parallel seems to exist between the proper names of species and of men" [Strickland 1835: 39].

with partially non-coincident taxa and categories for the same set of organisms. Accordingly, an array of such probabilistically interpreted taxonomic *N*-objects generates a certain probabilistic distribution of the potentially possible references of their *T*-designators. This may be observed, for instance, if the diagnosis and/or circumscription of a taxon are interpreted probabilistically as statements within particular taxonomic hypotheses with various degree of plausibility. Such *T*-designators with probabilistically defined references may also be considered conditionally rigid [Härlin 2005; Bertrand and Härlin 2008; Rieppel 2008; Pavlinov 2014, 2015a].

Lastly, the previous consideration allows us to suppose [Pavlinov 2014, 2015a] that, given the cognitive situation in general being shaped probabilistically, both *T*-designators and their references to taxonomic *N*-objects seem to be most adequately described by a probabilistic model of descriptive language developed by the Soviet logician Vasiliy Nalimov [Nalimov 1979]. Its important part is the *fuzzy logic* operating with the concept of "linguistic variable," the semantic and referential properties of which are determined contextually and probabilistically [Zadeh 1992; Kosko 1993; Dompere 2009]. According to this, in the probabilistically defined semantic space of systematics, each *T*-designator, treated as a specific "linguistic variable," as said above, is characterized by a certain probabilistic distribution of its possible references to the respective contextually defined taxonomic *N*-object. The choice of a particular meaning of the latter's designator basically depends on a particular taxonomic theory in which context this *N*-object is conceptually and operationally defined. An example is the variation in the meaning of the name *Reptilia* associated with a certain group of terrestrial vertebrates, which is interpreted in different ways in different classifications, whether typological or evolutionary-taxonomic or cladistic or something else.

Such a situation, in which the references of *T*-designators to respective taxonomic *N*-objects are determined by semantically varying contexts, is partly responsible for the instability of nominative nomenclature. This undesirable effect is minimized by "divorce" of the subject-area Codes ensuring the delineation of possible contexts ("worlds") in which *N*-objects (of whatever sense) are considered and named. This seems to provide a kind of "philosophical" explanation for the "Great Schism" of the nomenclature systems and Codes that occurred in the 19th and 20th centuries.

The *T*-designators perform three basic functions with respect to the taxonomic *N*-objects, viz., *designative*, *descriptive*, and *classificatory* [Pavlinov 2014, 2015a]. According to the former, a *designator* in its narrow linguistic sense is meant, presumed by its nominalist interpretation: it carries no semantic load and simply signifies a taxonomic *N*-object to individuate it lexically in the semantic space. In this case, the function of taxonym is expressed by the formula "a name is just a name," so any arbitrary symbol can be used instead of a traditional name. If the *T*-designator functions as a *descriptor*, it is semantically motivated to indicate certain features of the organisms allocated to the denoted taxon; so one may say it has a certain "sense" [du Plessis 2017]. Finally, classificatory function makes *T*-designator a *classifier*: it indicates, by specific linguistic means, the position of the denoted taxonomic *N*-object in a classification.[8] These three functions are of fundamental importance by underlying several basic concepts of taxonomic nomenclature, viz., nominalist, essentialist, and rational-logical (see Section 2.3). They are complemented by *mnemonic* (fixing the image of an object in the memory of the subject) and *communication* (as a means of communication between the subjects) functions. Although only taxa are usually considered in this respect, it is clear that the designators of other taxonomic *N*-objects (ranks, categories, etc.; see above) also perform these functions.

> In a recent publication by Antonio Valdecasas and colleagues, up to seven such functions are recognized: individuation, reflection of relationship hypothesis, information retrieval, explanatory power, testable predictions, conceptual power, and language [Valdecasas et al. 2014]. However, these authors seem not to distinguish between two "Linnaean foundations" of systematics, systematization and naming, so they mix classificatory and nomenclatorial considerations.

Descriptive taxonyms expressing certain characteristics of the denoted organisms comprise the following main semantic groups: *bionyms* indicate peculiarities of the way of life, *morphonyms* indicate peculiarities of morphology, *ethonyms* indicate peculiarities of behavior, *toponyms* indicate specific habitats, *chrononyms* indicate specific time. In addition, *T*-designators can be *onomatopoeic* (imitate certain sounds) and *metaphorical*; the latter include *iconyms* referring to certain magico-religious phenomena [Alinei 2021]. However, semantically motivated taxonyms are not all necessarily descriptive: examples are *eponyms* derived from the personal names of people or deities, including *patronyms* given in honor of them. Some ranknyms are semantically motivated in a hidden form: for example, taxonomic category of family was originally introduced in the 17th–18th century to indicate metaphorically the natural status of taxa of the respective hierarchical level [Magnol 1689; Adanson 1763]. Typonyms may also be considered semantically motivated, as far as respective prefixes in them (*holo-*, *para-*, *allo-*, etc.) indicate the nomenclatural status of the type specimens.

The classification function of *T*-designators is rarely considered, although it is implicitly presumed in most nomenclature systems, and in rational-logical nomenclature it is explicitly stated as the basic one. Thus, a taxonym-classifier can indicate that the referent taxon belongs to a certain taxonomic category or to a particular inclusive taxon. The classification function of ranknyms makes them semantically motivated, as it presumes indirect indication of the place of the corresponding taxonomic category in a fixed-rank hierarchy: e.g., in the contemporary Codes, the family is always ranked higher relative to the genus and lower relative to the order. The respective function of typonyms presumes indication of the nomenclatural status and function of nominotypes denoted by specific terms (holotype, paratype, etc.).

---

[8] A taxonym treated in such a vein can be likened to a museum locating label that bears information about the position of a specimen in the repository [Felt et al. 1930; Felt 1934].

The classification function of taxonyms is most diverse; it is provided by special ways of formation as group-specific and/or rank-specific *markers* using certain lexical means; there are several cases of such kind. One such case is the addition of a designation of the inclusive taxon to the complete taxonym of its subtaxon; the most typical example is the genus–species binomen. Another case is the formation of a whole taxonym or its constituent morphemes (stem, prefix, suffix) according to the position of the respective taxon in the classification, including its allocation to a certain category in the hierarchical system. Thus, formation of the names of suprageneric taxa (family, order, etc.) from the stem part of the generic name rigidly correlates with the respective including and included taxa in the classification. How the rank-specific suffixes of the names of suprageneric taxa are formed indicates their position in the ranked hierarchy: for example, in zoology, the suffix *-idae* corresponds to family, *-inae* to subfamily, and *-ini* to tribe. Suffixes can also be group-specific: for example, the family name ends with *-idae* in zoology, *-aceae* in botany, and *-viridae* in virology. In phylonomenclature, prefixes are used to mark group-specific names: for example, *Phyto-* for higher plants and *Phyco-* for algae. As is seen, all these lexical means facilitate recognition of allocation of the particular taxa to certain groups of organisms or to certain taxonomic categories.

Lastly, taxonyms may indicate a substantial status of taxa they denote, in one way or another, according to the provisions of particular theory-burdened nomenclature systems. Thus, some of the latter of the 18th–19th centuries requested verbal naming of supraspecific taxa of the main categories, whereas the taxa of subsidiary categories were admissible to denote with arbitrary symbols (e.g., [Linnaeus 1751; de Candolle 1819]). In phylonomenclature, the phylogenetic status of taxa is marked using specific prefixes in their phylonyms (*Pan-*, *Apo-*, etc.) [de Queiroz 2007; Cantino and de Queiroz 2010, 2020].

All nomenclature systems pay great attention to the lexical aspects of the structure of the professional language of systematics. One of the main issues here is the morphology of *T*-designators, which can be *verbal*, *symbolic*, or *combined*. These interpretations are governed by the corresponding principles of verbalness and symbolness. In the first case, *T*-designator is either a single word or a phraseme borrowed from a natural language or construed artificially in a similar manner. This approach is mandatory for all traditional and some other (such as phylogenetic) nomenclature systems. In the second case, *T*-designator is formed by a set of letters and/or symbols that may be combined arbitrarily (e.g., for a pure designation function) or designed to perform a certain meaningful (e.g., classification) function; the latter is most characteristic of rational-logical nomenclature systems. Certain combinations of these two basic morphological forms are presumed by some nomenclature systems, in which various symbols and/or numbers are included in the basically verbal taxonyms to provide their classification function (e.g., [Kluge 1999a,b, 2020]). The ranknyms are most usually verbal, but they take a digital form in numericlature, where they designate not ranks/categories proper, but rather non-strictly fixed levels of generality (e.g., [Hennig 1969]).

The *T*-designator, in any of these forms, may consist of one or several parts; in the second case, it is a *composite* designator (called *combination of names* in some nomenclature systems). However, the latter functions as a whole in denoting a particular taxonomic *N*-object, so a verbally construed *T*-designator is a *phraseme* understood as a stable multi-word unit [Cowie 1998]. Elements of a composite designator usually perform specific functions. Thus, different parts of a verbose essentialist *T*-designator may refer to particular features of named organisms. In binary nomenclature, one part of a phraseme corresponds to the generic name and another to the species epithet. In the above-mentioned nominal complex, one part refers to the name proper, and another contains a certain auxiliary information about it (authorship, etc.).

> Different nomenclature systems variously consider the use of punctuation and additional symbols in composite *T*-designators. Early systems embodying the essentialist concept allowed the use of punctuation marks in verbose taxononyms; they are prohibited in all contemporary Codes. The latter accept the use of parentheses to separate intercalary names in composite *T*-designators. Special characters (such as multiplication sign) are used to compose names for the taxa of hybrid origin (nothotaxa of botanists).

2.3 BASIC NOMENCLATURE CONCEPTS

As noted in Chapter 1, the descriptive language of systematics constitutes part of the epistemic component of a cognitive situation in which this scientific discipline operates. This component interacts in a complex way with the ontic one that provides specific understandings of "taxonomic reality" presumed by particular taxonomic theories. The language of systematics is elaborated to serve, in cooperation with other epistemic tools, as a reliable means of describing this "reality." This language, in turn, is structured by different nomenclature concepts that are developed to be most adequate to the conceptually depicted particular "realities," and they can be considered its specific "dialects." The overall structure of the latter's diversity is multifaceted, and can be represented by the following largely overlapping ways of categorization [Pavlinov 2014, 2015a, 2019].

From a theoretical standpoint, nomenclature concepts can be divided into two main groups depending on the method of their substantiation, *theory-neutral* (or *theory-free*) and *theory-dependent* ones; their names speak for themselves. Concepts of the first group are not based on certain explicitly formulated higher-order theories, their regulators are mostly "internal"; the empirical concept belongs here. Theory-dependent concepts are modeled by certain higher-order theoretical constructs, which are predominantly "external" in relation to language itself. Depending on the specific ways of their substantiation, some theory-dependent nomenclature concepts are *ontology-based* (essentialist, phylogenetic), while others are *epistemology-based* (rational-logical, pragmatic); a nominalist concept takes an intermediate position. The elaboration of theory-dependent concepts means rationalization of the whole of the nomenclatural activity, so the nomenclature they shape can be denoted as *rational* in a general sense.

According to another fairly general "division basis" focusing on different functionalities of *T*-designators, nomenclature concepts are divided into other main groups. *Descriptive* concepts imply or explicitly require that *T*-designators be descriptive, i.e., indicate somehow certain characteristic features of the denoted taxonomic *N*-objects. This group encompasses empirical and especially essentialist nomenclature, and also a significant part of the traditional nominalist nomenclature. *Designative* concepts develop *T*-designators, which are designed just to denote taxonomic *N*-objects and thereby to individualize (signify) them in a semantic space by certain designative (e.g., linguistic) means. This encompasses primarily some rational-logical and nominalist nomenclature concepts. *Classificatory* concepts imply that *T*-designators perform primarily a classification function, in some specific way indicating the position of taxonomic *N*-objects in a taxonomic system. Rational-logical nomenclature is mainly developing in this vein.

The distinction of nomenclature concepts depending on the method of "technical" formation of *T*-designators provides the following groups thereof. In *verbal* concepts, *T*-designators are mainly lexemes (words, phrasemes) of a certain natural language or some lexical structures similar to them; all descriptive and many nominalist nomenclature concepts belong here. In *symbolic* concepts, various numerical or letter combinations, "formulas," etc., function as *T*-designators, and the principles of the latter's formation are elaborated within certain artificial languages; they are characteristic of rational-logical and partly nominalist concepts. Symbolic *T*-designators can be descriptive (like "formular" designations in the Petit-Thouars system), classificatory (such as in the "philosophical" language of Wilkins, in numericlature, etc.), or designative (see Section 7.1). Some nomenclature systems allow *mixed* verbal-symbolic formation of *T*-designators (for example, in virology).

The classifications elaborated by systematics are normally hierarchical, with most contemporary ones being with fixed ranks/categories. The latter circumstance is reflected in taxonomic nomenclature by making some of its concepts *rank-dependent*. This means that certain principles and rules are connected, in one way or another, to the ranked hierarchy of a taxonomic system, viz., to the level of generality of taxonomic *N*-objects (taxa and categories). Such dependence may affect both nominative and regulative nomenclature. In the first case, rank dependence involves specific formation of *T*-designators (taxonyms, ranknyms), and in the second case, an application of certain nomenclature regulators (for example, typification) is rank-dependent. Accordingly, the absence of such connection makes nomenclature concepts *rank-independent*.

The most noticeable nomenclature concepts that deal with the taxonyms are considered below.

The **empirical concept** is fundamentally theory-neutral, semantically it is mostly descriptive, and semiotically it is verbal. It is shaped and developed spontaneously under the influence of certain implicit ("internal") regulators, among which the most significant are general cognitive and linguistic laws and principles of the functioning of natural languages. The most obvious variant of this concept is folk nomenclature, which is characteristic of pre-scientific systematics (see Section 4.2.1). However, it is important to note that folk nomenclature is not fully free from a metaphysical "load," which is provided by intuitive ontology [Cruz and Smedt 2007]. The latter plays the role of a kind of "proto-theory," on which the general structure of descriptive language of pre-scientific systematics depends to a certain extent; this is reflected by the above-mentioned concept of linguistic world picture.

The **pragmatic concept** is close to the empirical one. It does not appeal to any complex theoretical constructs, but subordinates the language of systematics to certain explicitly specified pragmatic regulators (such as principle of convenience) that dominate over others. All nomenclature systems, to the degree they are intended to resolve practical tasks, encompass certain elements of pragmatics.

The **essentialist concept** was developed by scholastic systematics as an ultimate variant of the descriptive nomenclature. It is characterized by the most pronounced semantic motivation for the formation of T-designators. Its key idea, which goes back to antique natural philosophy, is expressed by the formula "a thing and its name are the same" [Losev 1990]; something similar is stated by the Chuang-tzu aphorism "Things are so because they are called so" (cited after [Chuang-tzu 1964]). As far as the antique essentialist tradition is concerned, this is because both the "thing" and its name are but two different interrelated expressions of the same Platonic *eidos* (see Section 4.2.2), so "there is no separation between naming and defining" [Slaughter 1982: 66]. From this point of view, the recognition of the essence of an organism, its expression by the latter's "genuine" name, and finding its "genuine" place in the System of Nature constitute an inseparable trinity of the cognitive procedure. In the terms of systematics, this means that classifying and naming of organisms are closely intercorrelated as they are governed by the same general essentialist principles. The logical genus–species scheme developed by neo-Platonists and scholastics laid the foundation for the important principle of binarity: every notion corresponding to a particular essence (except for the highest one) must consist of two parts, "generic" and "specific" ones (in the logical sense). This is true, in particular, for T-designators denoting taxa as expressions of their *eidê* (essences) manifested in the essential characters of the organisms belonging to them.

The **nominalist concept** was developed by post-scholastic systematics; it was substantiated by the nominalistic natural philosophy of the New Time (F. Bacon, J. Locke), to which the basic idea "a name is just a name" corresponds. According to this, T-designators are simply "labels" of taxonomic N-objects: they perform primarily a designative function and do not necessarily have any semantic motivation. Therefore, the form of T-designators, generally speaking, is accidental with respect to the features of denoted taxonomic N-objects, so they may be of whatever form—either verbal, symbolic, or combined. The original principle of binarity was simplified to the principle of binomiality, according to which each of two parts of the complete genus–species binomen must consist of one word or a word-like lexeme. This concept is preferred to the essentialist one due to the fact that it imposes fewer conditions on semantic motivation and lexical regulation of taxonyms and, therefore, makes the entire nominative nomenclature potentially more stable.

This is an appropriate place to call attention to the different meaning of *binary* and *binomial* ways of the formation of taxonyms (see also Section 3.3). In fact, they are not synonymous: the second principle is a special case arising from the first, so they should be clearly distinguished [Stejneger 1924; Sprague and Nelmes 1931; Pavlinov 2013, 2015a]. Unfortunately, this distinction is not observed in most contemporary nomenclature systems [Hawksworth 2010].

The **rational-logical concept** of the 16th century was based on the general idea of philosophical rationalism presuming strict correspondence between both rationally and uniformly arranged Nature, knowledge about it, and the language describing it (see Section 7.1). This concept requires taxonyms to perform primarily or exclusively a classification function by indicating the position of taxa in a uniformly arranged overwhelming classification system. It is evident that T-designators thus formed are loaded semantically, and their semantics is determined primarily by their classification function. However, they may be formed to perform a partly descriptive function in nomenclature systems of a rational-logical kind. Within the framework of this concept, T-designators are regulated by certain rather strict principles and rules, according to which they must be formed based on a single scheme or a single "formula" by means of a certain strictly fixed combination of respective symbols. These principles and rules are specific to particular rational-logical nomenclature systems; the most developed variant is provided by numericlature considered most compatible with the contemporary digital technology.

The **phylogenetic concept** was developed within the framework of a cladistic version of phylogenetic systematics (see Section 7.2). It is one of the most expressively theory-dependent, with this dependence being expressly stated. Its metaphysical basis is composed of a specific understanding of "taxonomic reality" as the phylogenetic pattern shaped by the hierarchical network of kinship relationships between holophyletic groups (clades). Accordingly, phylogenetic nomenclature is presumed to be most "concordant with evolutionary [in fact cladistic, *IP*] concepts of taxa [and] more closely conform to the manner in which they are conceptualized" [de Queiroz and Cantino 2001: 269].

3 Basic Principles of Nomenclature

The nomenclature principles, understood in a general sense, are the main regulators of nomenclatural activity dealing with the solution of nomenclatural tasks (see Section 2.1.3). In traditional nomenclature systems, they are most often reduced to the rules of a rather practical kind, although some are called "laws" to emphasize their special significance and mandatory character (e.g., the law of priority). Because of this predominantly pragmatic focus, these principles remained until very recently without a solid integral concept ordering them in a unified system according to their functions and significance. In this regard, three cycles of works are to be mentioned in which attempts are made to develop such systems on different conceptual bases: one has a more applied nominalist character [Dubois 2005, 2006b, 2011a, 2015; Dubois et al. 2019], the second considers them in the context of phylogenetic theory [de Queiroz and Gauthier 1990, 1992, 1994; de Queiroz 1992, 1997, 2005, 2007; Cantino and Queiroz 2010, 2020], and the third considers nomenclature principles from a more general theoretical perspective [Pavlinov 2014, 2015a, 2019].

In this chapter, the author's version of the system of basic principles of taxonomic nomenclature is considered from a rather general perspective not limited to particular nomenclature systems. The regulative nomenclature is viewed in its entirety as an array of the norms, principles (provisions), and rules developed over 300 years of its history. They are considered in several basic aspects, viz., general regulative, cognitive, linguistic, juridical, taxonomic [Pavlinov 2014, 2015a, 2019].

The general regulative aspect of consideration of nomenclature encompasses the most fundamental principles that relate to the general organization of nomenclatural activity, including its functioning and development, i.e., they are a kind of "meta-principles" of nomenclature.

From a cognitive science perspective, under consideration is the general meaning of designations of the objects studied by systematics: why and how they could and should be designated (named, etc.). Here, the key point is a certain relationship between taxonomic objects and their designations, and one of the guiding ideas is monosemy, implying one-to-one correspondence between them.

From a linguistic perspective, the following main issues are dealt with. The first considers how grammar (mainly syntax) regulates the structure of the linguistic units (lexemes) that are used to designate taxonomic N-objects: whether they should be verbal or symbolic, how many separate morphemes they have to include, etc. Another issue is whether taxonomic designators should strongly comply with certain linguistic principles or may be arbitrary. The etymological issue involves semantic motivation of the designations of taxonomic objects.

The juridical consideration of nomenclature involves social regulation of its content and functioning by certain covenants. The most important issue here is analysis of the principles and rules that are designed to ensure the universality and stability of taxonomic nomenclature in all its meanings.

When considering nomenclature in the taxonomic aspect, the main issue is its relationship to the taxonomic theory, understood in a broad sense, i.e., including the ways of defining "taxonomic reality" studied by biological systematics and the principles of describing its structure depending on its particular definitions. This involves, above all, the ontological status (nature) of the objects of systematics, e.g., whether they are real or nominal, and if real, what kind of reality they are ascribed, and how this status may affect the principles of their designation. This kind of connection is omitted or even negated in the consideration of nomenclature in an empirical vein, but it is very significant as one of the general regulators of the descriptive language of systematics. In particular, it is this relationship that determined both the development of the essentialist concept of nomenclature in the 17th–18th centuries and its replacement by the now dominating nominalist one in the second half of the 18th century.

The pragmatic consideration of nomenclature may be added to these main groups of principles. It involves analysis of the regulators aimed at making the whole of taxonomic nomenclature as practical in its applications as possible.

The cognitive and linguistic principles can be, with some reservations, considered "internally" motivated: they are explications of certain primary "laws" of functioning of any linguistic system as part of the epistemic component of the cognitive situation. The juridical, taxonomic, and pragmatic principles are "externally" motivated: more special norms of a secondary nature are of key importance in their formation; they relate basically to the purposeful rationalizaton of taxonomic nomenclature.

The main objective of this chapter is to streamline the basic principles constituting regulative nomenclature, as it has developed to date, in a certain generalized hierarchically organized framework [Pavlinov 2014, 2015a, 2019]. The latter's highest level is set by five major blocks that integrate nomenclature regulators according to their basic functions and basic motivations. They correspond to the above-depicted main aspects of consideration of nomenclature.

> Alain Dubois and his colleagues suggested another block structure of nomenclature principles, which is laid on a mostly empirical basis to correspond to the steps of the nomenclature process recognized by them [Dubois et al. 2019]. Accordingly, the nomenclature principles were grouped as follows: (a) general; (b) regulating the availability of names; (c) regulating the taxonomic allocation of names; (d) regulating the validity of names; and (e) regulating the registration of names [Aescht 2018].

Basic Principles of Nomenclature

The hierarchical approach to arranging nomenclature principles adopted here is important in that it allows the identification and analysis of the main factors acting within each block and thus determines the general structure of a semantic space of descriptive systematics. The general principles perform mainly a framework function; they specify the nomenclatural tasks to be solved and offer their solutions in a general manner (as norms), so they are largely declarative. In the contexts shaped by them, the principles of more particular meaning are elaborated, which constitute a basis for practical realization. The latter is provided by the working rules of direct action designed to solve specific nomenclatural tasks. All this makes it possible to identify and formulate more clearly nomenclature regulators of different levels of generality and to indicate their relative importance, their interrelations, subordination, etc.

According to this approach, the nomenclature regulators are considered in this chapter following the hierarchical structure of the above-mentioned semantic space. Within each of the blocks outlined above, the order of consideration of regulators corresponds basically to the significance and sequence of resolving those nomenclatural tasks, to which they are applied (see Section 2.1.4). For each major task, the leading general principle is defined first, and the working principles/rules are then considered in its context as possible options for specific solutions of the respective task. With this, the nomenclature principles are considered, as far as possible and relevant, in their historical dynamics to demonstrate their roots and succession. These principles are usually considered as applied to taxa, but they are undoubtedly relevant to other taxonomic N-objects as well.

Most of the terms used here are borrowed from both the earlier, now acting, and draft Codes. A few terminological novelties are proposed mostly to denote the basic (general) principles introduced for the sake of a more rigorous and consistent arrangement of the whole of the system of regulators shaping nomenclatural activity. These basic principles/provisions are marked with **bold** to distinguish them from the working principles/rules, which are marked with *italics*.

3.1 GENERAL REGULATORY PRINCIPLES

The principles of this block are aimed at the general ordering of taxonomic nomenclature as a specific linguistic system. On this basis, as indicated above, they can be considered "meta-principles" of taxonomic nomenclature, the latter's general organizational foundation. Their action is all-encompassing, deeply mediated, and determines the whole of the structure and development of the language of biological systematics. Most of these principles are implicit in their actions, but at least two of them, universality and stability, are expressly declared basic in some Codes. Here, several "meta-principles" of this block are considered that seem to have the greatest relevance to the general regulation of nomenclatural activity.

The **principle of adequacy** is probably the primary and most fundamental in implying a certain correspondence of the language of description of "taxonomic reality" to the latter's structure. According to this, the main function of the principle of adequacy is linking the development of the language of taxonomic descriptions with the development of the general taxonomic theory outlining the "taxonomic reality" being described. Being basically cognitive, it reconciles more particular nomenclature principles, primarily from taxonomic and also partly linguistic blocks, with the basic provisions of taxonomic theory.

The **principle of rationality** of taxonomic nomenclature is of paramount importance for the latter's development and functioning. It explicates the general philosophical principle of rationality as an organizing force of the whole of scientific activity, including professional language as an important part of its epistemic component. This principle presumes elaboration of the explicitly formulated principles and rules, their systematization and arrangement into nomenclature systems and Codes, and development of mechanisms for their adoption, change, application, etc.

The **principle of systemity** requires that every particular nomenclature system be complete and consistent so that its nomenclature regulators be mutually compatible. This important condition, albeit without explicit reference to this principle, is mentioned in the introductory sections of many Codes. The main provisions of this principle make an array of nomenclature regulators a kind of non-formal quasi-axiomatic system, and its ideal is the nomenclature holo-system (see Section 2.1.1).

Systemity presumes that many nomenclature principles, especially working ones, do not act in isolation but in certain combinations. Some principles establish methods for solving particular nomenclatural tasks, while others specify conditions for particular decisions, including certain restrictions for them. For instance, the cognitive principle of designative certainty works in association with the taxonomic principle of taxonomic certainty; conditions for applying the principle of priority are specified by the principle of ranking; application of the principle of homonymy suppression is restricted by the independence of Codes, etc.

Another manifestation of systemity, as well as of rationality, is the clear structuring of nomenclature Codes. It involves proper delineation of the main groups of nomenclatural tasks and principles, correct subordination of general principles and working rules, etc. The current Codes substantiate this structuring as predominantly pragmatic rather than "theoretical" by its motivation, so they cannot be considered effectively structured. An arrangement of the nomenclature provisions (norms, principles, rules) outlined in this chapter illustrates another possibility of their rationally based systemic structure.

Theoretism *vs.* pragmatism. This pair of largely opposing principles indicates two general ways of laying the foundation for the development and structuring of nomenclature concepts and systems. The *principle of theoretism* means that taxonomic nomenclature should be developed on a rational basis according to certain general theoretical considerations; for example, by establishing essentialist or nominalist or phylogenetic principles of nomenclature systems. The *principle of pragmatism* implies that the latter should be developed to meet primarily the practical needs of both the developers and users of classifications, and therefore be simple and easy to understand and apply (the *principle of simplicity* of [Dubois 2005]). It is obvious that any nomenclature system represents a certain balance of theoretisms (at least in a hidden form) and pragmatisms (usually stated expressly).

The **principle of overall universality** requires that taxonomic nomenclature be unified in both its regulative and nominative functions. This is one of the most important drivers of development and the basis of the functioning of taxonomic nomenclature, therefore, it is proclaimed as one of the basics in Codes. It is presumed that the universality of nomenclature extends both to all taxa of living organisms and to all systematists dealing with them. The former ("object" universality) addresses "taxonomic reality" and its linguistic representation, implying a unified application of all nomenclature regulators to all *N*-objects. The latter ("subject" universality) addresses taxonomic communities and invokes a unified application of these regulators by all subjects of nomenclatural activity. Dubois' principles of *exhaustiveness* and *universality* proper [Dubois 2005] correspond to these two addresses.

The general requirement for the universality of the language of taxonomic descriptions is obvious, but in a certain "absolute" sense it is quite utopian. This language is universal in some basic (primarily cognitive) norms, but local in the application of many working rules. The latter are associated with the research and linguistic specifics of particular taxonomic communities, and are partly caused by the specifics of the classified objects. As a result, the unified language of systematics is split into already-mentioned "dialects" due to the divergence of particular nomenclature systems on conceptual, subject-area, regional, and other grounds. This may be formalized as the general *principle of locality* that was never acknowledged explicitly, but in fact operates implicitly in rather diverse ways.

In the second half of the 19th century, two main nomenclature "dialects," botanical and zoological, were legalized by the *principle of the independence* of Codes; they were added with several others in the 20th century. Although officially proclaimed as one of the key regulators in the preambles of Codes, it is in fact a working lower-order rule with no deep idea behind it. Its objective is to restrict application of the principles of suppression of homonymy and priority (fixing different starting dates), as well as other working principles/rules to certain subject areas of systematic biology. At present, the taxonomic community tries to overcome this diversification trend by developing unified nomenclature systems in the form of *BioCode* and *PhyloCode*, but with no evident success.

A special case of the action of the principle of locality is the restriction of application of certain important principles (suppression of synonymy, etc.) to special nomenclatural groups encompassing para- (= morpho-) and ichnotaxa [Sarjeant and Kennedy 1973; Melville 1981; Bengtson 1985; Rasnitsyn 1986, 2002; Greuter et al. 2011]. In fact, this means the appearance of particular nomenclature systems (paranomenclature, ichnonomenclature) for these specific kinds of taxonomic *N*-objects.

The newest system of numericlature should also be mentioned (see Section 7.1), which develops its own designation systems most suitable for digital technology, including Internet resources [Kennedy et al. 2005; Page 2006; Patterson et al. 2006, 2010; Pyle and Michel 2008; Schindel and Miller 2010].

The **principle of overall stability** requires nomenclature, in both its regulative and nominative meanings, to be globally stable, i.e., unchangeable both in its basic regulative propositions and *N*-objects regulated by them. It supplements the principle of overall universality; together they define a goal towards which taxonomic nomenclature should strive. With this, it shapes the context for the functioning of many other principles and, therefore, is systemically important for the whole of taxonomic nomenclature.

However, the whole history of the latter indicates that its overall stability is not only a non-attainable ideal but also "counter-scientific" to a degree. In fact, it contradicts the basic conditions of the above-considered principle of adequacy, according to which taxonomic nomenclature is doomed to change following the dynamics of the whole of systematics, including its theoretical part. Accordingly, it is quite reasonable, from a theoretical standpoint, to introduce the alternative *principle of overall instability* as an important general regulator responsible for the systemic development of the whole of taxonomic nomenclature. The above-mentioned and other particular nomenclature concepts and systems, appearing as answers to certain challenges posed by the developing cognitive situation of systematics, are good illustrations of its action.

This pair of associated but opposite principles has a specific implementation as the principles of partial stability and lability concerning T-designators only (see Section 3.4).

3.2 COGNITIVE PRINCIPLES

Considered from a cognitive science perspective, the general meaning of designations of particular taxonomic N-objects, viz., organisms, taxa, and ranks/categories, is investigated: why and how they need to be designated (named). This block comprises the principles which are associated with basic cognitive activity and deal mainly with the establishment of references between taxonomic N-objects and their T-designators. Therefore, they are primary relative to others (apart from the general regulatory ones) and belong to the "internal" regulators of nomenclature [Pavlinov 2014, 2015a]. They underlie the development and function both of empirical nomenclature of folk systematics [Atran and Medin 2008] and the rationally organized taxonomic nomenclature of scientific systematics. The main difference between them is that some of the latter's versions intentionally strengthen certain cognitive principles (such as designative certainty, semantic motivation, etc.).

The **principle of designation** is probably the most fundamental: it asserts that any cognizable object, when singled out, must be denoted by a specific designator assigned to it, viz., by a term, notion, name, symbol, etc. [Nikishina 2002; Kubryakova 2004]. There seems to be no reasonable alternative to it in the cognitive activity: any object studied by systematics, to become an element of the latter's semantic space, must be designated, otherwise it is simply absent in this space. The inclusion of a T-designator in the N-definition of any taxonomic N-object is the mandatory condition of its being nomenclaturally consistent. Therefore, this principle (also called the *principle of onomatophores* [Dubois et al. 2019]) belongs to the most important nomenclature regulators: according to *Linnaean Canons*, each taxon must be named immediately after it is recognized [Linnaeus 1736, 1737a, 1751].

In systematics, the designation can be *preliminary* or *final*. In the first case, so-called *open nomenclature* may be used to denote taxa with a not quite certain taxonomic allocation [Matthews 1973; Bengtson 1988; Barskov et al. 2004; Sigovini et al. 2016; Minelli 2019; Horton et al. 2021]. The respective composite T-designator includes specific abbreviations like "*ex gr.*" ("from the group"), "*aff.*" ("close to"), etc., added with the name of a taxon with which the one in question is preliminarily associated (e.g., *Homo ex. gr. sapiens*). In the second case, a unique taxonym is assigned to a taxon which gets the status of *nominal* (i.e., named) or *nomenclatural*. Obviously, insofar as judgments about taxonomic N-objects are always probabilistic, their designations cannot be final (ultimate) in an "absolute" sense.

A distinction should also be made between *original* and *subsequent* designations. The original designation of a new taxon means the allocation of a particular name to it in the original description. The subsequent designation refers to: (a) replacing the original name with another (substitution) one because of a revealed homonymy; or (b) correcting it because of a revealed error (misprint, etc.). In this regard, the original and subsequent taxonym spellings are distinguished in some nomenclature Codes. The terms "Section" of Tournefort and "Order" of Linnaeus can be considered original and subsequent designations of the same taxonomic rank/category. Another variant is change in typonym of a type specimen due to the change in its nomenclatural status, e.g., when syntype becomes lectotype.

The **principle of accentuation** (or *emphasis*) means that the more an object is distinguished as cognitively significant for a certain reason, the more distinguished in the "linguistic world picture" its designator tends to be to make it more recognizable [Atran 1990; Uryson 2003; Ellen 2008]. According to this, a general rule works in folk systematics: more significant objects are usually designated by shorter names to facilitate their memorization and recognition in the process of communication [Brown 1984; Ellen 2008]. Therefore, the folk taxa of higher ranks, as cognitively most accentuated, are most often one-word, whereas the names of lower-rank taxa are usually verbose. Scientific nomenclature implements this general principle in a specific manner: it is presumed that "natural" groups are to be mandatory designated by specific formally regulated T-designators (names etc.), while those of "non-natural" ones may be of any kind (typographic symbols etc.).

Designative certainty *vs.* uncertainty. The fundamental significance of designative certainty in cognitive activity is obvious: the more strictly (unambiguously) a notion is associated with (refers to) a certain object, the greater the likelihood of different subjects of nomenclatural activity meaning the same taxonomic N-object when using the same T-designator. This condition is implemented by the *principle of designative certainty* asserting the need for one-to-one correspondence between taxonomic N-objects and their T-designators. The development of the forms of implementation of this principle is one of the conditions for the practical realization of the general principle of rationality and, therefore, is one of the key factors that govern and direct the development of taxonomic nomenclature.

Contrary to the basic provision of designative certainty, in reality, many-to-many correspondence between taxonomic N-objects and their T-designators occurs. Its general cause is the above-mentioned fuzzy character of the descriptive language of systematics, which is inevitably and irreparably inherent in every natural linguistic system [Nalimov 1979]. It is manifested in the existence of two kinds of violation of one-to-one correspondence, homonymy and synonymy (see Section 2.1.2). Therefore, the *principle of designative uncertainty* is implicitly present in every nomenclature system, which impacts (contrary to aspirations of the nomenclature practitioners) on many aspects of nomenclatural activity.

The condition of certainty *vs.* uncertainty is most usually considered as applied to the taxonomic and linguistic *N*-objects, but it equally holds for the acting *N*-objects. This is formalized by the *principle of acting certainty*, according to which definitions, wording, rules, etc., should be formalized so clearly as to avoid confusions in their interpretations and applications. However, this end cannot be reached either because of the fuzzy character of the language of systematics.

According to the basic condition of the principle of designative certainty, every taxonomic *N*-object is to be denoted by a unique *T*-designator, so it is specified by the **principle of monosemy**; the latter is also called the *principle of univocality* [Dubois 2005] or *explicitness* of reference [Cantino and de Queiroz 2010, 2020]. By this condition, the principle in question presumes suppression of *polysemy*, i.e., multiple references between taxonomic *N*-objects and their *T*-designators. Thus, "another side" of the principle of monosemy may be represented by the *principle of polysemy suppression*. It is most usually considered as applied to taxa and expressed by the formula "one taxon—one name"; rationalization of taxonomic nomenclature began with it in the 17th century [Rivinus 1690]. The same is obviously true for designations of taxonomic categories/ranks. However, this requirement is not followed in many cases for various reasons, so it is reasonable to introduce the *principle of admissible polysemy* (see below).

The meaning of the principle of monosemy, in turn, is specified by the *principle of constancy of T-designators*, referring to the respective taxonomic *N*-objects: it presumes that, if the latter change for various reasons (e.g., if circumscription of a taxon is changed to a degree), their designators should remain unchanged. In the case of composite taxonyms, this principle is applied only to their essential parts. At the species-group level, it is abided by only for the epithet denoting "final" taxon (species, subspecies, race, etc.) and not for the name of the containing taxon (genus, species, etc., respectively): the latter's name changes if the "final" taxon is transferred from one higher-level taxon to another. If the *T*-designator is interpreted broadly as a nominal complex, including indication of authorship and the date of its establishment and/or change in itself or in its application, the principle of constancy is abided by only for the name proper and not for the other parts of the nominal complex.

The principle of monosemy presumes the need for selection among homonyms/synonyms that compete for the status of an official *T*-designator of a particular taxon. This selection is based on the differential assessment of taxonyms with respect to their *priorability* (preferability, seniority). A basic provision for this is provided by the *principle of unequivocality of names* [Ride 1988], and the regulation of respective conflicts between competing names by applying certain criteria of their assessment is generally considered by the *principle of zygoidy* [Dubois 2013; Dubois et al. 2019]. These criteria are specific to particular nomenclature concepts. In the essentialist one, the theory-based *principle of verity of name* is most significant, whereas in nominalist nomenclature, the respective criteria are provided by the pragmatically substantiated *principles (rules) of priority* and *usage* of taxonomic names (see Section 3.4).

The above-mentioned general principle of polysemy suppression is actualized by two working versions considering separately homonymy and synonymy; these versions are as follows.

The *principle (rule) of homonymy suppression* (usually called the *principle of homonymy*) excludes the use of the same *T*-designator to denote different taxonomic *N*-objects. In particular, in the case of taxa, it suppresses the use of the same taxonym as a valid name for different taxa. This principle is given a broader interpretation in some nomenclature systems: for example, one of Linnaeus' canons prohibits using anatomical terms (referring to partonomy) for the generic names of plants (referring to taxonomy). The rules for regulated changes in taxonyms violating suppression of homonymy are summarized by the *principle of neonymy* [Dubois et al. 2019].

Particular Codes differ markedly in the severity of conditions for the application of this principle; in particular, to what extent the names must be similar to consider them homonyms. Zoological Code is more liberal in this respect than botanical and microbiological ones. Generic taxonyms with similar spelling, which may be the subject of confusion, are suppressed as *parahomonyms* or *perplexing names* in botany and bacteriology (for example, *Asterostemma* and *Astrostemma*). On the contrary, the "*one-letter rule*" allows the use of generic names that result from the different romanizations of the same non-Latin word in zoology (for example, *Charonia* and *Charronia*, *Hydrothrix* and *Hydrotriche*), with different gender suffixes (for example, *Pica* and *Picus*), and some others. However, in the case of species-group names, etymologically identical epithets with variable spellings (for example, *sulphureus* and *sulfureus*) are considered homonyms in all Codes. In phylonomenclature, identical names referring to the clades of different phylogenetic meaning may not be considered homonyms.

Application of this principle is group- and rank-specific, therefore it is proposed to distinguish between *horizontal* and *vertical* homonymy [Kubanin 2001] (see Section 2.1.2). Horizontal homonymy of all species-group epithets is considered within a genus in zoology, while that of intraspecific epithets is considered only within a species in botany. Suppression of homonymy of the genus-group taxonyms is limited by the principle of independence of Codes, so the respective identical names are called *hemihomonyms* [Starobogatov 1991; Shipunov 2011]. In the case of vertical homonymy, genus–species tautonymy, i.e., coincidence of the parts of the same genus–species binomen, is suppressed in most of the contemporary Codes with the exception of zoological one.

Several other special variants of broadly understood horizontal homonymy of generic names are considered in the 19th and 20th centuries; they are prohibited in some Codes, but are not mentioned in others. *Pleonasms* were equated with homonyms in early nomenclature systems and therefore are considered undesirable, especially if they are parts of a genus–species binomen (as in the case of the domestic cow, *Bos taurus*). Cases of *omophony* are considered variants of parahomonyms, and their use is not recommended or even suppressed in some early Codes. For species-group and suprageneric taxa, primary and secondary homonyms are distinguished, but they are treated differently (see Section 2.1.2). Junior homonyms are always replaced for species-group taxa, while for suprageneric taxa, they are considered taking into account the spelling of respective basionyms.

> In some early nomenclature systems, the taxonyms that coincide in the Latinized spelling, but differ (including semantically) in their original

> language, were suggested not to be treated as homonyms. For example, in the Latin name *Neomys* assigned to two genera of mammals, the first syllable (*Neo*) has different Greek bases, viz., νέω (swim) in *Neomys* Kaup, 1829 and νέος (new) in *Neomys* Gray, 1873 [Palmer 1904]. They are considered undoubted homonyms in the contemporary Codes.

The *principle* (*rule*) *of synonymy suppression* (usually called the *principle of synonymy*) prohibits using different *T*-designators (names or terms) for an "official" denoting of the same taxonomic *N*-object in different classifications and eventually in all relevant publications.[1] It is most usually considered with respect to taxonyms: in most of the Codes, a taxon must be denoted by only one valid (correct) name under any circumstances. This requirement goes back to the earliest rulebooks of the 17th–18th centuries; it is strictly followed within the main subject-area Codes, whereas it is applied rather loosely in the nomenclature of cultivated plants.

Certain provisions restrict the application of the principle of monosemy: according to them, some names are conditionally not considered homonyms or synonyms. These restrictions are generally regulated by the above-mentioned principle of admissible polysemy that applies in certain special cases stipulated by particular Codes. One of these restrictions is imposed by independence of Codes. Polysemy is not suppressed for the names of taxa higher than the family-group category: it is permissible to use for them both new automatically typified and traditional descriptive names in botanical nomenclature, whereas such names are not regulated at all in zoological nomenclature. In some groups of fungi, its application was previously presumed to the particular life cycle phases of the same species, which were denoted by different non-competing taxonyms. The suppression of monosemy is partly weakened for para-, morpho-, and ichnotaxa that are based on different parts of fossil organisms or on their traces: they are allowed to be denoted by different taxonyms even if there might be a reason to suppose their belonging to the same orthotaxon [Anonymous 1999, 2012; Greuter et al. 2011]. In the nomenclature of cultivated plants, polysemy is admissible for local varietal names [Brickell et al. 2009].

> In order to make taxonyms easy to recognize in texts and therefore more "certain" visually, those designating the genus-group and species-group taxa are traditionally italicized. It is suggested that this rule should be extended to the taxa of any rank [Thines et al. 2020].

Semantic motivation *vs.* neutrality. This pair of regulators affects the mechanism of substantive justification of certain aspects of nomenclatural activity. According to the *principle of semantic motivation*, this mechanism is associated with the substantive

---

[1] Taking into consideration the temporal dynamics of treatments of both taxa and their name, this restriction is applied to simultaneous cases.

content of taxonomic research aimed at cognizing "taxonomic reality." On the contrary, the *principle of semantic neutrality* means the absence of such an association. The first principle is implemented by the descriptive (especially essentialist) and rational-logical concepts of nomenclature, and the second one by its nominalist concept.

Semantic motivation of descriptive taxonyms (morphonyms, biononyms, toponyms, etc.) goes back to folk nomenclature and is strengthened by its essentialist concept. This motivation for taxonyms at the levels from species to family rank-groups is largely discarded in most of the Codes implementing a nominalist concept of nomenclature. However, it is preserved in the nomenclature of viruses, where the names of species and genera usually indicate the characteristics of their biology [Anonymous 2013]. It is recognized as an admissible option for naming higher-rank taxa (of order and higher) in botanical nomenclature from its early versions to the present [de Candolle 1813, 1867; Anonymous 1909, 2011, 2012]. It is also sometimes applied informally for zoological higher-rank taxa (orders and higher); these taxonyms are not officially regulated by the Zoological Code.

Semantic motivation is implicitly presumed also by the classificatory function of taxonyms. Indeed, as far as the latter are construed in order to provide a kind of description of the position of taxa in a classification, they are not purely formal but meaningful, i.e., substantively burden.

The names of fixed taxonomic ranks/categories (ranknyms) are semantically motivated in early traditional nomenclature: their names are usually based on those of military (centuria, legion), social (family, tribe), or state (kingdom, empire) units. This terminology remains as a "relic" in contemporary rank-dependent nomenclature regardless of its nominalist character.

Nomenclatural acts as such are most often semantically neutral with regard to the substantive context of formulating and solving nomenclatural tasks. However, there may also be exclusions: an example is the case of the semantically motivated restriction of an application of the principle of priority when choosing valid species names of wild animals and their domestic descendants [Gentry et al. 2004].

Finally, the nomenclature regulators themselves may be semantically motivated. The clearest example is provided by some provisions of the rank-dependent nomenclature, which were implicitly based on the recognition of different ontologies of taxa of different levels of generality [Pavlinov 2014, 2015a] (see Section 1.2).

3.3 LINGUISTIC PRINCIPLES

The principles encompassed by this block, as its name suggests, are set mainly by the general (largely lexical) norms of the organization and functioning of the language systems. Considered from a taxonomic perspective, these principles are basically theory-neutral and "intrinsic" in their regulation regardless of substantive taxonomic context. And yet, some are linked to certain provisions of taxonomic theory (e.g., to ontology-based ranked hierarchy) and therefore are indirectly theory-dependent in their application. With this, it is to be remembered that most of these principles may be considered substantiated by a linguistic theory in its general understanding, so they cannot be treated as theory-neutral in an "absolute" sense.

The principles of this group are quite numerous. Beginning with Linnaeus, they were given special importance by "linguistic purists," who demand strict observance of Latin linguistic standards in the formation of taxonyms [Linnaeus 1736, 1737a, 1751; Saint-Lager 1880, 1881, 1886; Clements 1902; Stearn 1985]. In contemporary nomenclature, their importance is not as great because of its nominalist nature, but certain rules of correct formation of Latin or Latinized taxonyms are provided in most of the contemporary Codes.

Verbalness *vs.* symbolness of *T*-designators considers two different approaches to their formation, verbal or symbolic; they are fixed by two respective principles.

The *principle of verbalness* presumes that the designation of *N*-object must be a lexical unit of natural or similarly construed language, i.e., the *name* in the general sense. Accordingly, such a designator is generally denoted as a *lexeme*, which can be either a single *word* (also word-like lexeme) or a verbose *phraseme*, i.e., both semiotically and semantically a solid phrase. The morphology of *T*-designators is regulated by the principle of wordness (see below). The general nomenclature concept implementing this principle is called (in tautology) *verbal*. In systematics, the latter is historically primary: it is inherent in empirical folk nomenclature, it passed from there into scientific nomenclature, and became fundamental in its traditional nomenclature systems, as well as in some non-traditional ones (such as phylonomenclature).

According to the *principle of symbolness*, any symbols (signs, numerals) and their combinations not related to natural (and lexically similar) languages can be used as a *T*-designator. The *symbolic* nomenclature based on this principle underlies the rational-logical concept, which is implemented by the "philosophical" language of science, numericlature, and some other nomenclature systems (see Section 7.1). Currently, such nomenclature is used to denote taxonomic *N*-objects identified on the basis of cyto- and molecular genetic data (haplotypes, phylotypes, etc.). This non-canonical nomenclature develops in parallel with the traditional one [Ratnasingham and Hebert 2007; Morard et al. 2016], it is figuratively referred to as *gray nomenclature* [Minelli 2017, 2019], and its area of application was called "dark matter" of systematics filled with "dark taxa" [Page 2016; Ryberg and Nilsson 2018].

Both principles can be applied in combination to produce *mixed* nomenclature that was quite common in the 17th–18th centuries. In one of its variants, especially significant ("definite," "natural," etc.) taxa are designated verbally according to certain linguistic standards, whereas less significant ones are designated symbolically and rather arbitrarily. Another option is to insert certain symbols in the verbal designations of taxa. Currently, the use of such mixed taxonyms is officially admissible in the nomenclature of cultivated plants and viruses [Brickell et al. 2009; Anonymous 2013], and this was also suggested in phylonomenclature to distinguish between species with different phylogenetic status [Graybeal 1995]. In the "hierarchical" nomenclature system of the Russian zoologist Nikita Kluge, a mixed naming of taxa is employed to indicate their levels of generality [Kluge 1999a,b, 2020]. Suggested treatment of the composite taxonyms as "nominal complexes" encompassing some information about their publication (authorship, date, etc.) [Lanham 1965; Dayrat et al. 2004; Dubois 2012] also makes them a variant of mixed nomenclature.

The *T*-designators, either verbal or symbolic, carry out effectively two main functions, designation and (to a lesser degree) description. The symbolic form is more suitable for performing the classification function, whereas the verbal form is more suitable for the mnemonic function. With regard to the communication function, verbal taxonyms are more effective in traditional media, while symbolic ones are more adapted to contemporary digital technologies.

It should be noted that a certain arbitrariness of contemporary "taxonomic Latin" (see below) gives any taxonyms a shade of symbolism. For any formal word-like *T*-designators to be used in traditional nomenclature, the only linguistic restriction today is their readability and pronounceability (e.g., the generic name *Tadarida* in zoology); this provision goes back to M. Adanson.

The **principle of wordness** regulates morphology of primarily verbal *T*-designators regarding the number of separate words in them. If extended also to non-verbal designators, it regulates the number of separate (non-connected) symbols in them. This principle operates in combination with other regulators, primarily with the above-considered cognitive principles of accentuation and verbalness/symbolness.

In folk nomenclature, the wordness is regulated implicitly by the above-considered principle of accentuation (emphasis). In scientific systematics, it is initially shaped by the *principle (rule) of binarity* derived from the classification genus–species scheme (see Section 4.2.2). The latter principle states that the designation of any notion, except for that at the highest level of generality, must consist of two parts corresponding to "generic common" and "species particular." From this point of view, it is the composite "genus–species" nominal complex that functions as a solid meaningful (genuine, legitimate) *T*-designator of a taxonomic *N*-object, with the "generic" part taking leading position and the "species" part being just its epithet with no independent meaning.

In essentialist nomenclature, the lexical structure of a genuine name, which corresponds in its meaning to the taxonomic diagnosis, is determined by the need to reflect the essence of organisms belonging to the respective taxon. The length of such a name-as-diagnosis depends contextually on a particular classification, viz., on the nature (essential characters) of the "genus," number of "species" in it, and their own essential differences. In general, the more "species" in the "genus" are recognized and, therefore, the more features are needed to distinguish them, the more words their genuine specific epithets should include, so the *T*-designator becomes a solid phraseme (phrase name).

If a "proximal genus" is not further divided in the given genus–species scheme, it becomes "final species," and its name becomes a specific epithet. Therefore, in scholastic systematics, if only one "species" is distinguished in the "proximal genus," the former is usually not assigned its own epithet [Rivinus 1690; Tournefort 1694; Linnaeus 1751]; formally, the length of such a specific epithet is "null-word." If a few clearly different forms ("subaltern genera" of the next lower step of division, "final species") are recognized in their containing "intermediate genus," then one-, two-, or three-word epithets are enough to denote them. In a more complexly structured "genus," when essential features of its "subaltern genera" or "species" cannot be expressed briefly, their *T*-designators may include more than a dozen words. If they refer to different parts of organisms, they are often separated by punctuation marks according to grammatical requirements. Such long names can be assigned to any "intermediate" and "proximal" genera, but the most verbose ones are usually characteristic of "final species."

With the development of nomenclature, the change in wordness proceeded toward reduction of taxonym length; this can be seen as a manifestation of the *principle of brevity* of names (which implements a more general principle of linguistic parsimony). This reduction involved first and most fully the names of "genera": in the 18th century, they became one-part and often one-word at all levels of hierarchical division. As a consequence, the principle of binarity appeared limited to the level of "final species," so the latter's complete taxonyms remained two-part genus–species binomens. With this, the specific epithets were characterized by noticeable verbosity in essentialist nomenclature, so Linnaeus was forced to limit their length to 12 words because "the shorter the specific name the better" [Linnaeus 1751: § 249, § 291]. They were such phraseme names that Linnaeus called "genuine specific" [Linnaeus 1751].

In nominalist nomenclature, the reduction of wordness spread also to the species level: multi-word genuine specific epithets were replaced by one-word trivial ones; the latter were often obtained by uniting several words into one. However, as far as the principle of binarity was preserved, the complete species taxonym remained two-part (binary). As a result, the principle of binarity turned into the *principle (rule) of binomiality* postulating that the complete species taxonym is a genus–species binomen consisting of two separate words, viz., generic name and specific epithet.[2] Thus, these principles (or rather rules) of binarity and binomiality that govern the structure of taxonyms should be distinguished: the second is a special case of the first, and their identity in botanical and bacteriological nomenclature is incorrect [Sharp 1873; Stejneger 1924; Pavlinov 2014, 2015a] (see Section 4.2.2).

> The principle (rule) of binomiality is usually attributed to C. Linnaeus, although its introduction was not part of his reform (see Section 5.3). It is given such great importance in contemporary traditional nomenclature that the latter is often called binomial (or binary, which is only partly correct). This obvious "relic" feature is caused by the fact that the main objective of traditional systematics, at the time of its initial formation, was set as the elaboration of classification of organisms "according to their genera and species" [Bauhin 1596]. Therefore, according to the essentialist provision, each organism allocated to a certain genus and species had to be designated by the respective genus–species binomen.

In contemporary Codes, the principle/rule of binomiality is a rather powerful nomenclature regulator with retroactive action. Accordingly, if binomiality is not observed

---

[2] The term "binomi*n*ality" is often used, but this is grammatically incorrect [Stejneger 1924; Pavlinov 2015a; Aubert 2016].

Basic Principles of Nomenclature

in a work, both the latter and the names in it are declared nomenclaturally "void."[3] The mandatory nature of this principle caused a number of specific nomenclature problems actively discussed in the second half of the 19th century, and their resolution has required special considerations and reservations.

One of them concerned the necessity of nomenclatural legitimation of intraspecific forms as taxonomic *N*-objects. The reason was that incorporating their characteristic epithets into complete *T*-designators made the latter multi-word contrary to the key condition of binomiality. To overcome this controversy, it was proposed to supplement the principle in question with the official recognition of *trinomial* and *quadrinomial* nomenclature systems [Coues et al. 1886; Banks and Caudell 1912]. By this, the *principles of trinomiality* and *quadrinomiality* were implicitly introduced. These proposals were not formally adopted, but it was ruled that complete infraspecies names do not contradict the principle of binomiality, since the latter (in its strict sense) applied to the species only. To emphasize this, some Codes prescribe that rank (status) of the respective infraspecific taxon should be indicated when its own epithet is provided (which adds another word to composite taxonym).

Another issue concerns designations of taxa of additional categories (subgenera, sections, etc.) included in the complete species names that also make them not strictly binomial. To comply with the condition of binomiality, such additional names are called *interpolated (intercalary)* and considered not to violate the principle in question. To accentuate this, a word denoting the respective additional taxon is provided in the complete composite species-group taxonym in parentheses with, if necessary, indication of its rank.

A similar discrepancy is observed with the verbose designation of hybrids or graft chimaeras by combinations of the names indicating species allocation of their parents. According to the adopted rule, such composite taxonyms are declared not to violate the principle of binomiality because they are assigned not to taxa proper, but to organismal *N*-objects. In botany, the latter may be both hybrid forms and particular organisms; their multi-word designators are called *collective names* or *hybrid formulas* [Hawksworth 2010]. In contemporary zoological nomenclature, only stable hybrid forms are considered in this capacity, and their names are formed following the standard rules adopted for intraspecific taxa.

Many opponents of the principle of binomiality (and binarity in general) rightly regard it as an anachronism and point out the main practical problem caused by it, viz., the potential instability of genus–species binomens. Indeed, with the change in generic allocation of species, the latter's complete name changes in two ways: first, the generic part of a binomen is always changed; second, its species epithet must also be changed in the case where it becomes a secondary junior homonym.

The alternative *principle (rule) of uninomiality* is proposed as a radical solution for these problems. According to its main provision, the complete names of species-group taxa should be strictly one-word (even if composite) and remain unchanged under any taxonomic context (i.e., they become potentially rigid designators). This

---

[3] Contrary to this general provision, the use of polynomial taxonyms was suggested as a "new principle of nomenclature" [Kluge 1999a, 2020].

proposal was made several times in the 18th and 19th centuries [Adanson 1763; Amyot 1848; Reynier 1893]. In contemporary systematics, this principle is defended by some adherents of phenetics and cladistics [Cain 1959; Michener 1964; Lanham 1965; Dayrat et al. 2004, 2008].

> Distinguishing between taxonyms of different wordness cannot always be clear-cut. Indeed, as indicated above, a rigidly constructed genus–species binomen can be considered semantically a single solid lexeme, in which generic and species parts have no independent meaning. Such a binomen can be reduced grammatically to a single composite word with its parts written with a hyphen [Adanson 1763] or solidly [Michener 1964]. However, contemporary Codes do not go deep into the semantics of taxonyms, so binomens are always treated as two-word lexemes. With this, they do not (unlike *Linnaean Canons*) limit the length of either part of the composite genus–species binomens, so they may consist of several dozen letters [List 2020]. However, it is still strongly recommended, following these *Canons*, not to make them too long [Dubois 2011c].

The **principle (rule) of Latinization** means that all scientific taxonyms are Latin or Latinized (have the form of Latin). This practice was optional in herbalistics, where the "official" plant names might also be Greek; it was subsequently adopted as a general rule in early scholastic systematics and finally legalized by the *Linnaean Canons*. It is one of the main working principles in contemporary Codes. Accordingly, the names in any other languages are suppressed for use as valid (correct) *T*-designators in professional systematics, and they are used only as auxiliary ones to establish a connection between scientific and other texts. To make such names available (legitimate) taxonyms, they undergo Latinization (romanization), i.e., transliteration, such that the letters (letter combinations) of a non-Latin language become Latin letters (letter combinations).

In addition to names, Latin is used to formulate taxonomic diagnoses. In scholastic systematics, this practice was overwhelming. In zoology, it was partially preserved by the mid 19th century and then abandoned. In botany, the Latin diagnosis remained mandatory in the original descriptions until recently; it is considered optional in the last versions of the Botanical Code [Anonymous 2012].

To this principle (rule) is added the *principle (rule) of classicality*, which requires strict compliance of linguistic rules of formation of taxonyms with classical Latin. It was considered among the basic rules by "linguistic purists" of the 18th–19th centuries. Numerous paragraphs in the *Linnaean Canons* were devoted to this issue in regulating the grammatical, spelling, etymological, etc., rules for taxonomic names. Linnaeus himself divided them into linguistically appropriate and inappropriate and demanded that the latter be unconditionally rejected [Linnaeus 1736, 1751]. As a development of this idea, it was later proposed that only grammatically appropriate names be considered with respect to their priority and usage as valid (correct) names [Saint-Lager 1880, 1881, 1886; Clements 1902].

Basic Principles of Nomenclature   63

As an alternative to the traditional Latinization, the so-called *New Biological Nomenclature* was suggested; this was based on the use of the universal language Esperanto [De Smet 1973, 1991a,b]. The latter was discussed by the Association for the Introduction of New Biological Nomenclature (established in 1971, ceased to exist by 2009), but the professional community of systematists did not consider it, since its implementation would entail replacing almost the entire nominative nomenclature.

With the prevailing of the nominalist concept of nomenclature, linguistic "purism" was supplanted by the *principle (rule) of arbitrariness*, according to which many classical linguistic norms are just recommended or completely canceled. With certain restrictions, such arbitrariness is acknowledged by all contemporary Codes. In an extreme form, the arbitrariness is implemented by the above-considered principle of symbolness.

At present, undoubted traces of classicality occur in the form of Latin grammar and spelling rules in some nomenclature Codes. They refer, for example, to the rules of sequence and agreement in the grammatical gender of parts of the genus–species binomen, the ways of forming the names of suprageneric taxa, etc.

3.4 JURIDICAL PRINCIPLES

The principles of the juridical block regulate nomenclatural activity on a kind of lawmaking basis that is formed as a result of certain agreements among members of certain taxonomic communities. In this respect, this basis is analogous to the legislative one [Balch 1909]; the fundamental difference is that Codes of Nomenclature do not have such a full-fledged status as the Codes of Laws provided by their legislated lawmaking and law enforcement [Jeffrey 1992; Pavlinov 2014, 2015a]. The former are regulated by public organizations, the latter by state and inter-state authorities.

The most general principles of juridical block, though being declarative, are of decisive importance in shaping the very grounds for the functioning of rationally constructed nomenclature systems (codification, supremacy of Codes, mandatory status of their provisions, etc.). Therefore, it is necessary to be aware of them to understand the basics of the overall structure and proper functioning of the whole of nomenclature. Some of them are expressly formulated in the Codes just in this capacity. With this, certain juridical principles act as working rules that apply directly both to nomenclatural acts (publications, authorization, etc.) and to *T*-designators.

The below consecution of conceptual consideration of principles of this block reflects mainly their significance as juridical regulators, starting with the most general and finishing with the most particular in this regard [Pavlinov 2014, 2015a]. However, this order should be different from a pragmatic perspective: the most important and therefore taking the first place are those that provide for the stability of nominative nomenclature, whereas others are sound to the degree they serve to reach this goal [Lewis 1871, 1872, 1875; Greene 1896; Rickett 1953; van Steenis 1964; Cronquist 1991].

Naturalness *vs.* conventionality. This pair of principles roughly corresponds to natural and positive legislature [Azarkin 2003]. The *principle of naturalness* was formed in the initial period of the emergence of scientific systematics under the influence of the biblical world picture. Accordingly, its basic idea was quite clearly expressed by J. Pitton de Tournefort, who believed that the "Creator of all things Himself has endowed us with the right and ability to name species" [Tournefort 1694: 3]; echoing him, S. Willdenow wrote that the naming of plants is based on "fixed rules drawn from nature herself" [Willdenow 1805: § 145]. In contrast, the *principle of conventionality* implies that nomenclature systems, especially in the form of conventionally established Codes, are formed as a result of agreements among members of certain taxonomic communities. Typical examples are the working principles/rules that are considered basic, viz., linguistic (binomiality, Latinization, etc.), juridical (the ways to ensure stability of names), and taxonomic (the ways to ensure certainty of taxa and their names).

Codification *vs.* precedent. This pair of principles roughly corresponds to the statutory and case-law forms of legislature [Azarkin 2003]. The *principle of codification* implies that each nomenclatural act is to be based on the principles and rules officially adopted as generally applied on a mandatory basis: they are to be referred to by nomenclaturists dealing with the relevant tasks. The *principle of precedence* implies that a certain solution of a specific nomenclatural task, once proposed *ad hoc*, may subsequently serve as an argument for similar solutions to a similar task; it is this precedent that is referred to by other nomenclaturists.

In the history of systematics, almost all nomenclature regulators, including the most significant ones, were initially formed and fixed in practice by precedents, and then they were codified on a conventional basis. An example is the standardization of taxonomic hierarchy, wordness and typification rules, linguistic rules for the formation of rank-dependent taxonyms, etc. This general scheme is implied by the *principle of usage*, as applied to nomenclature acts, according to which mass precedents may become subject to codification. A specific argument from English judicial case law was given in favor of the precedence acceptance, which reads, "A common error makes law" (*communis error facit jus*) [Lewis 1871, 1872, 1875]; this corresponds to an informal notion of *established custom* [Hawksworth 2010].

Taxonomic nomenclature was purposefully codified from the mid 19th century, when the development of Codes came under the jurisdiction of special collegial bodies, viz., commissions and committees at scientific societies. All current Codes by default are built on the basis of the principle of codification: they contain fixed standard rules for solving typical tasks, which are mandatory for the members of a taxonomic community that adopted the respective Code. With this, the Codes most often have a negative attitude toward the precedent regulation of taxonomic nomenclature; in some, direct reference to precedents is suppressed [Anonymous 1999; Cantino and de Queiroz 2010, 2020]. Therefore, particular decisions by international commissions or committees regarding specific cases are not considered as the basis for the same decisions in similar cases in the future (although probably their latent influence is still present).

The **principle of supremacy of Code** is one of the guiding principles in contemporary nomenclature: it states that the codified principles and rules for handling *N*-objects of all kinds (taxonomic, linguistic, acting, see Section 2.1.2), as they are summarized by officially adopted and recognized Codes, have absolute priority over personal opinions. In this regard, this principle is tantamount to the juridical principle of supremacy of law. This general provision holds true also for international collegial bodies (commissions, congresses, etc.): they can make specific decisions that contradict certain codified provisions, but the limits of their activities are also ruled by the respective Codes.

It is to be emphasized that this principle is strictly local in its applications. As just noted, for each Code, it is actual only for that taxonomic community, whose members adopted it and agreed to follow its provisions. This general limitation is formalized by the above-mentioned "informal" principle of locality. From this it follows that a similar local status holds for the working principles/rules presumed by particular Codes.

The basic conditions of the principle of supremacy are specified by the *principle of legitimacy*: this implies that for nomenclatural acts and their results to be nomenclaturally consistent (legitimate), they must strictly comply with the provisions of the respective Codes. It is supported by the general juridical norm, according to which ignorance of a law does not relieve one from responsibility for its violation. Therefore, if this compliance is not met (even if unintentionally), the nomenclatural act and its result are recognized as nomenclaturally "void."

This general principle is reinforced by several working principles, including those of direct action and automaticity, considered below.

Mandatory *vs*. recommendatory status. This pair of principles complements the principle of supremacy of Code by setting out specific modalities for the officially recognized nomenclature regulators. According to the *principle of mandatory status*, the relevant principles and rules must be strictly obeyed by members of the respective taxonomic communities, and their disobedience or abuse entails punishment in the form of suppression of validity (correctness) of the particular nomenclatural acts and its results. One of the conditions for ensuring the mandatory effect of the nomenclature principles is their codification and collegial authorization presumed by respective principles of the juridical block. According to the *principle of recommendatory status*, at least some regulators are not so rigid: their applications are just recommended "whenever feasible."

Obviously, this distinction is not absolute. First, every nomenclature regulator is mandatory only for the members of a taxonomic community that acknowledges the respective Code containing it, whereas it may not be as powerful in other communities. Second, the modality of regulators can change over time as the Codes evolve. In both cases, the principle of priority serves as an evident illustration: the requirement for its mandatory execution was formulated only in the mid 19th century, but followers of the principle of usage did not agree with this. Finally, the same regulator may be mandatory in one application and advisory in another; an example here is typification of taxa of different ranks in botany: it is mandatory at levels from family and below, but recommendatory at higher levels of the taxonomic hierarchy [Anonymous 2012].

With all these provisions in action, an unavoidable "subjective component" occurs in the solution of many nomenclatural tasks, so the applications of respective "mandatory" clauses in specific situations depend on the opinions of specialists. So they become in a sense "higher" than the Code, and the latter's provisions turn out to be, in general, not so much mandatory as recommendatory.

The dilemma of mandatory *vs.* recommendatory status of the Codes' provisions is currently a subject of discussion sparked by descriptions of new species-group taxa based on phototypes (photos of live organisms) and genetypes (molecular genetic data) that do not strictly comply with standard rules for typification (e.g., [Chakrabarty 2010; Strand and Sundberg 2011; Dubois et al. 2014; Hawksworth et al. 2016]; see Section 3.5).

This dilemma becomes especially acute in the case of so-called *taxonomic vandalism*, which refers to the recent mass non-qualified descriptions of new taxa [Borrell 2007; Moore et al. 2014; Páll-Gergely et al. 2020; Wüster et al. 2021]. The new names they introduce are formally available, but in fact these taxa cannot be considered fully scientifically substantiated; therefore their names actually just litter scientific nomenclature. In this regard, some professional zoologists protest against an official recognition of such names and consider it possible "setting aside certain provisions of the Code [as] an effective last resort defence against taxonomic vandalism and enhance the universality and stability of the scientific nomenclature" [Wüster et al. 2021: 1].

The **principle of direct action** means that working nomenclature regulators are to be applied directly for the execution of nomenclatural acts without any additional "by-laws." Evidently, it holds for working principles and rules, but not for general "declarative" ones. It is complemented by the *principle of automaticity* (*unambiguity*) presuming that, for such nomenclature regulators to be directly applied, they should be formulated so clearly as to avoid confusion and diverse admissible interpretations. It is evident that the principle of direct action is one of the important prerequisites for the Code supremacy to be consistently and strictly observed and its provisions to be mandatory.

However, there are many problematic situations when, for various reasons, the condition set by this principle cannot be met; therefore, an opposite *principle of indirect action* comes into force. It implies an introduction of secondary regulators (recommendations, clarifications, "opinions," etc.), delimiting or specifying conditions for applications of the primary one. The latter principle is accompanied by the *rule of interpretation*, according to which provisions of at least some working principles may be officially recognized to be subject to particular interpretations. However, the latter are considered an exclusive prerogative of approved authorities by all contemporary Codes, whereas personal interpretations are admissible by some earlier rulebooks (e.g., [Banks and Caudell 1912]).

Pro- *vs.* retroactivity. The application of nomenclature regulators has a two-way "time dimension." Most of them are *proactive*, i.e., facing the future and regulating possible prospective nomenclatural acts and novelties resulting from them. Others are *retroactive*; they are turned to the past and affect the assessment of the previous nomenclatural acts and their results. Accordingly, the nomenclature systems declare explicitly or at least imply the *principles of proactivity* and *retroactivity*.

The principle of retroactivity is declared as one of the basic in botanical and bacteriological Codes [Anonymous 1909, 2012; Lapage et al. 1992]. It is lacking in other Codes, but retroactive action of certain important principles is presumed in them as well [Mayr 1969]. The application of certain regulators to previous and future acts and names is considered separately in some Codes, so their temporal application is "asymmetric." For instance, to ensure the stability of nominative nomenclature, *Code of the British Association* recommended preserving old names violating some (not too significant) provisions, but rejecting such names if they were proposed in the future [Strickland 1837; Strickland et al. 1843a]. Among the retroactively acting principles, those of priority and binomiality are of special importance when considered conjointly. According to them, nomenclatural acts and names are not considered available (legitimate) if they were established previous to a certain date and/or did not meet the condition of binomiality.

It is to be emphasized that the retroactive action of these and some other important nomenclature regulators has one somewhat strange feature. In jurisprudence, the general principle "law has proactive but not retroactive effect" (*lex prospicit, non respicit*) is among the most strictly observed. It admits a possibility of applying the principle of retroactivity mainly to mitigate sanctions imposed by the laws in force earlier [Azarkin 2003]. Accordingly, a possibility of establishing or increasing sanctions for past actions that were not provided for by the legislation in force at the time of their commitment is to be approved *ad hoc* for each case by a special verdict of the competent legal body. In taxonomic nomenclature, however, the principle of retroactivity establishes *a posteriori* penalty in a standard (non-exceptional) manner for past "non-justiciable" acts in a very harsh form. Based on this norm, the Codes that were adopted from the mid 19th century declared invalid the whole of the early nominative nomenclature due to its inconsistency with certain requirements of these Codes (such as binomiality), although there had been no such regulatory norms at earlier times, and this is anti-ethical [Gray 1821; Lindley 1832]. As a result, many "non-Linnaean" names widely used in zoology and botany during the 18th and early 19th centuries were excluded from scientific circulation. Such retroactive decisions look especially mocking given that the nomenclature of Linnaeus, with reference to whom this norm was proclaimed, was not strictly binomial and he did not consistently follow the priority when selecting preferable taxonomic names (see Section 5.3).

The **freedom *vs.* non-freedom of taxonomic decisions** concerns the results of conducting certain classificatory tasks, as far as they affect resolving certain nomenclatural tasks.

The *principle of taxonomic freedom* is declared as one of the basics in the introductory sections (preambles) of most of both early and contemporary Codes. It asserts that the latter's regulatory functions extend only to the solution of nomenclatural tasks associated with the naming as such, but do not apply to the solutions of classificatory tasks including recognition and ranking of taxa. However, this declaration does not actually hold in the practice of nomenclatural activity: indeed, as emphasized above, the Codes regulate manipulations not only with the designations of taxonomic *N*-objects but also with these objects themselves (see Section 2.1.4). The latter means that certain decisions regarding recognition of these objects must comply with the relevant prescriptions of the Codes.[4]

It is to be noted that at least some such prescriptions are based on certain theory-dependent substantive considerations of "taxonomic reality." Therefore, to the extent that nomenclature concepts imply such considerations in some way or another, they presume that the taxonomic decisions entailing nomenclatural ones should agree with the respective background metaphysics (see Section 1.2). This agreement inevitably makes nomenclaturally relevant taxonomic decisions not only non-free but also theory-dependent. The latter is implicitly present, for example, in a prescription to recognize and name only those taxa that are considered "real" within the corresponding conceptual contexts implicit in the respective nomenclature concepts and the Codes implementing them. For example, they can be "definite genera" in *Linnaean Canons*, "natural groups" in *Cadolle Jr. Laws*, clades in *PhyloCode*, "non-hypothetical" groups in zoological nomenclature, etc. Another case of a hidden but quite severe limitation of the free space of classification decisions is the officially recognized system of fixed ranks in traditional Codes. This system cannot be changed, and the taxa are considered nomenclaturally consistent (legitimate) only if they are allocated to the officially recognized rank categories. This provision is expressly stated in the virological Code; infra-subspecies taxa are allowed in botanical, but forbidden in zoological Codes.

So, the freedom in question declared by the Codes is merely a non-realizable "good wish" rather than a working principle. This indicates an implicit and at the same time very significant effect of the alternative *principle of taxonomic non-freedom* with its rather diverse manifestations, of which those just mentioned are but few instances.

And yet, generally speaking, systematists are really "free" to do whatever they please with the taxa without coordinating their actions with the particular Codes. But a refusal to comply with their provisions that regulate quite rigidly classificatory tasks entails an equally rigid penalty. Indeed, following the above-considered principle of supremacy, the taxonomic communities, within which these Codes operate, are "free" to recognize the respective nomenclatural acts inconsistent, and the names established by them unavailable (illegitimate).

With this, a real taxonomic freedom is quite easy to achieve by an admission that any group of organisms that might be recognized and ranked by systematists on whatever conceptual and other possible basis deserves to be considered a taxonomic *N*-object by the very fact of its recognition. Such a group is to be treated nomenclaturally consistent to the extent that (a) it is distinguishable and (b) its

---

[4] It is to be emphasized especially that I mean not personal opinions or treatments about circumscription, rank, etc., of particular taxa but just the decisions that are of nomenclature concern, as far as they yield certain nomenclatural actions and pretend to be reconized by the respective taxonomic community.

name agrees with certain mandatory provisions. This seems to be the only real prerequisite for making not only taxonomic decisions actually free from any "extra" non-nomenclatorial limitations but also, by this, the whole nomenclatural activity theory-free. However, it remains just to wonder if systematic biologists would agree to take seriously any "artificially" defined *N*-objects that have the same nomenclatural status as those they proudly recognize as "scientifically sound." The relevance of this issue is illustrated by the so-called "gray nomenclature," which frequently involves preliminary (hypothetically) recognized taxa (Minelli 2017; Williams 2021).

The norms of **partial stability *vs.* lability** implement the above-considered pair of overall stability/lability as being addressed primarily to *T*-designators. The condition of partial stability/lability presumes elaboration of the working principles of direct action, while the declarative condition of overall stability/lability does not; this explains their consideration within different blocks.

The *principle of partial stability* means the immutability of nominative nomenclature, so it is sometimes declared (in other terms) as the most fundamental for taxonomic nomenclature [Strickland et al. 1843a; Lewis 1871, 1872, 1875; Greene 1896; Rickett 1953; van Steenis 1964; Cornelius 1987; Cronquist 1991]. It is presumed that the main objective is the stability of names (their *robustness* according to [Dubois 2005]), while the stability of regulators (principles and rules) is intended to ensure it [Lewis 1871, 1872, 1875]. The alternative *principle of partial lability* of nominative nomenclature was never considered as its regulator, but it permeates the whole of nomenclatural activity as a consequence of the inevitable dynamics of regulative nomenclature. This entails instability not only of the names of particular taxa but also of the terms denoting *N*-objects in different nomenclature systems.

The multiplicity of competing synonyms made it necessary to develop a certain method for implementing basic conditions of the principles of partial stability and monosemy. However, efforts to achieve this goal were faced with a serious problem of multiplicity of working principles and rules suggested for this by different nomenclature concepts and systems. Their discussion yielded one of the most serious collisions in the development of taxonomic nomenclature throughout the whole of the latter's history. Their different treatments became one of the causes of the "Great Schism" of taxonomic nomenclature in the 19th century (see Section 6.3.2).

According to the essentialist concept, the use of genuine names that best reflected the principal essences of organisms was implied as the main means of ensuring the stability of nominative nomenclature. However, different systematists could understand the organisms' essences in different ways and reflected them in different descriptive names. Therefore, as noted above, this approach did not lead to the desired goal of stability of names. The adoption of the nominalist concept made meaningless all discussions of "genuineness" and "appropriateness" of particular names; this greatly facilitated the acceptance of a single and therefore stable taxonym for each taxon as its permanent "label."

Within the nominalist concept, two alternative working principles were elaborated to ensure the stability and monosemy of taxonyms, which were based on the specific criteria of selection of homonyms/synonyms to be used as valid (correct) names of the respective taxa. According to the *principle (rule) of priority* (often called "law," *lex prioritatis*), such a name should be the earliest one by the date of its publication; some more recent Codes replace terminologically priority with *precedence*. Such a name is called *senior*, and other relevant names are called *junior*; they are not used in such a vein. According to the *principle (rule) of usage* (*lex plurimorium*), this should be the name most commonly used as valid (correct) within a certain recent time interval. With this, an important reservation is made in both rules: the name in question must be initially available (legitimate) by being published in accordance with certain rules (see below). Thus, both working principles are time-based: publication dates of the names are considered by the principle of priority, whereas certain periods of time during which they are in official circulation should be taken into account by the principle of usage.

The *date precedence* is estimated based on the publication date of the respective nomenclatural acts. The precedence can be *absolute*, if the names are compared with no time limits and the earliest one is selected; or *fixed*, if a specified date is set as the starting point, from which the precedence begins to be considered. In the case of absolute priority, the problem is the absence of any warranty that a name once selected as the "oldest" one might appear to be a junior synonym of another, far "older" one. In the case of fixed priority, one of the main "technical" problems is selection of the particular starting point for applying the principle in question, which the whole of the taxonomic community would agree upon. The main arguments in favor of absolute and against fixed priority are: (a) the arbitrariness of any fixed dates, so disagreements on them entail instability; and (b) the desirability of compliance with certain ethical standards with respect to predecessors. An extreme instance of this consideration is the most recent appeal to use indigenous species names as having undoubted superiority over any scientific ones because of being both the oldest and ethically preferable [Evans 2020; Gillman and Wright 2020]. The main counter-argument is impossibility in many cases to correlate unambiguously the old names with currently recognized taxa because: (a) the authors of early encyclopedias (such as Theophrastus or Dioscorides) most often understood genera and species in a greatly differently way to modern systematists; and (b) they usually poorly defined the taxa and their names. Regarding the suggestion to use indigenous names instead of already established scientific ones, it does not take into consideration the problem of multiple local synonymy of widespread species and thus will plunge the whole of the "new" indigenous-based nomenclature into chaos.

The principle of priority, in its contemporary dominating treatment, implies *limited priority* in all Codes. First of all, as just said, it is limited "from below" by fixing a certain starting date (fixed priority). In botanical nomenclature, the application of this principle is limited to each of the taxonomic categories; it is not mandatory for the names of taxa higher than family rank; the names of fossil morphotaxa compete only within the same morphotaxonomic group. In zoological nomenclature, the application of this principle is limited to each of the rank-groups and is additionally regulated by the principle of rank coordination. Other restrictions on the application of this principle are caused by the need to preserve certain commonly used names (see below).

In the first Codes of the 19th century, it was proposed to define the starting point for considering fixed priority by the publication dates of specified major works, in which the principle of binomiality was followed quite consistently for the first time. Later, it was also suggested that the validity of classifications in these works should be taken into account. The selection of particular monographs (with corresponding dates) meeting these two primary criteria became the subject of considerable controversy that led to the multiplication of particular Codes in the second half of the 19th century; therefore, different works and dates are fixed in contemporary subject-area Codes. In some earlier nomenclature systems, the priority of names was considered, taking into account the contexts of their establishment and/or application. According to *Kew Rule*, this context for specific epithets was set by their combination with the names of "true" genera [Bentham 1858, 1879; Hiern 1878; Jackson 1887]. According to Saint-Lager Reform, priority had to be considered only for names that satisfied the linguistic principle of classicality [Saint-Lager 1880, 1881, 1886]. In the *PhyloCode*, priority (precedence) of phylonyms is determined by the date of their proper phylogenetic definition [Cantino and de Queiroz 2010, 2020].

Besides the date, *position precedence* may be in effect. When two synonyms are published in the same work, the one preceding the page or even the line is taken for the senior one. This rule was widely applied in the late 19th and early 20th centuries, but lost its significance subsequently: the decision on the priority of such names is made by the first reviser (see below).

According to the *principle of usage*, as applied to taxonyms, the latter not mentioned in the main systematic monographs during a given time interval are to be qualified as forgotten (aborted) and henceforth not considered as potentially valid (correct). This principle is applied in "*strong*" or "*weak*" formulations. In the first case, certain rules are applied for: (a) fixing the time interval within which the frequency of usage of taxonyms is considered; and (b) calculating this frequency [Anonymous 1999]. In the second case, the principle is clothed in a recommendation to "follow tradition": it is thus formulated, for example, in the contemporary Botanical Code [Anonymous 2012]. Another version of this principle takes form of the rule "*once [junior] synonym—forever [junior] synonym*," according to which a taxonym once listed as a junior synonym should not be revived again as a senior one [Coues et al. 1886]. In the second half of the 19th century, several fixed "non-usage" intervals were proposed to consider taxonyms forgotten according to the "strong" formulation: 30, 50, and 100 years. They were adopted in several earlier Codes and are considered in the contemporary Zoological Code [Anonymous 1999].

The merits and shortages of each of these working principles/rules implementing the general principle of stability were actively discussed throughout the 19th century and later, with all debaters emphasizing the obvious conventionality of whichever choice. The potential instability of nominative nomenclature due to rejection of currently used names in favor of old and now forgotten ones was indicated as the main drawback of the *lex prioritatis* [Lewis 1871, 1872, 1875; Robinson 1895]. The main charge against the *lex plurimorium* was that it presumed rather vague criteria of usage: this led to a certain arbitrariness and subjectivity in the selection of valid (correct) names and hence also to a violation of the stability of nominative nomenclature [de Candolle 1867, 1883; Britton et al. 1892; Dubois 2010b]. The supporters of the principle of usage, emphasizing the pragmatics of taxonomic nomenclature, considered it more reasonable because it is based on "practice," while the principle of priority established "dictate of theory" [Robinson 1895]. At present, ensuring the stability of taxonyms of species-group taxa by softening the requirements of the principle of priority is sometimes substantiated with reference to the needs of nature conservation [Heywood 1991; Dubois 2010c].

To maintain a balance between two ways of ensuring the stability of taxonyms, the *principle (rule) of preservation* is applied to protect junior synonyms and homonyms from being rejected in certain dubious situations. It was adopted by botanists in the mid 19th century, and by zoologists and others in the mid 20th century. On this basis, Official Lists and Indexes of *conserved* and *protected* names are maintained, which are not subject to the principle of priority and other standard regulators for the selection of valid (correct) names. An objection was raised against this principle that it contradicted the universal principle of the supremacy of law [Anonymous 1907], but it is nevertheless recognized in almost all contemporary Codes.

The *principle (rule) of rejection* complements the principle (rule) of preservation; it is opposite to the latter in meaning, but pursues the same objective of ensuring the stability of the existing nominative nomenclature. Taxonyms, to which this principle is applied, receive the official status of *rejected* names and are deprived of the possibility of further use for the purposes of nomenclature. This principle can be applied not only to the names but also to the entire works by excluding their further use in solving nomenclatural tasks. Based on it, Official Lists and Indexes of rejected (invalid, etc.) names and works are maintained.

In many cases, the preservation and rejection of taxonyms and works require the use of official powers of collegial nomenclature bodies according to the principle of authorization (see below).

In most contemporary Codes, preference is given to fixed priority in combination with the principle of conservation. The exception is Virological Code, which is based on the principle of usage [Anonymous 2013]. Both options are situationally allowed in the nomenclature of cultivated plants [Brickell et al. 2009].

The **principle (rule) of publication** states that, for every nomenclatural act or taxonym to be recognized as available (legitimate), it must be published according to certain rather strict provisions expressly formulated in the Codes. This principle, although rather general, is also a working rule of direct action; its importance is emphasized by the proposal to designate it as the *principle of nomenclatural foundation* [Dubois 2011b, 2013].

Publication generally means the appearance of a properly formatted and issued text or graphic illustration containing a certain nomenclatural act, such as an original description of a new taxon (with new name), or a change in its position in classification with change in its name, or change in spelling or application of the previously established name. The text or illustration that meets the criteria adopted by respective Codes receives an official status of *publication* (*published work*). In contemporary botanical and phylogenetic nomenclature, publication with the original description of taxon is referred to as *protologue*. In botanical and partially bacteriological nomenclature, two kinds of publication are distinguished, *effective* and *valid*. In zoology, publication by its set of criteria corresponds to the valid one in botany.

Recognition of a publication as valid entails recognition of the nomenclatural consistency (availability, legitimacy) of the respective nomenclatural act it contains; otherwise, this act is treated as nomenclaturally inconsistent (unavailable, illegitimate). The most significant "substantive" criteria of publication include the use of Latin alphabet and binomial nomenclature, coupled with the correct N-definition of both a new taxon and its name.

The set of more "formal" criteria varied with the development of nomenclature, and they continue to change. Thus, in botany until the end of the 19th century, one of the admissible forms of publication was the distribution of the *exicates* (authorized herbarium sheets) with the hand-written names and characters of new taxa on them [de Candolle 1867; Engler et al. 1897]; it was cancelled subsequently. The most significant (and even the only) criterion for the publication to be recognized as valid retained until recently its appearance in some replicated and sold scientific editions (books, journals). Currently, the list of admissible publishing methods is expanded by some Codes to incorporate modern information technologies. For example, digital publishing via the Internet, subject to certain restrictive conditions, is now legalized in zoology and botany [Anonymous 2012; Editorial 2012]. Contrary to this "liberal" trend, calls are made to select more severely the platforms, with competent peer-reviewers, for publications of new taxonyms to avoid possible incorrectness with their establishing [Measey 2013; Dubois 2017b; Bílý et al. 2018; Krel 2020].

In addition to publication, two other mandatory conditions for recognizing the validity of nomenclatural acts and/or availability (legitimacy) of taxonyms are introduced in some Codes. One of them is the *registration* of taxonyms in the corresponding officially approved International Registrars [Cantino and de Queiroz 2010, 2020; Greuter et al. 2011; Editorial 2012; Anonymous 2013; Parte 2014], usually with aids of advanced digital technology [Patterson et al. 2010; Penev et al. 2016]. Another is the *establishment* of taxonyms following certain specific prescriptions [Cantino and de Queiroz 2010, 2020; Greuter et al. 2011]. These additional procedures are intended to guarantee that new taxonyms meet the main nomenclature criteria. Probably, the reason for their adoption was a significant number of new names in publications based on the molecular genetic data that do not fit the current nomenclature standards [Minelli 2017].

The **principle of authorization** concerns certain aspects of nomenclatural activity conducted by the latter's subjects; it has two main interpretations, general and particular. In the first case, it regulates the nomenclatural activity by the collegial authorities dealing with the interpretation of other nomenclature regulators and particular nomenclatural acts. In the second case, it regulates the nomenclatural activity of individual systematists concerning certain nomenclatural acts and T-designators.

This principle in its *general* meaning implies reference to those collegial subjects of nomenclatural activity who are responsible for the introduction and/or approval and/or rejection of certain nomenclatural novelties, from the adoption of new Codes or amendments to acting ones down to making particular *ad hoc* decisions concerning concrete nomenclatural acts, publications, and taxonyms. This reference ensures fulfillment of one of the main conditions imposed by the principle of codification, viz., an official confirmation or rejection of the validity and nomenclatural consistency of the respective novelties. Thus, the contemporary Codes presume that their new editions are to be approved by the relevant international collegial authorities, viz., Commissions, Committees, Congresses, etc. This authorization also extends to the Official Lists of conserved and rejected names and works, to the respective official Registrars, etc.

Authorization in its *particular* meaning implies the indication (citation, ascription) of the personal authors of particular nomenclature novelties, including description of new or changes in the treatment of previously established taxa and their taxonyms, designation of or change in the nomenclatural status of nominotypes, etc. The authorship indication (citation, ascription) confirms that the given author(s) introduced the given nomenclatural act in the given work, e.g., established a new taxonym in the given spelling and application. In botany, rules about referring to authors of taxonyms are rather strictly formalized [Brummitt and Powell 1992; Hawksworth 2002]. In order to strengthen the link between the taxonym and its authorship, it is sometimes suggested that they should be considered parts of the same composite *T*-designator [Lanham 1965; Dubois 2000, 2012; Dayrat et al. 2004]. With this, it is assumed that such authorization actually means a link not to the author proper, but to the work in which the corresponding nomenclatural novelty appeared; it is proposed, therefore, to consider not persons' names that may occasionally change, but their signatures associated with the respective works as the subjects of authorization [Dubois 2008a, 2012].

Authorization thus understood solves at least two important nomenclatural tasks. First, it provides a more precise "individuation" of a particular taxonym among possible synonyms and homonyms. The second task involves, in combination with the above-considered principle of designative certainty, a more accurate indication of the meaning that is embedded in the application of the given taxonym. The Russian zoologist Andrey Kubanin attached much importance to this provision and considered the rules of taxon N-definition with or without reference to its authorship as setting two different "nomenclature paradigms" [Kubanin 2001].

This principle in its particular understanding implies two kinds of authorship associated with two forms of *T*-designation. The *original* (or *primary*) authorship corresponds to the original designation and refers to the originally established spelling and meaning (application) of the *T*-designator. The *subsequent* (or *secondary*) authorship corresponds to the secondary designation and refers to the change in the *T*-designator, both itself (change in spelling, replacement with another one), its nomenclatural status (recognition of its being unavailable, junior synonym, etc.), and application (due to revision of the denoted taxon's circumscription and/or rank). In phylonomenclature, two forms of authorship of phylonyms correspond approximately to the above-indicated standard groups, *nominal* (who first published a name) and *definitional* (who first associated this name with a clade) [Cantino and de Queiroz 2020]. Similarly, a distinction is proposed between *nomenclatural* and *taxonomic* authorship [Dubois 2012]. Indication of only original (primary) or both kinds of authorship is mandatory in different Codes.

Certain nomenclatural acts that are subject to subsequent (secondary) authorship are specifically regulated by the *principle (rule) of first reviser*. The latter status is assigned to the systematist who is the first to validly introduce a certain nomenclatural act associated with change in status of a taxonomic N-object (taxon rank or circumscription, nominotype status, etc.) and/or in T-designator (spelling, application, etc.). According to this principle (rule), the first reviser is endowed with the authorship rights with regard to these changes, and the principle of priority applies to the respective actions.

> Citation of the names of authors of taxonyms has an unpleasant side effect: it encourages dishonest systematists to describe a large number of new taxa (mainly species) in order once again to put their name next to another taxonym in pursuit of "cheap glory." Such specialists are called "species manufacturers" [de Candolle and Cogniaux 1876], and their pursuit of rampant descriptions of new taxa is called "nomenclatural mihi itch" [Needham 1930; Evenhuis 2008b], which causes the above-mentioned "taxonomic vandalism." To eliminate this effect, it is proposed to abandon mandatory indication of the authors' names accompanying taxa names [Dubois 2008b], although this contradicts the main purpose of the principle of authorization.

The so-called *qualifying clause* in phylonomenclature has a meaning similar to secondary authorization in its particular meaning. It has to do with identifying and/or modifying the circumscription of a monophyletic taxon in deciding which taxonym is to be allocated to it [Cantino and de Queiroz 2010, 2020].

Subject inequality *vs.* equality means the recognition or non-recognition of the prerogative rights of the subjects of nomenclatural activity with respect to both the nomenclatural acts they perform and the taxonyms they establish or change. The respective *principles of subject inequality* and *equality* operate in conjunction with the principle of authorization. They are stipulated in some Codes as a means of ensuring the stability of nominative nomenclature.

When considering these subjects in general, their inequality is established by the fact that the nomenclature acts carried out by the officially recognized collective authorities (bodies) are of greater importance than those of informal collective or individual subjects. However, since the official recognition of such decision-making collective authorities is always limited to particular scientific communities, this regulatory norm has a local significance.

As to individual subjects, the prerogative right of the original author of a taxonym or the first reviser on its spelling, application, and selection (something like "copyright") is recognized, providing that some other, more significant principles and rules are not violated. On the other hand, the author of a taxonym is denied any privilege to change it without sufficient grounds (e.g., just because of its "impropriety" or "disagreeableness"). As a result, the author is in a sense alienated from the taxonym immediately after its publication, so he/she is equated in this sense to any other nomenclaturists. This condition was being discussed from the early 19th century; it was stated clearly in one of the Codes of the early 20th century: a "name once published cannot be retracted, even by the author. Nor does the author of a name, after the name is published, have any more privilege than any other person with that name" [Banks and Caudell 1912: 8]. In this regard, the question arises about the possibility or impossibility of retraction of a work with the description of a new taxon by its author because of an obvious error it contains [Dubois 2020a, 2020b; Vlachos 2020].

The **principle of depositing** presumes the preservation of the type specimens of the species-group taxa in certain officially recognized natural science repositories (museums, herbaria, microbial culture collections, etc.). This principle is regulated by all current Codes in one way or another, and it is closely linked to the principle of typification; in fact, they were put into effect simultaneously. Generally speaking, it has a recommendatory character, but at the same time it is considered one of the key conditions for ensuring the correctness of taxonomic descriptions [Dubois 2017a]. Recently, this principle began to spread to the preservation of voucher (reference) specimens that serve as a source of molecular genetic data for the description of new taxa. With reference to the principle of authorization, it is required that the unique identifiers of type and voucher specimens be indicated in the respective publications. A new accent in the consideration of this principle is currently brought by the tendency to minimize the impact of collecting activity on the decline in number of rare species of animals and plants in Nature [Loftin 1992; Norton et al. 1994; Remsen 1997; Collar 2000; Donegan 2008; Winker et al. 2010; Pavlinov 2016].

3.5 TAXONOMIC PRINCIPLES

The nomenclature regulators of this block deal with the N-definitions of both taxonomic and linguistic N-objects; accordingly, they are twofold. Some are mainly associated with the ranks, taxa, etc., by regulating them in a specific manner. These principles refer (explicitly or implicitly) to or are derived from those, according to which some important properties are ascribed to the classifications and their elements (e.g., ranked hierarchy, substantive status of taxa, etc.). Other principles regulate the ways in which ranknyms, taxonyms, etc., are established and assigned to the respective taxonomic N-objects; as indicated above, at least some of these principles were originally substantiated by different ontologies ascribed to taxa of various levels of generality (see Section 2.1). Since the structure of classifications and the evaluation of the "natural" status of taxa and their ranks are based on certain substantive concepts developed by taxonomic theories, all nomenclature principles of this group are taxonomically theory-dependent to a greater or lesser degree. Therefore, these principles actually deserve to be called "taxonomic" [Rasnitsyn 1992, 1996, 2002; Pavlinov 2014, 2015a].

Ranking *vs.* non-ranking taxonomic hierarchy. All traditional Codes developed by nomenclature systems based on ranked taxonomic hierarchy regulate the rank structure of classifications in sufficient detail. Therefore, the *principle of ranking taxonomic hierarchy* plays an important role in them. It implies codification of the rank structure, i.e., fixation of the ranks as specific taxonomic N-objects by their specific N-definitions. The latter include a direct indication of the ranks recognized as "legitimate" in particular nomenclature systems, their subordination, and terminological denotations.

In zoology, the ranked structure is additionally regulated by the *principle of rank coordination* (= principle of coordination) The latter presumes that: (a) the basic (main, primary, key) ranks and the additional (subsidiary, secondary, auxiliary) ranks associated with them are united into the same *rank-groups* (there are three of them);[5] and (b) the taxonyms denoting subordinate taxa within the same rank-group are equivalent with respect to their nomenclatural status. Such a coordination is provided for by the automatic typification that links respective taxa by defining them with reference to the same nominotype. Both the ranks within each rank-group and the names of the respective taxa are called *coordinate*, and the change in rank of a taxon within the same rank-group does not mean the establishment of a new taxon. In addition, some Codes recognize *interpolated (intercalary)* ranks; the same term is used to denote the names assigned the respective taxa; their inclusion in the ranked hierarchy is subject to specific rules. Botanical and bacteriological nomenclature lacks such direct rank coordination, but it is presumed implicitly by the rule connecting the original (*basionym*) and derivative (*neonym*) taxonyms in the case of change of category allocation of the respective taxon.

One of the main manifestations of this principle is that, for a taxon to be acknowledged as validly (correctly) established in the original description, it is to be allocated to a certain rank category recognized by the respective Code. Accordingly, only taxonyms assigned to such "legitimate" taxa are considered available (legitimate); it was suggested that this restriction should be formalized as a general rule [Moore 2001]. For example, zoological nomenclature does not recognize infrasubspecific categories, so the taxa allocated to them and their taxonyms are treated as nomenclaturally inconsistent (unavailable, illegitimate). If such a taxon is subsequently allocated to a recognized category, its revalidation is required, which in fact means establishing a new taxon with a new name.

The alternative *principle of non-ranking taxonomic hierarchy* is adopted in nomenclature systems based on the rankless taxonomic hierarchy. It is inherent to scholastic systematics with its genus–species scheme, and it is embodied by *PhyloCode*.

Discreteness *vs*. non-discreteness. The general norm underlying this pair is also purely classificatory, as it is an extension of the same taxonomic norm regulating certain properties of classification units and ranks [Pavlinov 2018, 2021]. It is explicitly or implicitly present in all Codes, with some containing a direct indication that both the taxa and their ranks should be discrete. This requirement is formalized by the *principle of taxonomic discreteness*. Accordingly, the alternative *principle of taxonomic non-discreteness* presumes the possibility of the taxa being partially (probabilistically) overlapped and allocated to several categories. All these options are taken into consideration by extensional *N*-definition of taxa (see below).

In nomenclature, the principle of discreteness serves as one of the additions to the principle of taxonomic certainty in combination with the principle of monosemy. The

---

[5] Up to five *nominal-series* corresponding to the rank-groups are distinguished in the *Linz ZooCode Proposal* [Dubois 2000, 2005, 2006b; Dubois et al. 2019].

non-overlapping of taxa and their categories provides a more precise circumscription of each of the former, which serves as a prerequisite for the more rigid references of their taxonyms. This principle operates almost rigorously in most of the Codes, save for nomenclature of cultivated plants. The latter stipulates the possibility of cultivars belonging simultaneously to different cultivar-groups depending on the characters they are classified by [Brickell et al. 2009].

At the generic level, the principle of discreteness is manifested in the fact that the complete specific taxonym contains only one generic name; in the case of overlapping genera, their names may be included in the complete specific name [Tobias 1969]. At the species level, at least two options can be considered partial violations of this principle. One is the standard indication of several parental species in the names of hybrid forms. Another corresponds to cases when an unambiguous allocation of the specimen to certain taxa is impossible, so open nomenclature is used [Sigovini et al. 2016].

Rank dependence *vs*. independence of nomenclature is an attribute of either a ranked or rankless hierarchical arrangement of classifications, respectively. As emphasized earlier, both recognition or denial of ranks and rank dependence of nomenclature were substantiated initially by certain metaphysical considerations and therefore are theory-dependent to a greater or lesser degree. The contemporary rank-dependent nomenclature does not refer directly to any metaphysical background when substantiating the ranked taxonomic hierarchy, so the rank dependence of some of its regulators is retained in it as a kind of "relic." With this, the contemporary rank-independent nomenclature is explicitly justified metaphysically (ontologically) by reference to the nominal (subjective) character of fixed ranks.

The *principle of rank dependence of nomenclature* is a consequence of the principle of ranking taxonomic hierarchy, but it is not part of it: indeed, even in the case of clearly fixed ranks, the methods of designation of taxa may not be obligatorily associated with them. This principle is a part of traditional nomenclature; its assessment differs from its recognition as almost the main achievement of Linnaean reform [Ride 1988; Schuh 2003] to a very negative one [Griffiths 1976; de Queiroz and Gauthier 1990, 1992, 1994; de Queiroz 1997; Ereshefsky 1997, 2001a,b, 2002, 2007b]. It is suggested that the nomenclature system covering the entire ranked taxonomic hierarchy by respective regulators should be called *duplostensional* [Dubois 2015; Aescht 2018].

Traditional nomenclature is largely rank-dependent (ranked), which means that the application of many important principles and rules of N-definitions of taxa depends on the latter's ranks. Thus, the principle of priority applies only to the names of taxa of the same rank or rank-group, which also holds true for the suppression of homonymy in zoology. The suppression of the genus–species tautonymy in botany is also partly rank-dependent. The principle of typification is rank-dependent in the sense that the choice of nominotype (taxon or specimen) depends on the rank of a typified taxon.

This principle manifests itself most clearly in the *principle (rule) of rank specificity* (the *principle of homogeneity* of [Dubois 2005]), according to which more or less specific rules of N-definitions of taxa and their T-designators are applied at different levels of ranked hierarchy. Some rules are directly rank-dependent (e.g., the ways of taxonym formation), whereas for others this principle specifies certain conditions of application (such as monosemy, priority, typification). Such rank dependence may be considered a nomenclatural explication of the taxonomic *principle of rank equivalence* considered by some taxonomic theories [Pavlinov 2018, 2021]. To the extent that rank equivalence is not observed, the nomenclature system is called *pseudoranked* [Dubois 2007, 2008c; Dubois et al. 2019].

> Discussing rank-dependent nomenclature, A. Dubois strongly distinguishes in a specific manner between notions of rank and category, which are considered by him nomenclatural and classificatory (taxonomic), respectively, and he unites them by the notion of *taxonominal hierarchy* [Dubois 2007]. This means, according to the terms adopted herewith, that the former notion belongs to the sphere of nomenclatural tasks, while the latter, to the sphere of the classificatory tasks. But such distinction seems to be erroneous, for in fact, categories do not exist without ranks; their taxonomic treatment deals with their recognition, while their nomenclatural treatment deals with their designation [Pavlinov 2015a].

The rank dependence of taxonyms allows them to perform a certain classification function by indirectly indicating the position of taxa in a hierarchically arranged taxonomic system with specific lexical means; it was proposed that the ranked hierarchy endowed with such a property should be called *flagged* [Stevens 2002]. This dependence performs two basic functions: it is (a) *taxon-indicative* in marking the allocation of a named taxon to a certain containing taxon; and/or (b) *rank-indicative* in marking the allocation of a named taxon to a certain taxonomic category.

The rules for the formation of the rank-dependent names of taxa, to make them taxon-indicative, are themselves rank-dependent, and are as follows. The names of the higher rank-group taxa (from family and above) are formed by automatic typification (see below), which establishes a direct link between the nominotype and the inclusive taxon being typified. The complete names of the species-group taxa (species, subspecies, race, etc.), according to the principle of binarity, consist of two parts: one is the species/subspecies/race epithet and another is the name of containing genus/species.

The formation of rank-indicative taxonyms is regulated by the following rules. The names of the taxa of ranks from the genus-group and higher are one-word, while complete names of the species-group taxa are bi- and multi-word. With this, generic taxonyms are written in the singular, those of suprageneric taxa are written in the plural, and specific epithets are usually adjectives. The names of suprageneric taxa include rank-specific marker suffixes to differentiate those that are based on the same basionym; such suffixes may also be group-specific [Moore 1974; Greuter et al. 1996, 1998; Hawksworth 2010]; it is proposed that the taxonyms with the same rank-specific suffix are called *unified* [Kluge 2020]. This rule is observed for the taxa of the rank-groups of family in all traditional Codes, for the orders in botany, bacteriology, virology, and in bionomenclature, and for the classes in botanical and bionomenclature. Zoological nomenclature does not regulate officially the names of taxa higher than the family rank-group, though such regulation was applied to several grand taxa [Gadow 1893; Poche 1911; Berg 1932; Stenzel 1950] and it was suggested it should be considered as a general rule [Rodendorf 1977; Rasnitsyn 1982; Starobogatov 1991; Duboi 2000, 2005, 2006b, 2011b, 2015; Dubois et al. 2019].

The alternative *principle of rank independence of nomenclature* is developed based on the idea of rankless classifications. It means independence of any nomenclatural acts and their results from the hierarchical structure of classifications. Currently, this principle is officially declared in phylonomenclature; numericlature can also be considered rankless, as its digital coding of hierarchy does not imply recognition of fixed ranks. The "hierarchical" nomenclature system of N. Kluge does not presume fixed ranks, but special numerical markings are included in the taxonyms to indicate the particular levels of hierarchical structure of classifications, to which respective taxa are allocated [Kluge 1999a,b, 2020].

According to the **principle (rule) of group specificity**, certain group-specific morphemes are included in the designations of taxa to show their position in the global taxonomic system. This principle, by providing for the classification function of taxonyms, is an important part of some versions of rational-logical nomenclature, including numericlature; it is partly applied in contemporary virological, botanical, and phylogenetic nomenclature systems, as well as in *Draft BioCode*. This principle involves all supraspecific taxa in virology and the highest-rank taxa only in other Codes; group-specific markers are above-mentioned suffixes in the traditional Codes or prefixes in phylonomenclature. Taxonyms of the botanical hybrid forms (nothotaxa) are group-specific in indicating the latter's parent forms. The specific pre- and suffixes in the designators of oolithotaxa (fossil egg shells) and ichnotaxa of fossil traces [Vialov 1972; Simpson 1975; Mikhailov et al. 1996; Rindsberg 2018] also agree with this principle. Accordingly, the alternative *principle (rule) of group non-specificity* is relevant to all cases, in which group specificity of taxonyms is not observed.

The general **principle of nomenclatural verity** is based on a specific interpretation of the general concept of *verity* (truth) as applied to N-objects. It provides a stricter understanding of the latter's verity as the compliance of their N-definitions with certain conditions presumed by respective nomenclature systems. These objects can be either taxonomic or linguistic (acting are not considered here), so this principle is twofold in its particular interpretations and applications, taxonomic and nominative. One of the basic functions of this principle is that it specifies certain conditions for the application of some other important nomenclature regulators (principles of designative certainty, suppression of polysemy, etc.).

These two basic treatments of verity in nomenclature may be defined as follows: *taxonomic verity* refers to taxonomic N-objects (mainly taxa), while *nominative verity* refers to the linguistic N-objects (mainly taxonyms). Both were initially elaborated within the essentialist nomenclature; in particular, they took the form of definiteness of taxa and genuineness (legitimateness) of their names in *Linnaean Canons* (see Section 5.3). They were inherited from this concept, with certain modifications, by the nominalist one. For each of these treatments, particular principles (rules) are developed with respective *criteria of verity*, which are of two main kinds: the substantive ones are based on the ontological interpretation of taxa and/or their names, whereas the formal ones refer to nomenclature provisions as such. To the extent that the ontology issues are considered by taxonomic theories (see Section 1.1), this makes both substantive principles and criteria and the nomenclature systems elaborating them theory-dependent.

The *principle of taxonomic verity*, in a rather general form, asserts that only *"true"* taxa, however understood, deserve to be recognized and *N*-defined, including their naming. Accordingly, the taxa not fitting certain criteria of "trueness" are implicitly considered "untrue," and hence nomenclaturally inconsistent. The most significant substantive understanding of taxonomic verity is characteristic of systematics of a realistic kind: a "true" taxon is supposed to exist in "reality" and not just in the minds of systematists. By this, the principle under consideration is evidently theory-dependent: it is the taxonomic theory that provides both particular understandings of "taxonomic reality" itself and particular criteria of "reality" (verity) of taxa as its elements [Pavlinov 2018, 2021]. Therefore, nomenclature concepts derived on the basis of substantively different taxonomic theories imply different understanding and criteria of taxonomic verity.

Thus, *Linnaean Canons* presumed an essentialist understanding of the "definite genera," whereas the nomenclature systems of the 19th century referred to "natural groups" as consistent *N*-objects (e.g., [de Candolle 1867; Dall 1877]). In *Kew Rule*, the concept of verity is applied explicitly: it says that a species unit makes sense only when it is associated with the "true genus" [Bentham 1858, 1879]. In phylonomenclature, "true" taxa are defined as monophyla to be recognized and named [de Queiroz 2007; Cantino and Queiroz 2010, 2020]. The suppression of "hypothetical" (presumed, "unreal") taxa in favor of "real" ones in Zoological Code [Anonymous 1999] seems to also follows from this provision.[6]

In rank-dependent nomenclature systems, the taxon's verity depends also on its position in the taxonomic hierarchy: only those taxa are considered "true" and deserving of naming that are allocated to officially recognized categories. Thus, in some nomenclature systems of the 19th century, it was proposed that taxa of the main ranks were designated by standard names, while those of subsidiary ranks may be designated by arbitrary symbols (e.g., [de Candolle 1813]). In contemporary nomenclature systems, the taxa of nearly all secondary ranks (except for subfamily) are not acknowledged in virology [Anonymous 2013], and the infra-subspecific taxa are prohibited in zoology [Anonymous 1999].

A more formal understanding of the taxon's verity is tied to its operational *N*-definition as a taxonomic *N*-object. This means that a taxon is considered nomenclaturally consistent ("true") and deserves to be named if it is established according to certain codified working principles/rules. The latter are formulated by the principle of certainty (see below), which implies that, if the respective clauses are not observed (for example, if characters or nominotypes are not indicated), such

---

[6] It is curious enough that Zoological Code, with this, acknowledges ichnotaxa based on traces of the "hypothetical" (presumed) extinct organisms [Anonymous 1999].

taxon is nomenclaturally inconsistent ("untrue") and therefore its name is unavailable (illegitimate).

The *principle of nominative verity* concerns *T*-designators (mainly taxonyms); it has two meanings. According to one of them, a taxonym is "genuine" (available, legitimate) if it is applied to a correctly *N*-defined "true" taxon, so this treatment directly refers to the taxonomic verity: one does not work without another. Thus, a taxonym is *empty* (*nomen nudum* or *vanum*) if it refers to a taxon that is not properly *N*-defined by the relevant specifiers; and it is *dubious* (*nomen dubium*) if it does not refer explicitly to any "real" taxon at all or refers to an "imaginary" taxon. According to the second treatment, the taxonym's verity is considered regardless of the evaluation of the taxon's verity. It is the second version that sets the main content of principle under consideration: only properly *N*-defined taxonyms can be considered "genuine" (available, legitimate) and therefore can be used as valid (correct) to denote taxa.

In the case of the substantive treatment of this principle, the taxonym's verity is linked to its ability to reflect properly the essence of organisms belonging to the named taxon. The specific name that met this criterion was called *genuine* (legitimate) by Linnaeus. This criterion was applied to trivial names in the majority of nominalist Codes of the 19th century, where such "true" names were termed *appropriate* [de Candolle 1813; Strickland et al. 1843a]. With the increase of influence of the nominalist accent in nomenclature, this criterion of taxonym's verity became of secondary significance. However, it was retained for the names of higher-rank taxa in botany [Sprague 1921, 1928]: e.g., it was stated that "The names of divisions and subdivisions, classes and sub-classes are taken from one of the main characters" [Briquet 1912: 15]; however, at present this provision is not followed so literally.

According to the nominalist interpretation of this principle, taxonym's verity is based on its predominantly pragmatic, rather formal interpretation. In this regard, the principle of nominative verity turns into the working *rule of availability* (*legitimacy*) *of taxonyms*. Following it, for the names of taxa to be nomenclaturally consistent (available, legitimate), they should be established according to certain formal criteria; it was proposed that such names should be called *hoplonyms* [Dubois 2000, 2012, 2015; Dubois et al. 2019]. Three major kinds of such criteria are developed to make taxonyms available (legitimate). Some are predominantly linguistic: they request the taxonyms to be Latin or Latinized, species names must be binomial, etc. The provision of strict compliance of taxonyms with the norms of classical Latin was among the most fundamental in *Linnaean Canons* and stimulated Saint-Lager Reform in the second part of the 19th century. Several criteria are borrowed from the juridical principles, and are based on the principle (rule) of publication. Finally, the availability of taxonyms depends on how correctly the taxa to be named are established based on the principle of taxonomic certainty (considered below).

According to the provisions of the principle of nominative verity, taxonyms that do not meet the above-mentioned and other relevant criteria are declared unavailable (invalidly published) and are removed from scientific circulation as "untrue." In particular, they are not considered with respect to homonymy/synonymy and cannot be used as valid (correct) names of taxa. With this, not only particular names but also entire works, in which such criteria are not observed, may be considered nomenclaturally inconsistent. Many classical works of the 17th–18th centuries were rejected for this reason after adoption of the retroactive principles of binomiality and Latinization in the 19th century. Currently, it is sometimes suggested that this practice is to be followed in order to de-validate certain recent "doubtful" systematic publications and complete editions [Kaiser et al. 2013; Measey 2013; Rhodin et al. 2015; Bílý et al. 2018; Troncoso-Palacios et al. 2019; Hołyński 2020; Krell 2020].

The **principle of taxonomic certainty** (definiteness) represents a particular interpretation of the fundamental cognitive principle of designative certainty as applied to the taxonomic *N*-object. It requires a strict operational *N*-definition of each taxonomic *N*-object as the main means of its fixation (individuation) in the conceptual space of systematics. So this principle is one of the most important in taxonomic nomenclature, and its possible alternative is hardly meaningful in practical application; however, probabilistic interpretation of the objects in question makes provisions of this principle not so rigid.

In the case of the type specimen as a specific *N*-object, its certainty is provided by a direct indication of its nomenclatural status (holotype, type species, etc.) in combination with the name of the typified species-group taxon.

The condition of the rank/category certainty (definiteness) is a characteristic feature of all Codes implementing the rank-dependent nomenclature concept. It is provided by a direct indication of the position of each particular rank in the fixed-rank hierarchy and denoted by a standard ranknym. The latter ensures nomenclatural comparability of the identically denoted ranks recognized in different taxonomic systems, regardless of the fundamental issue of their comparability considered metaphysically (e.g., see [Simpson 1961; Van Valen 1973; Holman 2007; Pavlinov 2018, 2021] on the latter). Rank certainty serves as a basis for the ranking of taxa by their allocation to certain rank categories, which is part of their circumscriptive *N*-definition (see below). According to the *rule of rank-category homonymy*, a ranked taxon is ascribed the same ranknym that denotes the rank/category proper. An indirect indication of the category allocation of taxa is provided additionally by their rank-specific names.

The methods for ensuring taxon certainty are regulated in most detail in all sufficiently developed Codes, which suppress any uncertainly established taxa. Their certainty is provided by their operational *N*-definitions by means of certain *specifiers*, viz., circumscription, diagnosing, and typification (see below). These specifiers correspond to different "logical" definitions of taxa as *N*-objects presumed by their particular substantive interpretations. In addition, some more formal requirements must be met, e.g., those presumed by the principle of publication. A taxon, whose certainty is provided by correct *N*-definition by applying the rules and criteria supposed by the respective Codes, is considered nomenclaturally consistent.

The theoretical burden of particular versions of the principle of taxonomic certainty implying different specifiers may be considered in different ways. According to one point of view, the diagnosing is "conceptual" by implicit referral to the essentialist interpretation of taxa, while the type is theory-neutral, being simply the "namebearer" [Ogilby 1838; Rasnitsyn 2002]. According to the opposite point of view, the fixation of nomenclatural type implies a certain typological idea, while indication of diagnostic characters is not associated with any substantive theory [Cook 1900; Dubois 2008a].

The above-considered principle of authorization in its particular sense can be considered an auxiliary means of implementing the requirements of the principle of taxonomic certainty. By this, the indication of primary and/or secondary authorship, including reference to the work with respective nomenclatural act, though not being a specifier proper, makes it possible to indicate more precisely the taxonym reference (application). This is emphasized by the introduction of the already mentioned notion of nominal complex that includes both the name proper and its authorship [Dubois 2000, 2005, 2012; Dubois et al. 2019].

The three main operational versions of the principle of taxonomic certainty based on different basic specifiers, as applied to the taxa, are considered below in more detail.

The *principle (rule) of circumscription* presumes indication of composition and/or boundaries in N-definition of a taxon; it corresponds to the latter's extensional definition. It was proposed that the nomenclature system based on this principle should be called *circumscription-based* (extensional), or *circumscriptive*. In this case, the most important specifier of a taxon is its circumscription (composition) provided by the list of its subtaxa (e.g., the genera in a family). An additional clause of taxon circumscription includes an indication of its rank in rank-dependent nomenclature systems. For a taxon characterized by temporal and/or spatial boundaries, their indication can also be considered its auxiliary circumscription. In addition, circumscription may be understood as establishing the limits of variation of the characters of a taxon [Rickett 1959]; thus treated, this principle partially overlaps with the principle of diagnosing.

The "*list*" circumscription in its standard interpretation is based on an understanding of a taxon primarily as a set of particular units (subtaxa, organisms); it may be *internal* (basic) and *external* (auxiliary). Internal circumscription presumes indication of those classification units that are included in the taxon being defined, while the units not included in it are indicated by its external circumscription. The latter is potentially inexhaustible and therefore usually not operational; but it is useful in some cases to specify the subtaxa (organisms) that certainly do not belong to the taxon being defined. These two modes operate independently in most cases, but inclusion of the reference phylogeny in the clade N-definition in phylonomenclature provides them simultaneously.

If the principle (rule) under consideration is taken as a sole (or at least basic) element of the N-definition, the change in taxon circumscription yields allocation of a new valid (correct) name to it. The circumscription was given great importance in the 19th century; in particular, reference to this norm was decisive in justifying the adoption of the first edition of Linnaeus' *Species plantarum* as a starting point for generic nomenclature in botany. There were objections to the fact that plant genera were not diagnosed in *Species plantarum* [Müller 1884; Kuntze 1900], but this argument was dismissed on the grounds that a taxon is defined primarily by its circumscription and not by characters [Anonymous 1909].

The "list" circumscription is basic in *PhyloCode*: a clade is defined theoretically as including the ancestor and all its descendants, and operationally by indicating its membership and position in the reference phylogeny. Accordingly, the clade may receive another name if its circumscription is changed significantly (though not specified to what extent) [Cantino and de Queiroz 2010, 2020]. Among the other contemporary Codes, the "list" circumscription is held for the botanical taxa of higher categories. In all other instances, circumscription is replaced by typification: accordingly, any changes in taxon circumscription do not entail alteration of its valid (correct) name, if its nominotype is preserved.

The "*rank*" circumscription means that a taxon is considered correctly N-defined and its name is available (legitimate) if it is allocated to a certain codified rank. Accordingly, it is presumed by contemporary traditional Codes that the rank allocation must be expressly indicated in the original description of a new taxon. In botanical nomenclature, the change in rank allocation is considered as establishing a new taxon with its name being changed according to rank-specific rules.

The principle of circumscription, combined with the principle of monosemy, is manifested in the practice of compiling the lists of junior synonyms. Such a list provides a clearer delineation of the given taxon recognized in the given classification: synonyms indicate previously recognized groups that are included in the respective taxon in this classification. This practice goes back to herbalists; in *Linnaean Canons*, a special chapter is devoted to the rules of compilation of synonymy lists. With an expanded interpretation of the latter, they also comprise *chresonyms* [Smith and Smith 1972], i.e., the names (combinations of names) under which the given taxon appears in different classifications.

The *principle (rule) of diagnosing* corresponds to the intensional definition of a taxon: it means the need to indicate the latter's diagnostic features in its N-definitions to make it distinguishable from the related taxa of the same rank (e.g., genera of the same family). The corresponding nomenclature system can be designated *diagnosis-based* (intensional). The intrinsic properties of organisms (traits, ontogenetic patterns, some ecological features, etc.) are the most usual elements of the differential diagnoses as the intensional specifiers. Certain characteristics of supraorganism aggregates are sometimes used in this capacity; this is exemplified by the diagnosis of one of the subgenera in the zoological genus *Gazella* by significant sexual dimorphism in the formation of horns [Ellerman and Morrison-Scott 1951].

The indication of the diagnostic characters of a taxon in its original description has been considered an important condition of its nomenclatural consistency since the very beginning of systematics. Its origin can be found in essentialist nomenclature, in which a taxonym ("genuine" name) coincides with the taxon diagnosis, so the changes in characters automatically entail a respective change in the name. To the extent that the diagnostic features of taxa are context-dependent,[7] this may occur, for instance, if a species is transferred from one genus to another or due to the change in species composition of the inclusive genus. Notwithstanding that the contemporary nominalist nomenclature is free of essentialist interpretation of taxa, indicating characters in the taxon N-definition is considered necessary to make it

---

[7] This was expressed, for instance, by Linnaeus' assertion that "the genus makes the character" [Linnaeus 1751: § 169].

nomenclaturally consistent and its taxonym available (legitimate) in all traditional Codes. However, they disassociate the name and diagnosis of taxon, therefore, any change in the latter's diagnostic characters (all things being equal) does not entail the establishment of a new name for it. Accordingly, different names allocated to a taxon due to changes in its diagnosis are considered objective (homotypical, nomenclatural) synonyms.

The main purpose of including diagnoses of taxa in their N-definition is to make them both comparable to and distinguishable from other described taxa [Pavlinov 2015a; Dubois 2017a]. In this regard, a certain problem arises with the most recent research on species systematics based on the analysis of molecular genetic data (DNA barcoding, turbo-taxonomy, metagenomic analysis, etc.) (e.g., [Ficetola et al. 2008; Brower 2010; Cook et al. 2010; Sunagawa et al. 2013; Renner 2016; Dadi et al. 2017; Davis and Borisenko 2017; Thines et al. 2018; Dupérré 2020; Ahrens et al. 2021]). The new taxa characterized in them by specific molecular sequences only (e.g., [Riedel et al. 2013a; 2013b; Meierotto et al. 2019; Sharkey et al. 2021]) turn out to be incompatible with those identified by "traditional" characters. This problem is partly solved both by preserving voucher specimens, in which such characters are available, and analysis of "historical DNA" extracted from old museum specimens (see below). However, "molecular diagnosing" of taxa is not the only novelty in contemporary descriptive systematics; another example is provided by using bioacoustic data for this purpose; this is critically important in some animal groups for identifying species, including new ones (e.g., [Sangster and Rozendaal 2004; Somervuo et al. 2006; Mielke and Zuberbühler 2013; Tishechkin 2014]). Advocates of new approaches suggest that the "molecular diagnosing" of new taxa be mandatory in contemporary Codes [Riedel et al. 2013b] and high-resolution images of type specimens should replace traditional descriptions [Renner 2016; Miguel 2020].

The *principle (rule) of typification* corresponds to the ostensive definition of taxon: it presumes: (a) indication of the nomenclatural type (N-type) in a taxon's N-definition in its original description; and (b) consideration of the N-type when taxon interpretation is subsequently changed. Accordingly, the nomenclature system based on this principle can be referred to as *type-based* (ostensional). The development of this principle, also known as the *method of type*, is specifically characteristic of the nominalist nomenclature of a traditional kind [Whewell 1847; Parkes 1967; van der Hammen 1981; Petersen 1993; Witteveen 2014; Pavlinov 2015a].

The application of this principle is rank-dependent: the N-type of a taxon of a certain rank-group is the taxonomic N-object belonging to the next lower rank-group. In the species-group, the N-type is a particular collection (herbarium) specimen or a strain (live culture). In nomenclatural groups of higher ranks, the N-types are the taxa, viz.: a species in the genus-group, a genus in the family-group, a family in the order-group, and an order in the class-group. With this, the N-types of sequentially subordinate taxa may be considered to compose a certain "cascade" beginning with the type specimens, so the latter are suggested to consider the N-types of all taxa regardless of their position in the ranked hierarchy [Dubois 2005; David et al. 2011]. It was suggested that there should be a terminological distinction between *classification* and *collection* types: the former are particular taxa, whereas the latter are particular specimens [Farber 1976; Moore 1998]. It is also suggested that the general word "*type*," because of its metaphysical connotation, should be replaced by a series of terms that are composed of the prefix "*nucleo-*" added to terms denoting taxa of various ranks, viz., nucleospecies, nucleogenus, etc. [Dubois 2005; David et al. 2011].

The typification of suprageneric taxa by reference to their genera was first suggested in the second half of the 18th century (Adanson), and it became commonly recognized as *automatic* from the early 19th century (Candolle Sr.). For genus-group taxa, indication of their type species first appeared at the beginning of the 19th century (Latreille), and it became a mandatory norm in its second half. For species-group taxa, typification by particular specimens was the latest to be recognized, being first proposed at the end of the 19th century (Coues). At present, the principle of typification presuming mandatory or highly recommended direct indication of *N*-types extends: (a) to taxa ranging from species-group to family-group in all traditional Codes; (b) also to order-group taxa in virology and partly (as an option) in botany; and (c) up to class-group taxa in bacteriology.

Theoretical consideration of the principle (rule) of typification involves an important question about the function performed by the *N*-type in descriptive systematics. According to the essentialist interpretation, the *N*-type of taxon is interpreted (somewhat metaphysically) as a bearer (incarnator) of its characteristic features, so it is indirectly related to the taxon intensional definition. The nominalist interpretation implies that the *N*-type simply represents or exemplifies the respective taxon, so it "bears" the latter's name rather than characters and serves as a means of ostensive definition of the taxon. More formally, this issue may be set as follows: with which particular *N*-object is the *N*-type tied, either taxonomic (taxon as such) or linguistic (taxonym as such) [Pavlinov 2015a]? According to the concept of semantic triangle (see Section 2.2), the type should be tied to the taxon, while from a purely nomenclatural viewpoint it is usually tied to the taxonym. The latter interpretation is expressed aphoristically as follows: "Taxa have circumscriptions but no types while names have types but no circumscriptions" [Nicolson 1977: 569]. According to this vewpoint, the *N*-type is considered a taxon's *name-bearer* (nomenifer), and it is denoted accordingly as *nominotype* or *onomatophore* or *onymophoront* [Simpson 1940; Bolton 1996; Dubois 2005; Dubois et al. 2019; Sluys 2021].[8] A lower-rank taxon involved in the same automatic typification with the inclusive higher-rank one becomes *nominative* of the latter: e.g., tribe and family typified by the same genus.

The "name-bearing" function performed by the type is most evident in the case of automatic typification, when the names of higher-level group taxa are derived from

---

[8] The name-bearing function of the *N*-types seems to make up a certain connectivity of classifying and naming inherited from essentialist nomenclature. As the British entomologist Harold Oldroyd suggests, "It is this type-concept that binds taxonomy to nomenclature [so that] taxonomy and nomenclature are interlocked [...] We have sacrificed taxonomy to nomenclature by ruling that the specimen that happens to be selected as type by the procedure of the Code must ever afterwards typify the species" [Oldroyd 1966: 254]. On this basis, he believes that "The type-system should be abandoned entirely" [Oldroyd 1966: 260].

the names of their nominotypical genera. However, in the case of species-group taxa, the *N*-type seems to mean something more than just a "name-bearer." In fact, a simple pointing at certain specimens (say, lying next on a desk) without providing information about which particular features of particular taxa should be observed in them does not make any sense when practically distinguishing these taxa [Pavlinov 2015a]. This is especially evident when the numerical methods are applied to the samples, in which all specimens, including types, are employed as just "character-bearers." So, one may conclude that, of the functions performed by the *N*-types of the species-group taxa, "character-bearing" is primary and "name-bearing" is secondary.

Functioning of the *N*-type as a name-bearer is realized in the following nomenclatural tasks. First, when the original name is to be replaced by a new one, the replacement name is associated with the same type as the replaced one. Second, when the taxon circumscription is changed, the original name is preserved for that redefined taxon, to which the respective nominotype is allocated (this condition is not mandatory in bacteriological nomenclature).

Since the *N*-type participates in both the ostensive *N*-definition of a taxon and in "bearing" its name, a quite remarkable and somewhat unexpected question arises, namely, whether the *N*-type belongs to the typified taxon of necessity or with contingency [Levine 2001; LaPorte 2003; Haber 2012; Witteveen 2014, 2021]. A clear-cut answer is not always obvious from a purely "philosophical" standpoint; for example, a negative resolution about necessity follows from possible (and real) cases of "misidentification" of the type specimens. This means that the type of a certain taxon in its well-established, commonly accepted understanding is proved to be a member of another species. The most famous among such cases is the Linnaean type of the Indian elephant that was shown subsequently to be a specimen of the African elephant [Witteveen and Müller-Wille 2020]. However, taking into consideration that the basic functions of the nomymotype (nomenifer) is to represent a particular species as both its name- and character-bearer, it is evident from an "empirical" perspective that the type has to be the member of the species it typifies [Heise and Starr 1968; Starr and Heise 1969; Pavlinov 2015a]. The contemporary Codes stipulate such a collision and resolve it by means of certain secondary regulators, such as *ad hoc* decisions and opinions of the respective authoritative bodies.

The typification can be *direct* (explicit) or *indirect*. In the first case, it involves the direct indication of the corresponding taxonomic *N*-object chosen as the nominotype (type taxon, type specimen). In the second case, fixation of the *N*-type is achieved by forming the name of the taxon from the stem part of the name of its nominotype; such typification in botanical and bionomenclature systems is called *automatic*, and the name of a nominotypical taxon is called *basionym* [Hawksworth 2010; Anonymous 2012]. In the case of species-group taxa, a variant of indirect typification is the publication of an image of a real specimen: it is presumed that it is the latter and not its image itself that serves as the actual type. As noted above, direct typification of newly established taxa of all or most of the officially recognized ranks is mandatory in contemporary traditional Codes. Accordingly, its absence makes the respective taxon undefined and its taxonyms unavailable (illegitimate); the same is entailed by the unavailable (illegitimate) status of the respective basionym from which the neonym of a newly described taxon is derived.

Typification may be *primary* or *secondary*. The former means an indication of the nominotype in the original description of a new taxon; the latter refers to a subsequent fixation of the nominotype. For example, for species-group taxa, the primary typification corresponds to the fixation of holotype in the original description, while the secondary typification means fixation of neotype or lectotype (see below) in the subsequent taxonomic revision. The secondary typification began to be routinely applied to the previously established non-typified genera and species since the end of the 19th century: the designation of their nominotypes is carried out to ensure the fulfillment of the principle of typification. In particular, a special project for the typification of Linnaean plant genera is currently underway [Jarvis 1992].

The development of the principle of typification as applied to species-group taxa appeared to be rather intricate. In fact, no original specimens were consistently indicated and preserved by early nomenclaturists, who admitted the possibility of replacing them in case they lost their diagnostic value (damages, decolorized, etc.); this is still allowed in contemporary bacteriological nomenclature. Distribution of the exicates by botanists among their colleagues may be considered an early version of typification [de Candolle 1867; Engler et al. 1897]. At the turn of the 19th–20th centuries, the need for fixation (indication) of the types in original descriptions and their preservation in collections was adopted by most American nomenclaturists and their followers [Coues et al. 1886; Britton et al. 1892; Bather 1897; Schuchert 1897; Thomas 1897; Marsh 1898; Cook 1900]. A recommendation to typify newly described species-group taxa was adopted in all Codes in the mid 20th century; nowadays it is mandatory [Winston 1999; Pavlinov 2015a].

Initially, this typification implied fixation of a single type specimen according to the rule "one taxon—one type"; however, designation of *type series* (*hypodigm* according to [Simpson 1940, 1961]) with several specimens subsequently became popular. Although considerable objections were put forward against such a multi-typification [Merriam 1897], this practice was justified empirically by the distribution of additional types (cotypes, paratypes, etc.) among different institutions to lessen the danger of them all kept together being incidentally destroyed [Chamberlin 1952]. The need to reflect intraspecific variability by means of the type series was supported at a theoretical level by so-called "population thinking" [Simpson 1940; Mayr 1942, 1969; Whitehead 1972]. Accordingly, several additional type categories (isotype, cotype = paratype, topotype, allotype, etc.) were distinguished along with the "main" types (type, holotype), and their number grew from initially several [Schuchert and Buckman 1905; Burling 1912] to several dozens by the mid 20th century [Frizzell 1933; Evenhuis 2008a].

The following main kinds of the type specimens are officially recognized in contemporary traditional Codes [Hawksworth 2010]: *holotype* (the only specimen fixed in the original description), *syntypes* (several specimens listed or presumed in the original description with no holotype being fixed), *lectotype* (a specimen selected and subsequently fixed from the series of syntypes), *neotype* (a specimen subsequently fixed in the absence or loss of the original specimens). A special kind of type series, *hapantotype* (equivalent to holotype), is recognized in zoological nomenclature as representing different phases of the complex life cycle in certain protists.[9] In addition, *paratypes* and *paralectotypes* may be fixed to supplement the holotype and lectotype, respectively; the holotype may also be complemented by *isotypes* and *epitypes* (in botany) and *allotypes* (in zoology). In bacteriology and virology, the type can be not only a micropreparation but also a *strain*, i.e., a living culture. An image (drawing, etc.) of the real fixed and preserved (in a museum, herbarium) specimen, on which the original description is based and to which a reference is provided, is allowed as a specific *N*-type in all nomenclature systems; however, as noted above, it is presumed that that specimen and not an image *per se* is the nominotype. Accordingly, this provision does not apply to lifetime photographs of living organisms, if the latter are not preserved in a collection (see also below). A more neutral notion of *voucher specimen*, along with the standard term "type" (without specifications), is used in the nomenclature of cultivated plants.

The just-listed nominotypes of the species-group taxa are of different importance as particular taxonomic *N*-objects, and they can be ranked accordingly as follows [Seberg 1984; Pavlinov 2015a; Vasudeva Rao 2017]. In general, they are designated informally as *primary*, *secondary*, *tertiary* (and occasionally even quarternary) types, respectively (not in a more general sense indicated above). Two kinds are officially recognized in zoological nomenclature, the main *name-bearing* types (holotype, syntype, lectotype, neotype) and "*other*" types without such function (paratype, paralectotype, allotype). In botanical and bionomenclature, all nominotypes formally have the same status, although they may also be divided into the same two groups, "main" and "auxiliary" (including isotype, epitype, etc.).

Unlike species-group taxa, nominotypes of the taxa of higher ranks normally do not have specific terms, though the term *nomenspecies* is suggested to denote nominotypes of genus-group taxa in some Codes [Hawksworth 2010]; this term is sometimes used in the sense of nominal species (e.g., [Hoffmann and Roggenkamp 2003]). Some terms applied to the type specimens in species-group taxa were proposed to be used for the nominotypes of genera [Schuchert 1897], but this proposal was not accepted.

For the species-group taxa, of certain nomenclatural significance is the *type locality*, i.e., the geographical and (if relevant) stratigraphical place of collection (capture, etc.) of a specimen that becomes the nominotype. An auxiliary specimen subsequently collected (caught) in this place is termed *topotype* or *adelphotype*.

---

[9] It seems reasonable to apply this type category also to other organisms with a complex life cycle (e.g., holometabolous insects) [Pavlinov 2015a].

A new emphasis in the typification of the species-group taxa combined with the principle of depositing is now brought by the problem of nature conservation that appeals to the minimization of the impact of the collecting activity on endangered animals and plants [Loftin 1992; Norton et al. 1994; Remsen 1997; Collar 2000; Donegan 2008; Winker et al. 2010; Pavlinov 2016]. In this regard, it is suggested that the possibility should be considered of using photographs of living organisms taken in nature without killing them as nominotypes in the case of a description of new species-group taxa [Donegan 2008; Marshall and Evenhuis 2015; Krell and Marshall 2017; Shatalkin and Galinskaya 2017]. Such name-bearing photographs are proposed to be called *phototypes* [Hawksworth 2010; Maxwell 2019].[10] However, adherents of traditional understanding of the principle of typification equate this practice to the description of *typeless species* making their names unavailable (illegitimate) [Timm and Ramey 2005; Dubois and Nemésio 2007; Nemésio 2009; Amorim et al. 2016; Santos et al. 2016; Dubois 2017c; Pine and Gutiérrez 2018]. One of the recent issues of the journal *Bionomina* (https://doi.org/10.11646/bionomina.12.1) is devoted to this problem.

One intriguing point of the latter is that the reference in the original description to a "physical" type specimen, preserved some way in a repository, is not strictly mandatory under zoological nomenclature [Wakeham-Dawson et al. 2002]. In fact, it is deemed that "*Whenever feasible*, new species-group taxa should be established on the basis of at least one preserved type specimen" ([Declaration 45], italics added). With this, it is important to keep in mind that the absence of "physical" specimens makes at least some results of the research on species systematics non-reproducible, so they appear to cease to meet one of the important criteria of scientificity [Pavlinov 2015a, 2016; Krell 2016; Dubois 2017c; Löbl 2017]. In particular, the case of "*Bicingulatus*" [Maxwell 2019] clearly indicates that the phototypes can be subject to certain falsifications, and certain technical tasks concerning approval of the identity of such "phototypes" should be considered in this regard [Aguiar et al. 2017]. Thus, this nomenclature problem seems to have no simple solution yet [Dubois and Nemésio 2007; Pavlinov 2015a; Cianferoni and Bartolozzi 2016; Raposo and Kirwan 2017].

To the issue of using phototypes as nominotypes replacing "physical" specimens partly relates that of using digital images (photographs) of museum (herbarium) type specimens [Häuser et al. 2005; Wheeler et al. 2012]. Such images are suggested to be referred to as *E-types* [Speers 2005; Abebe et al. 2014]; it is believed that their use, instead of (or in addition to) the specimens themselves, may facilitate research on taxonomic diversity and nomenclature in many species-rich groups on an international scale.

One of the "hot spots" in the application of the principle of typification to the species-group taxa has been highlighted recently by their descriptions based on the molecular genetic data shortly considered above. Suggestions to codify the use of this category of data in the typification have been discussed [Reynolds and Taylor 1991; Chakrabarty 2010; Santos and Faria 2011; Bull et al. 2012; Jörger and Schrödl 2013; Federhen 2014; Hawksworth et al. 2016]. Molecular sequences typifying

---

[10] Not in the sense of [Jørgensen 1997].

newly described species and subspecies are proposed to be designated as *genetypes* [Chakrabarty 2010].

In this regard, considering an essential difference between the primary and secondary information used in taxonomic studies [Pavlinov 2016, 2021], it is important to distinguish between the traditional type and voucher specimens, from one side, and the decoded molecular sequences, from another side. The former contain primary information, while the latter contain secondary information and are similar to digital images in this respect. Accordingly, one of the prerequisites for making typification by such non-traditional data compatible with the above-mentioned criterion of scientificity is that it is to be considered mandatory to preserve the voucher (reference) specimens, from which DNA/RNA samples were taken as the nominotypes of the new taxa, in certain repositories [Will et al. 2005; Rowley et al. 2007; Pleijel et al. 2008].[11] To become officially recognized, they can be termed *hologenophores* and *isogenophores* [Astrin et al. 2013] and reasonably considered the parts of the composite holotypes and paratypes (isotypes), respectively. In some cases, the vouchers with the molecular genetic data associated with them can be ascribed the status of *epitypes* in order to make them comparable with the old type materials [Evans and Mann 2009]. On the other hand, an analysis of the so-called "historical DNA" extracted from the old type specimens allows the latter to be effectively incorporated into modern total molecularization of taxonomic research [Mulligan 2005; Ellis 2008; Knapp and Hofreiter 2010; Rowe et al. 2011; Särkinen et al. 2012; Nachman 2013; Costa and Roberts 2014; Choi et al. 2015]. All this should provide future revisers with the possibility of checking species allocation of the molecular sequences kept at Internet repositories such as GenBank, BOLD, etc. [Weir 2014; Meiklejohn et al. 2019; Pentinsaari et al. 2020].

The principle of typification is not observed in phylonomenclature; however, indication of a hypothetical closest (direct) ancestor of the clade in the latter's *N*-definition may be considered a specific analogy of typification. This provision is therefore suggested to be formalized as the *principle (rule) of ancestration* [Pavlinov 2015a]. Based on the central assumptions of cladistics, only populations and species, and not higher-level clades, are considered in this capacity; with this, the term "ancestral species" is not used in *PhyloCode* as non-operational, and a certain *reference point* (corresponding to the clade basal node) is taken for the respective specifier [Cantino and de Queiroz 2010, 2020]. If some apomorphies characterizing the clade are attributed to its closest ancestor, the latter is designated as *cladotype* [Béthoux 2007].[12]

To conclude consideration of the principle of typification, it should be emphasized that it had been developed at a time when only morphological data were available capable of being "fixed" in some way or other with dead specimens. Currently, the kinds of data by which species-group taxa can be identified are more diverse, and at least some of them (such as ethological) can in no way be fixed with museum

---

[11] In this regard, the task of the proper museification of the genetic material becomes especially urgent [Martin 2006; Mandrioli 2008; Rowe et al. 2011; Jackson et al. 2012; Puillandre et al. 2012].

[12] This term also has other meanings [Aref'ev and Lisovenko 1995; Hawksworth 2010].

specimens in any form; accordingly, the voice records and videotapes fixing them should be considered as the valid type materials. Besides, certain ecological features of whole populations can be occasionally used for the recognition of new species-group taxa (e.g., [Nonveiller 1963]).[13] So, these (and others not mentioned here) new challenges facing the principle of typification clearly indicate that the latter is partly outdated in its traditional provision and application, and the time comes when its content must be seriously revised.

3.6 PRAGMATIC AND OTHER PRINCIPLES

The overall array of nomenclature regulators is hardly exhausted by those discussed in previous sections. In addition, the regulatory functions are performed by certain other principles and rules, which are assigned different meanings in different nomenclature systems, from basic to purely secondary. Below are briefly characterized two of their groups, which should probably be allocated to separate blocks, viz., *pragmatic* and *ethical*. All are supposed to be theory-neutral.

The **pragmatic principles** provide a certain general context for the functioning of the nominative nomenclature conditioned by its applied nature. The latter means that its principles and rules are good to the extent that they facilitate the solution of two main tasks fulfilled by practical systematics: (a) description of taxonomic diversity; and (b) communication between members of the taxonomic community, as well as between them and the users of various kinds.

From this quite understandable practice-aimed standpoint, the general *principle of pragmatism* should be considered the basic norm of taxonomic nomenclature. One of its key requirements is that the latter in both its meanings (regulative and nominative) is to be practical and easy to use. Accordingly, this principle usually appears in the form of a more particular *principle of convenience* (*practicality*), which is sometimes called *expediency* of nomenclature [Watson 1892; Rickett 1953]. In A. Dubois' nomenclature system, this requirement is formalized as the *principle of simplicity* [Dubois 2005].

Such a standpoint means recognition of the primacy of its "practical" over "theoretical" justification: the former refers to convenience and expediency, the latter to some "legitimacy" [Robinson 1895]. As noted at the beginning of this chapter, such a pragmatic context provides a specific rating scale for the evaluation of significance of many nomenclature rules considered above (priority, usage, etc.). In fact, their choice is dictated by their contribution to the implementation of the condition of expediency as a principal means for ensuring the stability of names [Lewis 1871, 1872, 1875; Greene 1896; Rickett 1953; Cronquist 1991]. Among contemporary Codes, a special emphasis on pragmatics is placed in the nomenclature of cultivated plants [Ochsmann 2003].

The primacy of convenience and expediency over "legitimacy" is criticized on the grounds that it allows too much room for the arbitrariness of particular "convenient" decisions. Its opponents believe that "the only way to make nomenclature really

---

[13] I thank Frank T. Krell (Denver Museum of Nature & Science, USA) for providing me with the reference to this article.

stable is by an unfailing adherence to a rule. If one exception is admitted another will be" [Greene 1891a: 286–287].

The **ethical principles** (most of them are rules rather than principles) are connected in different ways with the personification of the nomenclature activity and its results. They are found in all Codes starting with *Linnaean Canons*; in particular, the last edition of Zoological Code includes *Code of Ethics* [Anonymous 1999]. This connection is twofold.

First, this means authorization (in its particular sense) of certain nomenclatural acts and their results, viz., the establishment of new, or changes in the interpretation of previously established, taxa and their names. The ethical aspect of the principle of authorization is that it pays homage to the relevant subject of the nomenclatural activity, so it is sometimes referred to as the *principle of literary justice* [Greene 1891b]. The latter is often violated by the retroactive application of some principles (binomiality, priority, etc.), which led to the rejection of many pre-Linnaean works and names. Accordingly, nomenclaturists who are especially inclined to respect professional ethics protest against the retroactivity thus implemented (e.g., [Adanson 1763; Gray 1821; Lindley 1832; Douvillé 1882a,b]).

Second, this refers to the linking of taxonyms with certain people through the establishment of the former as patronyms dedicated to the latter. Such names can bear both "positive" and "negative" connotations, according to which their regulation has a double meaning. On the one hand, it is considered honorable to name (with a benevolent connotation) taxa in honor of the persons who, in one way or another, contributed positively to the development of science, culture, politics, etc. This standpoint was expressed by a classical "formula" by the British zoologist William Swainson, who once wrote following Linnaeus that "The highest regard of a naturalist is to have a genus called after his name" [Swainson 1836: 235]. On the other hand, it is considered unacceptable that the names might have a derogatory and even more insulting meaning: this is specifically mentioned as a recommendation in all Codes.[14]

Most recently, ethical rules for the formation of taxonyms related to ethnoethical [Evans 2020; Gillman and Wright 2020], ethical-political [Shiffman 2019], and even geopolitical [Anonymous 2005] considerations began to be discussed. The most radical suggestions presume replacement of at least certain "awful" (from certain ethical standpoints) scientific names with more "pleasant" indigenous ones; a more moderate attitude supposes a fruitful interrelation between folk and scientific systematics and nomenclature (e.g., [Agrawal 2002; Beaudreau et al. 2011; Barron et al. 2015]).

---

[14] Examples of a kind of "nomenclatural warfare" can be found in some Internet resources [Davis 2015; Issak 2020].

Historical Part

4 An Overview and the Beginning

The historical development of biological systematics as a scientific discipline studying "taxonomic reality" inextricably involves the development of its professional language to describe this "reality." An evident reason was that the descriptive language is an important part of the specific toolkit that systematics operates with, so it should be constantly improved to make its construction as adequate as possible to the structure of the described "reality." So, this general dynamics inevitably involves the development of taxonomic nomenclature, being an important part of the language of systematics, following the development of the latter's both theoretical foundations and pragmatic needs.

The inevitability of and necessity for such a development of nomenclature was noted by many nomenclaturists [de Candolle 1867; Saint-Lager 1880, 1886; Coues et al. 1886; Jordan 1911; Mayr 1969; de Queiroz and Gauthier 1994; Bowker 1999; Ereshefsky 2001a, 2007b; Schuh 2003; Knapp et al. 2004; Dubois 2011a; Pavlinov 2015a, 2019]. The nomenclatural means, with which the diversity of organisms is described, cannot help but change following changes in the ideas about the structure of this diversity, as well as about the methods for describing it. Therefore, defending an immutability of nomenclature by advocates of its empirical concept (e.g., [van Steenis 1964; Greuter 2004; Chaikovsky 2007]) is hardly a good answer to the challenges that change over time following the changes in understanding of what and how biological systematics investigates.

This chapter begins with a brief outline of the main trends and stages in the historical development of taxonomic nomenclature considered from a conceptualist perspective. Then the initial stages of the development of the language of taxonomic descriptions associated with pre- and proto-systematics are considered. This chapter, as well as the following chapters on the history of nomenclature, are based largely on the author's publications [Pavlinov 2013, 2014, 2015a, 2015b, 2019]. Some notable details of the history of nomenclature in botany and zoology can be found in a number of monographs and review articles [Sachs 1906; Green 1927; Heller 1964; Heppel 1981; Nicolson 1991; Malécot 2008; Dayrat 2010].

4.1 MAIN HISTORICAL TRENDS AND STAGES

The historical development of taxonomic nomenclature proceeded in an orderly fashion as a result of certain general regular factors. The most principal drivers (fundamental regulators) of its development can be somewhat conventionally classified as "external" and "internal" ones, which basically correspond to the ontic and epistemic components of the cognitive situation in which systematics operates (see Section 1.1). In the first case, the need to develop the language as a means of adequately describing "taxonomic reality" is of prime importance. In the second case, the main factor is the need of the taxonomic community for this language to be universal, solid, and easy to use.

To understand an overall trend in the development of taxonomic nomenclature, it is useful to consider the professional language of systematics a non-equilibrium complex linguistic system [Susov 2006]. The development of such kind of systems, considered in general, is subject to the following general regularities [Prigogine and Nicolis 1985]. Firstly, it is directed towards complication, so the language becomes complicated partly following a more complicated understanding of the structure of "taxonomic reality" and partly caused by some intrinsic systemic mechanisms, and this complication entials differentiation of language into certain "dialects." Secondly, as far as language is a social phenomenon, its development is profoundly motivated by the historical changes in the social functions including evolving human cognitive intentions [Kubryakova 2004; Gontier 2009; Allard-Kropp 2020]. And lastly, its development involves drastic revolutionary changeovers that make it largely unpredictable, which is evident from the history of nomenclature: the second half of the 18th century promised to be a triumph of the essentialist nomenclature personified by Linnaeus, but Adanson proclaimed the nominalist concept which became dominant from the early 19th century.

The development of scientific systematics involved *rationalization* (in the general sense) of its professional language, and this was associated with the purposeful development of the regulative nomenclature as an ordered array of certain principles and rules for handling taxonomic names. This dominant trend was part of the overall rationalization of the cognitive activity in science that distinguished it from non-scientific modes of cognition [Gaydenko 2003], so it was basically driven by the overall trend of the scientification of the whole of systematics [Pavlinov 2018, 2021]. This rationalization culminated in the codification of nomenclature that involved ordering of its initially scattered principles and rules into nomenclature rulebooks officially recognized by the taxonomic community.

The general need for the rationalization of nomenclature set a fairly powerful *integration* trend in its development directed by the search for a unified means for standardized solutions of standard nomenclatural tasks. On the other hand, different understanding of the principles and methods of systematics, as well as different interpretations of the means sought, set simultaneously the opposite *diversification* trend. It led to the emergence of different nomenclature concepts and systems based on different ideas about what and how systematics explores and describes. The primary role in the shaping of the second trend was played by the differentiation of the subject areas of systematics affecting the language of taxonomic descriptions: first were botany and zoology and then also microbiology. Regional differentiation of the international community of taxonomists also took place, which was provoked by centrifugal geopolitical processes in the 18th and 19th centuries [Lindbeck 1975; Wimmera and Feinstein 2010]. The divergence of differently motivated "dialects" of nomenclature (conceptual, subject, regional, etc.) became the main result of the realization of this diversification trend.

When considering the historical development of taxonomic nomenclature, it is crucial to understand that it is a harmonious combination of the stability and lability of its concepts and principles that made their totality quite efficient over a very long period of the developing and functioning of descriptive systematics [Dayrat 2010; Dubois 2011a; Pavlinov 2015a]. In fact, nomenclature regulators should maintain its reasonable stability at a certain stage in the development of descriptive systematics, but they cannot remain totally unchanged as the latter develops. In the absence of such historical dynamics, a gap would inevitably arise and widen between the conceptualization of "taxonomic reality" and the structure of the language designed to describe it. This means that the history of taxonomic nomenclature, notwithstanding the latter's being designed primarily to solve practical tasks, was conceptually driven (see Section 1.2). Its principal driver was the change in the theoretical (ontic before all) foundations of systematics, which was superimposed by the pragmatics of taxonomic descriptions.

The rationalization of taxonomic nomenclature was accompanied by its abovementioned complication due to developments of particular principles and rules. This is caused by complication of the ideas about the structure of "taxonomic reality" requiring a more complex linguistic means of its description. With this, the increasing complexity of the rationally organized language involves an increase in the number of regulated tasks and the tightening of their regulation. As far as ideas about both "taxonomic reality" and the principles of its exploration and description are elaborated by particular taxonomic theories, this provides a prerequisite to the development of particular nomenclature concepts and systems.

In the history of systematics, the maturation of taxonomic nomenclature was preceded and accompanied by the formation of descriptive languages in other main divisions of natural history. One of them is anatomical nomenclature, which is quite obvious and indicative: indeed, the invention of descriptive names of plants and animals that focus on their individual parts (e.g., cornflower, hornbill) requires prior systematization and naming of these parts (flower, bill). The same is true for the names of organisms that refer to their habitats, which should also be distinguished and named in advance (valley mayweed, mountain cedar, field sparrow, sea urchin).

There were several versions of the main stages in the history of nomenclature that highlighted different moments in it corresponding to the particular views of their authors (e.g., [Linnaeus 1751; Sprengel 1807–1808; Sachs 1906; Greene 1909; Malécot 2008; Dayrat 2010], etc.). Considered within the framework of the conceptually interpreted history of natural science, the main stages in the historical development of taxonomic nomenclature seem to correspond to those distinguished in the conceptual history of the whole of systematics [Pavlinov 2015a, 2019] (see Section 1.1). This is an inevitable consequence of the above-emphasized connection between "content" and "form" of taxonomic knowledge (see Introduction): they develop in conjunction and, therefore, rather synchronously (on a global time-scale).

At the pre-systematic stage, the prerequisites were laid for the formation of the descriptive language, as part of the classification method of cognition, on an empirical basis. The development of proto-systematics was associated with the early shaping of nomenclature by its initial rationalization. The latter resulted from an increasing understanding of the role of language as a means of adequately describing "taxonomic reality."

The subsequent development of scientific systematics promoted the rationalization of taxonomic nomenclature and, at the same time, proceeded with its fragmentation caused by changes in the cognitive situation of systematics. At the scholastic stage, the essentialist concept dominated, according to which the descriptive names of taxa should reflect the essence of organisms. At the post-scholastic stage, this concept was replaced by the nominalist one with its formula "a name is just a name." The rationalization of taxonomic nomenclature with the progress of post-scholastic systematics led to the development of Rules/Codes with strictly formulated principles and rules that were "dictating" to systematists how to designate properly the taxa recognized by them.

This general integration trend of the development of nominalist nomenclature was superimposed by the diversification one leading to its split into "dialects," which was driven by several particular causes. One of them was different interpretations of the ways of solving the same nomenclatural tasks: examples are type-based and circumscription-based nomenclature systems presuming different specifiers for N-definitions of taxa and their names (see Section 3.5). Another was the above-mentioned subject-area and regional differentiation of the whole of the taxonomic community, each "subcommunity" with its own particular nomenclature preferences. This diversification of nomenclature was strengthened by its rationalization: elaboration of more strictly defined regulative principles and rules led to a widening of the gaps dividing the particular "nomenclature subcommunities."

The most recent development of nomenclature seems to witness strengthening of its conceptualization due to certain shifts in understanding of both the structure of "taxonomic reality" and the linguistic means describing it. The most indicative examples are phylogenetic and new rational-logical nomenclature concepts.

4.2 THE EMPIRICAL ROUTE

The early forms of describing the diversity of manifestations of the surrounding world in general and organisms in particular are rooted in primordial classifying, constituting the basis of pre- and proto-systematics. Its immanent part is a rather simply organized descriptive language, whose main regulators are certain implicit cognitive and linguistic principles and laws of the functioning of natural languages. Being "internal," they shape its grammatical and lexical structure, as opposed to the rationally organized taxonomic nomenclature of scientific systematics with its powerful "external" regulators (see Section 2.1.3). At the same time, the empirical nature of early nomenclature does not mean its freedom from a certain prototype of the theoretical knowledge that acts as its "external" regulator. Indeed, general structure of the descriptive languages of empirical classifiers depends to some extent on certain ontologies presumed by respective cognitive situations.

4.2.1 Folk Nomenclature

Humans inherited classification activity from their biological ancestors to make it one of the primary and most fundamental forms of conscious cognitive activity [Lévi-Strauss 1966; Foucault 1970; Atran 1990; Ellen 2008]. A principal pivotal point differentiating humans' cognition from that of other animals is that it involves not only recognition of certain classes of objects but also their designation by certain linguistic means, which is stimulated by a kind of "language instinct" [Pinker 1993]. In the course of early human evolution, classification activity, supplemented by linguistic designation means, gave rise to folk systematics as a specific way of generalizing certain ideas of indigenous people about the structure of the world around them [Atran 1990, 1998; Brown 1991; Ellen 1993; Atran and Medin 2008; Pavlinov 2018, 2021]. This classification activity involved both the recognition of certain groups of organisms and allocation of certain names to them.

Thus, the development of *folk nomenclature* as a prototype of future taxonomic nomenclature began together with the development of the prototype of future systematics, i.e., together with folk systematics. The very fact that the generalizing names (notions) of organisms occur in indigenous languages evidences the existence of their classifications [Greene 1909; Bartlett 1940; Berlin et al. 1973]. Indeed, "ethnobiological nomenclature represents a natural system of naming that reveals much about the way people *conceptualize* the living things in their environment" ([Berlin 1992: 26; italics in the original]). Therefore, the folk taxa distinguished in folk classifications are given for the "external" observers (such as scientists) mainly through their names, i.e., linguistically, so such classifications were termed "biolinguistic" [Chamberlain 1992].[1] Accordingly, in recent studies, analysis of the[1] nominative folk nomenclature serves as a means of clarifying how indigenous knowledge reflects the structure of biological diversity [Atran 1998; Minelli et al. 2005; Lampman 2010; Ulicsni et al. 2016]. More particular issues refer to the structure and etymology of folk names of animals and plants, general and specific features of their formation in indigenous languages, etc. [Berlin 1973, 1992; Berlin et al. 1973; Brown 1984, 1986; Atran 1990, 1998; Taylor 1990; Ellen 1993, 2008; Markova 2008; Kolosova 2009].

Folk classifications are most often hierarchical, which is caused by the hierarchical structure of both Nature and human cognitive means [Atran 1990, 1998, 2002; Lyubarsky 2018]. However, folk nomenclature does not recognize fixed ranks and their special designations. An idea of the ranked hierarchy of these classifications was introduced by the "Linnaean" scholars [Berlin 1973, 1992; Atran 1990, 1998] as a kind of common "template" to compare folk and scientific classifications. The experts in folk nomenclature recognize and designate folk ranks based on modern classification terminology; as a result, such standard terms as "kingdom," "life form," "genus," "folk species," etc., appear in their descriptions.

The folk names of animals and plants are usually semantically motivated in a way "to mark natural kinds with names that reflect their *inherent nature*" ([Berlin 2007: 19; italics in the original; Caprini 2007]). They are divided into *primary* and

---

[1] This circumstance had been already understood by Theophrastus (3rd century BC), who noted that local people distinguished between plant kinds by giving them different names (Theophrastus 1916).

secondary according to their nomenclatural status, which partly has a certain onto-epistemic background. The primary folk names are assigned to the most cognitively distinguished folk taxa, viz., high-rank groups (kingdoms, life forms) or especially distinct folk genera. Such names usually serve as easily identifiable designators and are most often monoverbal (e.g., plant, animal, tree, worm, fish), which is also true for the designations of clearly distinguished folk "generic species" (e.g., stoat, weasel, mink, ferret). If intraspecific forms (as they are now understood) are especially distinguished for certain reasons, they are also assigned monoverbal primary names: examples are age groups in insects with complete metamorphosis (e.g., caterpillar and butterfly), sex and age groups in domestic animals (e.g., stallion, mare, and foal; bull, cow, and calf). The secondary folk names are assigned to less significant folk taxa of lower ranks: they are most often descriptive and act as two- or multi-word qualifying epithets (e.g., river otter, green-winged butterfly, red pine).

Folk nomenclature is characterized by polysemy: the languages of most local ethnic groups possess a more or less rich synonymy. One of the reasons is the borrowing of different names referring to the same folk taxa of animals and plants as a result of contacts between different tribes [Atran 1990, 1998; Atran and Medin 2008]. With this, ethnolinguists themselves often generate a specific synonymy by treating lower-rank folk taxa as manifestations of intraspecific diversity according to their contemporary interpretations. However, broad treatment of the particular species based on the biological concept of Dobzhansky–Mayr may be erroneous in the original contexts of folk classifications, in which the monoverbally denoted forms can indeed be classified as "species" by folk systematists (e.g., the Russian names plotva, chebak, vobla, soroga all refer to the same "scientific" fish species, *Rutilus rutilus*). On the other hand, coincidence of both ethnolinguistic and scientific classifications in the recognition of the same lower-rank taxa is sometimes considered as an evidence of the latter's taxonomic distinctness [Mayr 1969, 1988; Cotterill et al. 2014; Ludwig 2017]. It may be true in certain particular cases; but in general this is hardly correct, as such interpretation presumes an identity of "linguistic world pictures" based on substantially different systems of conceptualization, viz., pre-scientific and scientific [Pavlinov 2015a].

Due to the conservatism of folk classifications and folk nomenclature, the names of animals and plants unknown to local residents or early settlers are usually formed in such a way as to lexically correlate them with the already known taxa. For example, the pigs brought to the New World by Spanish conquistadors were called "village peccary" by Maya Indians due to their similarity to the native "forest peccary" (genus *Dicotyles*) [Atran 1999]. Many examples of this way of naming can be found in the well-studied ancient Greek folk systematics: for instance, the Indian rhinoceros was called "Indian one-horned donkey" (*indikos onos monokeratos*) [Atran 1990; Bodson 2005]. The first white settlers in Australia designated the local marsupials on the basis of their habitual similarity to the European placentals: marsupial wolf, marsupial marten, marsupial mouse, etc. The names with such etymology, in addition to their descriptive function, partially perform a classification function: they "inscribe" newly discovered creatures in the respective stereotype folk classifications on a typological basis.

Folk nomenclature left a significant trace in the regulative taxonomic nomenclature. This is expressed primarily in the fact that nearly all of the latter's versions

An Overview and the Beginning

(save for the logic-rational one) are shaped on the basis of the *principle of verbality* inherited from folk nomenclature. In particular, a noticeable part of this heritage is monoverbality of the primary and bi- or multiverbality of the secondary names of plants and animals. Semantic motivation of descriptive names in the essentialist nomenclature is also a heritage, though substantially developed, of the folk classification language. Recently, the possibility of combining in some way indigenous and taxonomic systems of nomenclature started to be discussed, with the main emphasis on the need for the latter to comply with certain ethno-ethical norms (e.g., [Evans 2020; Gillman and Wright 2020]).

However, there is an important difference between folk and scientific nomenclature: the former is much more flexible and may have significantly broader implications than the latter. Its general terms (notions, signifiers) may function not only as the names of concrete groups of organisms but also as metaphors that presume wider meanings by referring to the more general phenomena [Brown 1991; Pinker 1993; Hart and Long 2011]. An extreme variant of the metaphoric usage of such terms seems to be their functioning as totemic signifiers [Lévi-Strauss 1966; Revel 2007; Ellen 2008]. Contrary to this, scientific nomenclature is aimed at as strict a usage of *T*-designators as possible according to the *principle of designative certainty* (see Section 3.2). This difference is probably caused largely by the fact that folk nomenclature is irrational, while the taxonomic one (as part of the professional language) is rational in its basic motivation and regulation.

4.2.2 Language of Proto-Systematics

Proto-systematics covers an extensive period of the development of cognitive activity and the language of describing the diversity of Nature, during which the beginnings of their rationalization were set. This period starts with Antiquity, continues with scholasticism, and ends with herbalistics.

The main contribution of antique (primarily Platonic) natural philosophy to the future development of the taxonomic nomenclature is that it established a close interconnection between two "Linnaean" foundations of systematics, viz., the recognition of particular groups of organisms (classifying) and the allocation of particular names to them (designating). Their interconnection was determined by a unified essential interpretation of both.

Indeed, every designation, if it is a notion and not a proper name, expresses not a particular individual object (a given horse, a given tree), but a class of objects (a horse in general, a tree in general). In the Platonic world picture, such a "thing in general" itself and a notion (name in general) designating it are inextricably linked as particular manifestations of the same *eidos* (idea). Therefore, such a name is meaningful by being a verbal expression of the *eidos*: it is not accidental in relation to the given "thing in general," but expresses its essence—of course, if it is a genuine, or appropriate, name [Losev 1990]. Consequently, comprehending the genuine (true, appropriate) name of a "thing" is equivalent, in a certain sense, to the comprehension of its essence, and thus its place in Nature. On the contrary, a non-genuine (inappropriate) name can only disguise the *eidos* (essence) of a "thing" and prevent its comprehension.

Such natural-philosophical Platonic understanding of the genuine names as an expression of the essences (ideas) of the "things in general" will provide one of the most fundamental features of essentialist systematics and nomenclature of the 16th–18th centuries. In fact, (a) as far as the comprehension of genuine names of organisms yields the comprehension of their essences, and (b) the comprehension of the latter is a necessary condition for the comprehension of the place of organisms in the System of Nature best represented by natural classification, then (c) the classifying merges with the naming, according to which (d) systematics to a certain extent boils down to nomenclature and one of its main objections becomes the search for genuine names [Pavlinov 2015a, 2019]. This is expressed rather aphoristically by the "founding fathers" of early systematics as follows: "knowing plants is equivalent to knowing their names" (Tournefort) and "if you don't know the names, then you also lose things" (Linnaeus) (see Chapter 5). With this, every genuine name is ascribed a universally valid function of an essentialist descriptor [Stearn 1959; Foucault 1970; Slaughter 1982]: the more fully it functions in this vein, the more precisely and completely it "grasps" and expresses the essential properties of the organisms grouped in a taxon denoted by this name.

The antique proto-systematics did not concern the rational rules for naming organisms; their names were borrowed largely from folk nomenclature. An illustrative example was provided by *Enquiry into Plants* by Theophrastus (Greek Θεόφραστος; c. 371–c. 287 BC), which contained descriptions of the plants of the entire Hellenic ecumene (see the contemporary edition: [Theophrastus 1916]). Especially distinguished groups ("genera") were usually designated by him monoverbally, while the less distinct ones ("species") were more often biverbal (hereinafter in their later Latin transliteration) [Váczy 1971]. According to the folk tradition, one-word could be not only the names of "genera" (for example, *iris*, *drypis*) but also those of most notable "primary species" (for example, cornflower species were designated *tetralis* and *kentaurion*) and even varieties or "secondary species" (wild and cultural forms of fig trees were called *epineos* and *syka*). If Theophrastus distinguished several "species" in a "genus," then the most typical was designated identically with the latter, i.e., without a specific epithet, while others were assigned such epithets. With this, some of these epithets were monoverbal, but different from the generic name: for example, different species of willows were called *itea*, *itea leuke*, *itea melana*, *aelichi*. The "folk" character of Theophrastian nomenclature was also manifested in the way that he associated exotic plants with local forms and denoted them accordingly. For example, he classified the exotic banyan tree as an Indian variety (*syka indikos*) of garden fig tree (*syka*) familiar to him. These features of Theophrastian nomenclature will later be reproduced in the Renaissance herbals, as well as in some early scientific classifications.

Of the later most remarkable monuments of the antique naturalistic tradition, three should be mentioned that came out within a short time interval. They were subsequently rewritten and reprinted repeatedly and appeared to be of great importance for shaping the descriptive language of proto-systematics by becoming a standard for representing information about plants in the herbals [Osbaldeston and Wood 2000; Ogilvie 2003].

One of them was *De plantis libri* by the Greek historian Nicolaus of Damascus (Greek Νικόλαος Δαμασκηνός; born c. 64 BC), which was in fact an exposition of Theophrastus' *Enquiry into Plants* (then attributed to Aristotle). It was translated into Arabic and then into Latin in the early 13th century to become very popular, even to the point of being included in the list of compulsory studies at some universities [Long 1996].

The second monument was *Naturalis Historiae* by the Roman encyclopedist Gaius Plinius Secundus, or Pliny the Elder (23–79). This *Historiae* was largely a compilation of its predecessors addressed mainly to the laymen who did not have any special education and interests (see the contemporary edition: [Holland 2015]). According to Latin grammar, the names of many "species" were two-word, with similar organisms being designated by a common "generic" name followed by "specific" epithets: for example, *Salix amerina*, *Salix candida*, *Salix nigra* [Vázcy 1971]. This characteristic feature of Latin will determine the lexical structure of binary (two-part) taxonyms in the subsequent works of systematists.

The five-volume *De Materia medica* (Greek Περί ύλης ιατρικής) by the Greek physician Pedanius Dioscorides (Greek Πεδάνιος Διοσκορίδης; pp. 40–90) was more specialized, being addressed mainly to practical physicians. It was divided into books with their titles indicating the characteristic features of the plants described in them: for example, Book I was called *Aromatics* (see the contemporary reprint: [Dioscorides 2000]). Each book was composed of a series of descriptions, each entitled with one-, two-, or three-word names of the described organisms or products thereof: for example, *Iris*, *Nardus oreine*, etc. Some accounts contained something like a prototype of contemporary synonymy. For instance, the description of the iris was supplemented by the following synonymous names (in their subsequent Latin transliteration): Greek *iris illyrica*, *thelpida*, *urania*, *catharon* or *thaumastos*; Latin *radix marica*, *gladiolus*, *opertritis* or *consecratrix*, Egyptian *nar*.

The most important contribution of late Antique and subsequent Medieval scholasticism of the fourth to the 10th centuries to the subsequent development of both the method and the language of systematics involved elaboration of the logical *genus–species scheme* illustrated graphically by the so-called *Tree of Porphyry*. Its development implemented the general condition of the principle of onto-epistemic correspondence (see Section 1.1). Its ontic basis was shaped by Platonic natural philosophy, according to which Cosmos was the result of a sequential emanation of *eidê* from the highest to the lowest levels of generality. A similar cascade of epistemic causation must correspond to this descending cascade of ontic causation: some general essence must be first comprehended, which provided a key for the understanding of its various particular detalizations.

A well-organized deductive hierarchical classification method followed from this onto-epistemic consideration, and it was construed as a sequential division of the notions corresponding to the essences of different levels of generality. According to this method, each sequence began with the fixation of the highest "genus" (*genus summum*), proceeded with the recognition of intermediate "genera" (*genera intermedia*), and the last of the latter (*genus proximum*) was divided into "species" (*species infima*). Thus, the scheme in question introduced the logical notions of "genus" and "species," with setting their subordination, that would become fundamental for future systematics. It yielded a rankless classification, in which the levels of generality of the logical "genera" were not fixed, and the "species" corresponded to the last (also non-fixed) step of division.

This scheme is both a method of classification and a method of definition [Pellegrin 1987]. Accordingly, each essence outlines a set (logical "class") of the "things" endowed with it, and this set is designated by an appropriate notion that expresses its essence and becomes its name. A definition of such a "class" is provided based on the logical formula "*Genus proximum et differentiam specificam*" (generic common and species particular) [Heller 1964; Pellegrin 1987]. This means that a notion, to become the definition of both the respective set ("class") and the "things" belonging to it, must be descriptive and consist of two parts, "generic" and "specific." The first part corresponds to the essence of a "genus" to which the given notion related, and the second part corresponds to a particular manifestation of this essence that distinguishes this "species" from others in the same "genus." And as far as such a notion becomes a name assigned to the respective set (group of "things," including organisms), the same binarity becomes a fundamental property of all descriptive names generated by the classification genus–species scheme. This general *principle of binarity* of the formation of essentialist names will be adopted and implemented by the future essentialist nomenclature developed by scholastic systematics; and it will be transformed into the *principle binomiality* characteristic of nominalist nomenclature developed by post-scholastic systematics (see also Section 3.3 on binarity/binomiality).

The next important step in the development of the language of proto-systematics was the *Herbal epoch* with time boundaries nearly coinciding with the Renaissance of the 14th–16th centuries [Arber 1938]. It basically continued (or rather reopened) the antique empirical tradition of describing plants and animals, a part of which was the lack of interest in the regulative nomenclature. However, it acquired some rational elements of the nominative nomenclature having been developed by scholastics.

The Herbal epoch (herbalistics) advanced in the descriptions of living matter in two important issues [Ogilvie 2003, 2006; Pavlinov 2013, 2015a]. A significant expansion of knowledge about the diversity of plants and animals, having been brought by the Great Geographic Discoveries, led to a complication of both their classifications and, consequently, descriptive language to represent this suddenly "expanded" diversity adequately. For this, the basic classification categories of Genus and Species were borrowed from scholasticism to become actively used in the hierarchization of classifications, which were added with several other auxiliary categories (Class, Order, etc.). The descriptive language of the later herbals, that appeared to be most advanced in this respect, became the direct predecessor of the nominative nomenclature of scientific systematics.

Within the herbalistics, two principal ways of ordering and describing the diversity of plants and animals, inherited from Antiquity and partly from the Middle

Ages, could be recognized. They correspond roughly to the descriptive and analytical traditions of antique phytographys and took shape as predominantly "list" or "systematic" ways of arrangement of the diversity of the described objects, respectively.

In the first case, prevailing among herbalists, the descriptions of organisms were ordered linearly in alphabetical order (as in Dioscorides), with the first letters of the generic taxonyms serving as a unified basis of ordering. Accordingly, the followers of Dioscorides were focused on descriptions of the individual "species" without any explicit attempts to classify them into "genera" of whatever meaning. From the point of view of the future development of scientific systematics, with its dominant idea of the hierarchical arrangement of the System of Nature, the main disadvantage of such a way of representing the diversity of organisms was that none of their major groups appeared distinguished.

In the second case, an ordering was based on relationships between not the names of organisms, but between the organisms themselves, estimated by the similarities in their features and their habitats; this resulted in the hierarchical classifications pretending to be natural (as in Theophrastus). The striving for a hierarchical systematization was first manifested in arranging herbals into "books," and those into "chapters" and "sections." Their titles indirectly served as the collective names of the organisms described in them. The most advanced herbals employed certain elements of the scholastic classification genus–species scheme with the notions of "genus" and "species" individuating these divisions.[2]

The herbalists were not interested in the elaboration of regulative nomenclature as a more or less ordered array of rules for the naming of "genera" and "species" [Slaughter 1982; Atran 1990; Pavlinov 2013, 2015a]. When classifying flora and fauna, be they local or foreign, they relied mostly on antique and medieval sources in identifying new plants and animals and, whenever feasible, borrowed the most "appropriate" names from there. Due to this, an important part of the herbal tradition became compiling the lists of names of plants and animals mentioned in classical and sometimes local sources. Such lists allowed navigation of previous and current literature and the association of different names from different authors with certain organisms. They became the prototype for the subsequent standard for compiling lists of synonyms and their authorship in taxonomic descriptions.

The descriptive language used by the actors of the Herbal epoch was predominantly Latin inherited from the scholastics and recognized as an official language in the universities from which most advanced herbalists graduated. It served as the unified means for communication between physicians and natural scientists and provided them with certain standard terminology. However, the language of herbals was not that strictly canonized, with Greek being used not infrequently along with Latin. This practice was caused by the fact that the treatises of the majority of the antique predecessors of the herbals were originally written in Greek (Aristotle, Theophrastus, Dioscorides) and only a few texts were in Latin (for example, Pliny). With this, many

---

[2] These terms are given here in parentheses to indicate their logical status differing from the contemporary biological.

herbals were published first in local languages—German, French, English—and only thereafter were reprinted in Latin. However, regardless of the languages in which the main texts of the herbals were written, the title names of plants and animals were most usually classical Latin or sometimes Greek. This laid the standard for the future nominative nomenclature of "genera" and "species" with its principle of Latinization. However, in some local herbals, the basis of nomenclature was formed by the native names, while the names from antique sources appeared only in synonymy lists.

If "genera" and "species" without fixed ranks were recognized in the herbals, they were usually denoted as "primary" and "secondary." Their names were mono-, bi-, and multiverbal, but in many cases it was not evident to which particular levels of generality the respective parts of the phraseme taxonyms belonged. In some herbals, respective parts of the binary names might differ lexically: generic parts were Latin, while specific ones were Greek. In other cases, it is usually supposed that generic parts of multiverbal names were usually monoverbal, and their specific parts were multiverbal [Sachs 1906]. However, one cannot exclude that such consideration might be dictated by the later tradition developed within the essentialist route of the history of nomenclature.

The Herbal epoch may be quite naturally divided into three main stages [Pavlinov 2013, 2015a].

The *first stage* (the 13th–15th centuries) belonged to the turn of the Middle Ages and the Renaissance; it was marked by the handwritten reproductions of the treatises of Pliny, Theophrastus, and Dioscorides, with certain additions and explanations. They first appeared in Italy, so the epoch in question emerged there as part of the Renaissance epoch [Janick 2003].

The *second stage* (the 15th to the early 16th centuries) was associated with the beginning of book printing in Europe that, due to the new technical facilities, stimulated the rapid development of reference books. This not only simplified their production but also partially changed their structure by providing their authors with the possibility of compiling indices and cross-references. The most active actors of herbalistics of that time were physicians who continued Dioscorides' tradition, so their herbals were in many ways not so much systematic as medical references. The most noticeable among them were the "German fathers" of botany called thus by the German botanists Kurt Polycarp Joachim Sprengel (1766–1833) in his fundamental *Historia Rei Herbariae* [Sprengel 1807–1808]. Indeed, their contribution to the development of both proto-systematics and proto-taxonomic nomenclature was quite significant, though the works of other actors of the second stage of the Herbal epoch were no less impressive [Sachs 1906; Greene 1909; Ogilvie 2003; Pavlinov 2013, 2015a].

The first person usually mentioned among the "German fathers" of botany was Otto Brunfels (1489–1534) with his three-volume *Herbarum vivae eicones*. He used the terms "genus" and "species" in their general sense throughout to denote classification units of different levels of generality: for example, "*Genera eius ex Dioscoride*," "*De Artemisia octava specie*," etc. All clearly distinguished plant forms were usually ascribed generic status by him: for instance, two forms of the plantain were designated as *Plantaginis duo sunt genera, Maior & Minor* [Brunfels 1530]. Since *Herbarum vivae* was written in Latin, all title names of the plants described in it were Latinized, including those from earlier Greek and Arabic sources or from German folk nomenclature. Of the synonyms to be found in the literature, he preferred those which seemed most significant to him or were most often used by his contemporaries. The description of each "genus" and "species" began with a small section "*Nomenclaturae*" containing synonyms: first were Greek names, then Latin, and at the end German names ("Teutonic," according to Brunfels) [Sprague 1928].

The German-written *Neu Kreuterbuch von Underscheidt* by Hieronymus Bock (better known by his Latin pseudonym Tragus; 1498–1554), unlike most herbals of that time, was not just a compilation of previous authors, but contained original detailed descriptions of native plants [Bock 1546]. Tragus emphatically rejected the alphabetic arrangement of plants, which he thought to hinder understanding of their natural arrangement. He emphasized in the preface: "In describing things, I try to approach as close as possible to how the plants were apparently united by the similarity in their form by Nature" (cited after [Greene 1909: 239]). When denoting the plants, Tragus gave preference to the most commonly used names. Their structure and length depended on how well they reflected the essential characters of the plants and how many subaltern "genera" or "species" Tragus distinguished in a containing one. Thus, in the "genus" of hellebores (*Elleborus*, now *Helleborus*), he denoted "lower genera" *Elleborus albus*, *Elleborus niger sylvestris*, *Elleborus niger adulterinus hortensis*. With this, it is not possible to establish definitely which parts of his taxonyms, generic or specific, were mono- or multiverbal; in fact, contrary to the established tradition, his generic names might be more often multiverbal, while the specific epithets were monoverbal [Larson 1971]. Tragus paid great attention to synonyms: his descriptions of particular genera were supplied with long annotated lists of the names used by other authors. Very significant was Tragus' contribution to the initial development of botanic organography: he established a fairly strict terminology for more than a hundred anatomical elements of plants [Sachs 1906; Greene 1909; Larson 1971].

The undoubted "classic" of this stage of the Herbal epoch was *Trium priorum de stirpium historia* by the Flemish physician and phytographer Rembert Dodoens (Lat. Rembertus Dodonaeus; 1517–1585). Its first edition was an herbal almost literally: each page contained the large image of an individual plant entitled in Latin with a short description and synonyms in other languages [Dodoens 1553]. In its French edition [Dodoens 1557], the descriptive texts took no less place than the plant images, the title names of "genera" were in French only (e.g., Du Nenuphar), while the names of "species" were provided in the figure captions also in Latin (e.g., *Nymphea alba*). The Latin names of "genera" were not always constant for particular plants denoted by the same French names (e.g., *Camomille* and *Cotula* for Le Camomille), and those of "species" were one- and two-word (e.g., *Orontium* for the Le petit Antirrinum, above-mentioned *Nymphea alba* for Le Nenuphar blanc). It is to be noted especially that an introductory section of the 1583 reprint of *Trium priorum* was entitled *De Stirpium Generibus*: this was (one of the first) quite unambiguous indication of the key significance of the generic division of the plant kingdom, no matter how the "genera" were understood.

Another variant of the nominative nomenclature could be found in *Historiae stirpium* by the German physician and herbalist Valerius Cordus (1515–1544), who is rated above his colleagues by some botanical historians [Greene 1909; Larson 1971; Ogilvie 2006]. His posthumously published fundamental treatise was based mainly on that by Dioscorides, as it was specified as *"annotationes in Pedacij Dioscoridis"* on its title page [Cordus 1561]. Like other herbalists, Cordus used the term "genus" throughout to denote the classification category higher than "species." As to the latter, he often designated the typical "species" of a "genus" by a descriptive epithet, whereas others were just given numerals. Thus, in the "genus" *Ranunculi*, he listed the first species as *Ranunculus palustris*, and the rest (more than ten) as *Secunda* species, *Tertia* species, *Quarta* species, etc.

The *third stage* of the Herbal epoch (the mid–late 16th century) followed more the Theophrastus tradition, so its actors published naturalistic and even systematic rather than medical textbooks. One of its key figures was the Swiss physician and encyclopedist Conrad Gesner (Lat. Gesnerus; 1516–1565).[3] His early botanical tractats were of quite typical herbal kind [Zoller 1967], but the four-volume zoological encyclopedia *Historia animalium* (1551–1558) differed significantly from them in systematic content and structure. Gesner's contemporaries assigned him the informal title of "Swiss Pliny" for the completeness of his treatises, and J. Cuvier would call his *Historia* "the beginning of the history of zoology" [Fischer 1966; Gmelig-Nijboer 1977; Ogilvie 2006]. In a separately published *Nomenclator aquatilium animantium*, the chapters were called Orders—for example, *Ordo III. De Pisciculus*, *Ordo XII. De Cetis* [Gesner 1560]—so this important taxonomic category appears in the systematic literature for the first time.

A notable detail of Gesner's *Nomenclator* was that the hierarchical classification of the Order *Mollibus* was illustrated graphically by the Tree of Porphyry. In the latter, each division step was indicated by the characters distinguishing the respective "genera," so these characters actually figured as the latter's descriptive multiverbal names. This practice of naming "genera" in hierarchical classifications will be adopted and developed in essentialist nomenclature. This was not an isolated case of using the Tree of Porphyry in the last stage of the Herbal epoch: it could also be found in several other herbals of that time, e.g., in *Stirpium Adversaria Nova* [Pena and Lobelius 1570]; see also [Legré 1897].

A specific manner of recognizing and designating groups of various levels of generality developed by late herbalistics could be found in *Rariorum Plantarum Historia* by Charles de l'Écluse (Lat. Carolus Clusius; 1526–1609). This *Historia* was divided into books expressing a certain system of plants, each book with chapters describing particular "genera" and "species," whereas the "classes" were recognized in some of the chapters as reflecting variants of the leaves and flowers [Clusius 1601]. The plants were designated by Latin names, some borrowed from classical sources; others were Latin translations of the native names. The plant names appearing as the chapter headers could be mono-, bi-, or multiverbal: the former were certainly generic, while the latter probably denoted species. The specific epithets provided in the figure

---

[3] His name is sometimes spelled Gessner [Pyle 2000].

An Overview and the Beginning

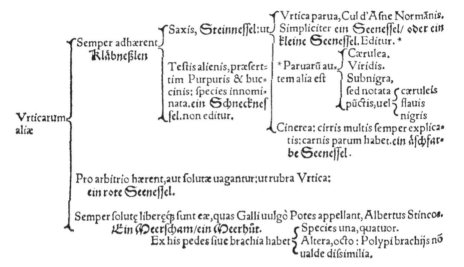

FIGURE 4.1 Tree of Porphyry on page 196 in Gesner's *Nomenclator* of 1560. This was probably the first instance of its usage in natural history. (From the Ernst Mayr Library of the Museum of Comparative Zoology, Harvard University.)

captions were quite diverse in form: they might be mono- or biverbal (*Hyocyanus Aegyptius* and *Hyocyanus niger vulgaris*), descriptive (*Myrtis domesticus fructis albo*) or numerical (*Cistus mas I, Cistus mas II*, etc.).

A remarkable example of quite an advanced stage in the development of the epoch in question was *De differentiis animalium* by Edward Wotton (1492–1555). It was divided into numerous chapters, each containing a detailed description of particular animals. The chapter titles included the latter's names in "vulgar Latin," and in addition, their one-word "official" Latin generic names were given in the margins. For example, Capitum LXXXIIII was entitled "De hystrice, erinaceo, & scyuro," and the generic names *Hystrix, Erinaceus*, and *Sciurus* were provided in the margins in the appropriate places [Wotton 1552]. This illustrates a specific kind of information retrieval system that had been developed by that time to navigate the voluminous books and will be widely used up to the early 19th century, including the works of the "founding fathers" of taxonomic nomenclature.

The completive and most advanced step of the Herbal epoch was marked by the works of the Swiss botanist Caspard Bauhin (1560–1624): their style was transitional between the herbalist and early essentialist routes [Larson 1971; Atran 1990;

De hyſtrice,erinaceo,& ſcyuro.   Cap. LXXXIIII.

HYſtrix etiam latet,totidémque diebus fert utero, & reliqua facit per- *Hyſtrix.*
inde ut urſa.Hyſtrix in Æthiopia frequentiſſima: eam generat & India & Africa.Eſt autem ex erinaceorum genere, & erinaceis ſimilis: ſpinis tergum hiſpida, (aculeos nanque pilorum uice gerit, ut erinacei, ſed longiores) quas plerunque cum intédit cutem,laxatas iaculatione emittit uoluntaria,ut aſſiduis aculeorum nimbis canes uulneret ingruentes. Magni-

FIGURE 4.2 Fragment of page 66 in Wotton's *De differentiis animalium* with the monoverbal generic name in the margin next to the main text. (From the Smithsonian Libraries.)

Cain 1994; Kupriyanov 2005; Pavlinov 2013, 2015a]. Among his works, the most significant were Φυτόπιναξ *seu Enumeratio plantarum* [Bauhin 1596] and its more advanced edition, Πιναξ *theatri botanici* [Bauhin 1623].[4] Bauhin's priorities were underlined by the latter's full title, including "*Methodice secundum earum & genera & species proponens* [...] *ab ipsis exhibitarum nomina cum earundem synonymiis & differentiis.*"[5] This work was divided into numbered books and then into named sections. The latter's titles referred to the plants considered in them, sometimes by one-word names (e.g., *DE GRAMINIBUS*), sometimes by the lists of "genera" in them (e.g., *LYSIMACHIA, VERONICA, TEUCRIUM, SCORDIUM*); most of these sections could be correlated in their rank with later orders and families. They were next divided into groups without indication of their status, and were denoted by one- or two-word names (e.g., *BRUNELLA; HEDERA TERRESTRIS*) and could be roughly identified with later genera.[6] If these "genera" contained a lot of "species," the latter were grouped according to their habitats, and these groups were denoted by subheadings (e.g., *Gramen paniculatum pratense* and *Gramen paniculatum arvense* in the "genus" *Gramen paniculatum*); they probably corresponded to the "proximal genera" of scholastics. Specific descriptive epithets were given in Latin in one- to multi-word forms (e.g., *Gentiana cruciata*; *Gentiana maior pallida*; *Gentiana pratensis floreintus lanuginoso*); less often they were in Greek (e.g., *Solanum μελανοκέρασον*). The rather informal category of section used by Bauhin will be fixed later as one of the main classification categories by P. de Tournefort.

C. Bauhin considered some rules for the formation of names in the preface to Πιναξ. In particular, he indicated that the "species" of the same "genus" must begin with the same generic name.[7] This rule will become one of the most important in taxonomic

---

[4] These books are usually referred to by replacing Latin words for the Greek ones in their titles, i.e., *Phytopinax* and *Pinax*, respectively.
[5] This could be read as "Methodical description [of plants] according to their genera and species [...] providing correct names and differences."
[6] Although C. Bauhin indicated "genera" and "species" among his priority in the title of Πιναξ, he mentioned the former term only in comments to some of the respective groups, while the latter term was sometimes indicated together with the naming of "genera" (e.g., *CYPERUS et eius Species*).
[7] The "invention" of this rule is usually attributed to Bauhin, but in fact it is a norm of Latin grammar presented, for example, in *Naturalis Historiae* by Pliny (see Section 4.2.2).

nomenclature, but Bauhin himself often departed from it: for example, in the "proximal genus" *Clematitis Indica quae lignum Colubrinum* he designated the following "final species": *Clematitis Indica spinoia foliis luteis*; *Clematitis malabarensis foliis Vitis, colore dracunculi*; *Colubrini ligni tertium genus in eadem provincia, vasta arboris magnitudine*; *Radix quaedam in Malaca, quaeadversus vulnera sagittis toxico illitis facta*. In addition, Bauhin paid great attention to the etymology of the names based on their descriptive interpretation. His generic names usually referred to the significant parts of specific names: e.g., the "genus" *Cyperus esculentus* included "species" *Cyperus rotundus esculentus angutstifolius* and *Cyperus rotundus esculentus latifolius*. Subsequently, this technique will be formalized as a general rule, according to which the genuine name of the "genus" should not disagree with the characters of its "species"; it will be followed not only by such essentialists as Tournefort and Linnaeus but also (with reservations) by some developers of the early version of nominalist nomenclature (e.g., [Jussieu 1789; de Candolle 1813]). Many generic names were proposed as new by Bauhin himself, usually through Latin transliteration of the original Greek or vernacular names, while the old classical names often appeared only in the synonymy lists. By this, he laid a tradition of replacing the taxonyms proposed by the predecessors with new more appropriate ones; this tradition will then be developed by essentialist nomenclaturists with their guiding idea of "genuine" names (Morison, Tournefort, Linnaeus).

Summing up the principal results of the development of proto-nomenclature during the Herbal epoch, the following important features should be highlighted. The works of early herbalists were mainly of applied character, so they mostly lacked any concept of natural systematization and an evident hierarchical arrangement of the groups of organisms described in them. At the same time, the scholastic classification categories of "genus" and "species" were quite consistently used in many works of this epoch, although not strictly delineated by their ranks and with no certain biological content. At a later stage in the development of this epoch, the books became arranged more systematically, accompanied by distinguishing and designating classification categories in a way other than in scholasticism: Order (Gesner), Class (l'Écluse), Section (Bauhin). In some of them, the scholastic classification Tree of Porphyry appeared for the first time to demonstrate the hierarchical subordination of "genera" of different levels of generality, with multi-word names referring to their diagnoses. Besides, the synonymy lists and references to the sources appeared in the herbals to lay the respective tradition for future taxonomic nomenclature.

5 The Essentialist Route

The first stage in the history of scientific systematics was marked by the development of a scholastic research program [Pavlinov 2018, 2021] (see Section 1.1). The ontological basis of its natural method was shaped by essentialist natural philosophy supplemented at the methodological level by the logical genus–species scheme.

A quite specific conceptualization of the descriptive language of early systematics became an important part of this research program. Its natural-philosophical foundation presumed comprehension of the proper places of organisms in the System of Nature corresponding to their essential properties interdependently with the comprehension of their genuine names reflecting those properties. As a result, the essentialist concept of nomenclature was developed based on the essentialist interpretation of descriptive taxonomic names.

It is important to emphasize that this concept appeared as a purposefully elaborated part of the professional language of scientific systematics. Therefore, it should already be called *taxonomic nomenclature* with sufficient reason, in contrast to the predominantly folk nomenclature of herbals.

5.1 MAJOR FEATURES

The most important characteristic feature of the essentialist understanding of classifying and naming of organisms is that their onto-epistemic premises largely coincide. Indeed, as just noticed, comprehending the essences of organisms is considered a necessary precondition of both their arrangement into natural groups and the formation of their genuine names. Due to this, systematization merges, in a sense, with naming, and systematics becomes boiled down, to a certain extent, to nomenclature. This allows one to understand the reason for such great attention that both "founding fathers" and finishers of scholastic systematics (such as Jung, Rivinus, Tournefort, Linnaeus) paid to elaboration of the rules for naming organisms. In particular, Linnaeus identified the nomenclature as the "second foundation" (along with the arrangement) of botany; in fact, it would be more correct to say—of the whole of systematics. Due to this, the development of the latter's natural method inevitably involved the purposeful development of the regulative nomenclature for the proper treatment of genuine names of plants and animals.

With this, the principles of semantic motivation and binarity become of prime importance for the developing essentialist nomenclature. The first principle presumes the essentialist interpretation of the taxonyms assigned to the essentially defined taxa: the latter must be given *appropriate*, or *genuine* (legitimate) names properly reflecting the essential features of the organisms allocated to them. Such descriptive names serve in many ways as taxonomic diagnoses, so they contain as many words as are necessary to reflect most adequately the presumed essences; hence the verbosity of the essentialist names. The second principle is based on the hierarchical genus–species scheme built into the natural method of scholastic systematics, so it is scholastic in its root. According to it, each taxonomic *N*-object is designated following the logical formula for the definition of an object (or a notion associated with it) through "generic common and species particular" (see Section 4.2.2).

Two important rules follow from the conjoint implementation of these two principles so that taxonym formation meets the provisions of the essentialist concept of nomenclature. First, according to the genus–species scheme, the complete name (taxonym) of each taxon in the classification hierarchy, except for the "higher genus," must be *binary*, i.e., consisting of two parts. The first part of a taxonym thus construed corresponds to the "generic common," and its second part corresponds to the "species particular," allowing for each "species" to be distinguished from others in the inclusive "genus."[1] Second, the essentialist name of "species" is context-dependent. This context is determined both by the "species" composition of the inclusive "genus" and the characters by which all its "species" are distinguished [Sharp 1873; Svenson 1945; Pavlinov 2015a]. This important circumstance is clearly expressed by Linnaeus, who asserted "that a character does not make a genus, but the genus makes the character" [Linnaeus 1751: § 169], therefore, the genuine specific name "should contain nothing else but only the feature by which it may be distinguished from others [species] of the same genus" [Linnaeus 1751: § 293]. As a result, the changes in either "species" composition of the "genus" or reconsideration of their essential features yield changes in their names.

This general norm affects the lexical structure of the names of "genera" and "species" as follows. The more "species" (subtaxa) are distinguished in a certain "genus" (inclusive taxon), the more traits are needed to distinguish them, therefore, the longer their complete name-diagnoses turn out to be. This rule applies primarily to specific epithets, but generic names can also be biverbal (e.g., *Primula Veris* of Tournefort). Thus, in *Species plantarum* by Linnaeus, the genuine name of "species" consists of three or four words in a small "genus," while it is longer and bulkier in a large "genus." The examples are, respectively, *GNIDIA foliis oppositis lanceolatis* and *VERONICA floribus solitariis, foliis digitato-partitis pedunculo longioribus* [Linnaeus 1753]. In such long composite taxonyms, the lexemes referring to different parts of plants are usually separated by a comma.

> It should be noted, however, that in scholastic systematics, just as in herbalistics, not all names were descriptive by their etymology. An example was a small

---

[1] The notions of "genus" and "species" are given in quotes here, as in the preceding chapter, to indicate their formal (non-biological) interpretation presumed by the logical genus–species scheme.

book *Nova plantarum americanarum genera* by the French botanist Charles Plumier (1646–1704) with descriptions of New World plants: he established the generic names as eponyms (*Plinia, Pittonia, Brunfelsia, Cordia*, etc.) or Latinized indigenous names (*Arapabaca, Nhandiroba*) [Plumier 1703].

The essentialist interpretation of taxonyms entailed a potential threat to the instability of nominative nomenclature. The reason was that systematists might disagree about particular properties of organisms to be considered essential, so they elaborated classifications that differed not only in the composition of "genera" and "species" but also in their names. For instance, gypsophila was assigned the following descriptive specific epithets in the works of the 17th–18th centuries [Smith 1821]: *Polygonum erectum angustifolium floribus candidis* (Menzelius), *Caryophyllum saxatilis foliis gramineis umbellatis corymbis* (Bauhin), *Lychnis alpina linifolia multiflora perampla radice* (Tournefort). It is currently known under the trivial name *Gypsophila fastigiata* (Linnaeus).

Almost all the works of scholastic systematists were written in Latin. Accordingly, the bilingualism of descriptions and taxonomic names inherent in some herbals nearly disappeared. Thus, the linguistic principle (rule) of Latinization became dominant.

The development of descriptive language of scholastic systematics in an essentialist way inevitably affected the development of detailed anatomical nomenclature: it began with J. Jung and continued with C. Linnaeus. A clearer formulation of the language of partonomy (i.e., the theory of parts), which went back to the herbalist H. Bock (Tragus), became an important prerequisite for the elaboration of the descriptive language of systematics itself, including rules for the formation and assignment of diagnostic taxonyms. For this reason, many of the writings of early systematists provided anatomical dictionaries with standard anatomical terms. In this regard, it is worth noting that the development of post-scholastic systematics also began with a more detailed development of anatomical descriptions, which "logically" preceded taxonomic ones [Pavlinov 2018].

The history of the development of rules for the formation and assignment of taxonomic names by scholastic systematics was quite gradual, spanning almost three centuries. It was closely linked to the development of the natural method of this research program, so the two main stages can be clearly distinguished in its history [Pavlinov 2015a, 2019]. The first stage covered the 16th–17th centuries, and was associated with the introduction of the first suggestions concerning the development of the unified nomenclature rules. At the second stage covering the end of the 17th and nearly the whole of the 18th centuries, key ideas were consolidated and strengthened; it ended with Linnaean reform. Throughout this long history, there was a progressive accentuation on the essentialist interpretation of the names of "genera" and especially "species," so the early scholastic systematics was less essentialist in the interpretation of taxonyms than later ones. At the same time, the development of the nominalist interpretation of names, which were used simply as "labels" of taxa, took place against a background of the predominant trend.

5.2 PRE-LINNAEAN STAGE

The first taxonomists, with whose works the scholastic stage in the development of systematics began, were essentialists in the interpretation of taxa. With this, paradoxically enough, they were not especially inclined towards the essentialist interpretation of taxonomic names. On the contrary, their nominative nomenclature systems contained noticeable nominalist elements that will form the basis for the future nomenclature of post-scholastic systematics. By this, they appeared to be pioneers in the initial development of contemporary binomial nomenclature.

Thus, in *De plantis libri XVI* by the Italian Aristotelian logician and systematic botanist Andrea Cesalpino (1524–1603), the Index contained all monoverbal names of "genera" and all biverbal names of "species," the latter with one-word specific epithets [Cesalpino 1583]. In *Methodi herbariae libri* by the Bohemian physician and systematic botanist Adam Zalužiansky (1558–1613), a kind of "double" nomenclature could be found. The multiverbal essential names of plant species were provided in the main text, and accompanied by trivial names in the margins as elements of an information retrieval system. Those of "genera" were monoverbal (e.g., *Panicum, Robur*), and those of "species" were biverbal (e.g., *Arundo vulgaris, Ochrus sylvestris*) [Zalužiansky 1592]. This "double" nomenclature system will be applied consistently by C. Linnaeus (see Section 5.3).

German philosopher, mathematician, and physician Joahim Jung (Lat. Jungius; 1587–1657) was probably the first to pay special attention to nominative nomenclature. His main ideas, relevant to the subject of this book, were presented in the posthumously published *Doxoscopiae physicae minores* [Jungius 1662] and especially *Opuscula botanico-physica* [Jungius 1747].[2] He made an important attempt to develop clear formulations of the *differentiae* of plant "genera" by elaborating formalized descriptions of unambiguously identified and terminologically fixed anatomical parts of plants. His *Opuscula* included a section *De Nominibus Plantarum* with consideration of certain rules for the formation of "true" generic names and specific epithets on morphological, geographical, eponymous, and other grounds. Jung had a negative attitude to the interest of his botanist colleagues in the compilation of extensive lists of synonyms; he declared many of them "false" and to be rejected and consigned to oblivion. In another section, *Homoida in Quibusdam Plants Neglecta*, Jung provided an extensive alphabetical list of generic names. The latter were almost constantly monoverbal (*Aconiti, Centaurii, Fumariae*), while rare biverbals were stable word combinations (e.g., *Dentis Leonis*). The complete designations of "species" began with generic names and were almost always two-word (*Gratiola Minor, Hyacinta Poetici*), less often three-word (*Jacea Lutea Spinosa*). It is noteworthy that mono- or biverbal specific epithets were most often the fragments of multiverbal phrasemes of those used by herbalists that Jung considered most appropriate.

Unlike their closest forerunners, scholastic systematists of the next generation developed clearly essentialist nomenclature, with its characteristic verbose descriptive

---

[2] However, his ideas about classifying and naming plants outlined in *Opuscula* became available much earlier in hand-written form; in particular, J. Ray was familiar with and influenced by them [Raven 1986].

(a)   *pingvifq,. Folia triangula. Fructus juba paniculacea minuti seminis.*
      Effig: apud Dod: 3. 1. 14.

Differentiæ Cyperi plures, fed vulgatiſſimæ duæ, cyperus longus & rotundus.

1. **Cyperus longus radice oblongiore, geniculata, et**
Longus. *graminis modo ferpente, caule altiori, folio majori, ab imo duntaxat, & in cacumine hærente, plura in orbem diſponuntur. Effig: & apud Math: lib. 1. Dioſc: cap. 4.*

Papyrus.   **Similem ej naturam habet papyrus Ægytia,**
Prawy   *quam demonſtrat Pena inter juncos & juncea.*
papir.   **Cyperi rotundi radix pixidulæ, vaſculive, aut**
2.   *poculi puſilli effigie habet, & inde κυπαιρον ſeu κυπειρω apparet dictam,*
Cyperus   *caule & folio minori.*
rotudus.

Dulcichinum eſt cyperus dulcis. In conſ: par:
Traſi Ve   *vere germinat: radice ori grata & eduli, inde traſus ſive tragus apellata:*
ronenſi-   *exotica eſt Italia & agri Veronenſis duntaxat, autore Mathi: ſeritur*
um.   *radice. Math: lib: 2. Dioſ: cap. 138. Dod: 3. 1. 15.*

(b)   ## MONENDVM XXVIII.

Denique, de herba qualibet in hac Iſagoge quod notatur obſeruatum, id requirenti vt ſit in promptu, en exhibetur Index Alphabeticus herbarum omnium hic commemoratarum cum adſcripto numero capitis, paragraphi, & lineae vbi reperire de ea liceat. Inſerta eidem Indici ſunt etiam alia pauca, quae non ſuo loco ab Auctore tradita poſſunt videri, vt argumentis capitum cum indice coniunctis totius operis materia ſit conſpicua. In indice tamen hoc Synonyma coniungere nunc non licuit. Reperientur ergo ſaepe obſeruationes diuerſae de eadem planta ad diuerſa eius nomina. v. g. Eiusdem Plantae alia monſtrabit Equiſeti, alia Hippuridis vocabulum. Quod vni appellationi Auctor non inhaeſit, voluit ille procul dubio plura appellamina vſu auditoribus reddere familiaria.

A.
Abies ix, 7, 2. 14, 4. x, 5, 2.
Abrotonum foemina xix, 8.
Abſinthium xxii, 19, 18.
   vulgare xxiv, 9, 6.
Acanthus x, 1, 3. xvii, 12, 1.
   ſatiuus xvii, 6, 7.
Aconita xvii, 14, 3.
Aconitum iii, 19, 5.
   corniculatum xxviii, 8, 2.
   hyemale viii, 9, 4.
Adiantum v, 6, 2. x, 2, 4. xxviii, 6, 3. 4.
Admirabilis Peruana ix, 13, 1.
Admorſa xix, 5, 2.

Æthiopis xxiii, 19, 2.
Ageratum xxii, 19, 14.
Agrimonia iii, 26, 2.
Alſine vii, 9, 3. xvi, 15, 3. xviii, 3, 4.
Amygdalus vi, 5, 4.
Anagallis aquatica xii, 9, 1.
Anchuſa viii, 6, 6.
Anemonae xxii, 21, 6.
Anemone viii, 9, 3. xi, 2, 2. xxi, 2, 2.
Anethum xi, 10, 2.
Anſerina vi, 5, 2.
Aparine ix, 7, 1. 14, 4.
Apium xi, 10, 2. 19, 10.
Apium hortenſe viii, 5, 3. 6, 8.
                                    Aphyl-

FIGURE 5.1 Examples of usage of different kinds of names by early systematists: (a) fragment of page 130 in Zalužiansky's *Methodi herbariae* of 1592 showing "double" nomenclature with descriptive multiverbals in the main text and designative mono- and biverbals in the margins; (b) fragment of page 60 in Jungius' *Opuscula botanico-physica* of 1747 showing consistent use of binomial nomenclature in the Index. (From the Smithsonian Libraries.)

names of "genera" and "species." With this, the last two decades of the 17th century appeared to be amazingly productive with regard to the development of the foundations of taxonomic nomenclature. At that time, the most significant works of Ray, Rivinus, Magnol, Tournefort, etc., were published with the first formulations of certain basic principles of nomenclature. They clearly marked the maturation of its essentialist concept, and most of them, after their consolidation by Linnaeus, will be inherited from it by the nominalist one.

A striking example of such consistently essentialist nomenclature was *Plantarum umbelliferarum* by Robert Morison (1620–1683). This book was abundantly illustrated with the Trees of Porphyry, and not with one or two, as in some advanced herbals or in Jungius' *Opuscula*, but for virtually all major groups [Morison 1672]. It was arranged into a rather sophisticated hierarchy of chapters, capitals, tables, sections, etc. Plant groups classified under the Caput category were denoted by one- or two-word names (e.g., *Angelica, Paftinaca aquatica*) or by other variants. For each such group, numbered lists of species were provided, with their taxonyms beginning with generic names followed by verbose specific epithets (e.g., *Oenanthe maxima Virginiana Paeoniae feminae foliis*; *Oenanthe Apii folio minor, caule firmiore*).

The scholastic nature of nomenclature of one of the leading taxonomists of the epoch under consideration, the English naturalist John Ray (Old Engl. Wray, Lat. Johannis Raj; 1627–1705), was manifested in the following features. He called all major taxonomic categories "genera," ranking them as "primary," "secondary," and "tertiary." He illustrated classifications of higher "genera" by the Trees of Porphyry, with indication of characters of the respective subaltern "genera" on them. In *Methodus plantarum nova* [Ray 1682]), the "higher genera" were designated by indicating their essential features (e.g., *Herbae Flore Stamineo*) or monoverbally (e.g., *Fungi, Litophyta*). The "lower genera" were designated by mono-, bi-, or three-word names (e.g., *Porus, Fungi lapidei, Musci marini lapidei*). It is noteworthy that in later *Synopsis methodica animalium quadrupedum*, Ray often provided two taxonyms for each "genus": one was a multiverbal phraseme characterizing its position in the classification and its essential features, while another was its monoverbal "proper" name. For example, he designated the genus of deer as "*Quadrupeda Ruminantia bisulca cordibus deciduis ramosis, seu Cervinum Genus*" [Ray 1693].

A very important step in the rational development of taxonomic nomenclature was taken by the German physician and phytographer Augustus Quirinus Bachman (Lat. Rivinus; 1652–1723). In botany, he became best known for his fundamental three-volume work *Ordo Plantarum* (1690–1699). He presented his systematic method, including a pretty detailed consideration of nomenclature, in its introductory section published separately as *Introductio generalis in Rem herbariam* [Rivinus 1690]. Rivinus seemed to be the first to contribute expressly to elaboration of the regulative nomenclature by suggesting a set of several important rules for the formation and assignment of taxonomic names. The first and most important of them sounds like this: "there are as many distinct of plant generic names as there are distinct genera" [Rivinus 1690: 9]. This clause evidently meant that each genus should be designated by a unique name, which was certainly the first formulation the principle of monosemy. With this, Rivinus stated the discriminating character of the "generic" names as a "fundamental rule: *Those different in seeds or flowers, should be indicated by different names.* And vice versa. *Those similar in seeds or flowers, should be indicated by similar names*" [the same page, italics in the original]. The next followed two important rules for the selection of names: according to the *rule of homonymy*, of all usages of a given name to designate different "genera," the antecedent one should be retained; according to the *rule of synonymy*, of all names used to designate a given "genus," the most appropriate should be preferred. Thus, the first of these rules anticipated the nominalist principle of priority, while the second one fitted entirely into the essentialist concept. The main provisions for the names of taxa formulated by Rivinus were as follows: universality (*universalis*), expressiveness (*clara*), singularity (*distincta*), and constancy (*constans*) [Rivinus 1690: 18–19]. It was also suggested that all names should be used in a Latinized form only, even if they were borrowed from other languages (Greek, Arabic, native).

Rivinus took a decisive step towards the constant use of monoverbal designations of "genera." Criticizing the then frequent use of multiverbal generic names by his predecessors and contemporaries, he suggested using only monoverbal names. So, commenting on the above-mentioned Morison's *Plantarum umbelliferarum*, Rivinus established *Lupulus* instead of Morison's *Convolvulus perennis heteroclitus floribus...* (nine words in total), and reduced his *Melitotus siliculis pendentibus curtis...* (seven words in total) to just *Melitotus* [Rivinus 1696]. According to Rivinus, the generic names could be (in modern terminology) morphonyms (*Bicapsularium, Umbelliferarum, Unifolium*), topo-, epo-, zoonyms (*Armeniaca, Asclepias, Pedicularis*), analogical (*Cruciata, Sagitta*), or biologically indicative (*Convolvulus, Filipendula*).

Several others of Rivinus' clauses concerned "species." Thus, he suggested making mandatory a commonly accepted rule, according to which the complete name of "species" must begin with the name of its "genus," which made it binary. Following another rule, the only "species" in a monotypic "genus" should be listed without a specific epithet. In a polytypic "genus," it was also proposed that the first "species" in the list should be listed without a specific epithet, while other species should be assigned monoverbal epithets, and not necessarily descriptive ones (for example, *Solanum fruticosum, Amomum Plinii*). Thus, his species nomenclature was rather consistently binomial, and only partly essentialist.

An important contribution to the initial development of the ranked hierarchy was made by the French botanist Pierre Magnol (1638–1715): he introduced the category of Family in his *Prodromus historiae generalis plantarum* [Magnol 1689]. However, the hierarchical structure and nomenclature in this work remained in general quite traditional for that time. Like Bauhin, he divided his work into numbered Sections, while the families subordinated to them were designated by diagnostic features (e.g., *Plantarum Bulbosarum*; *Florem imperfecta, stamineo, & cum parvis*; etc.). The hierarchical structure of classifications of each family was represented traditionally by the parenthetical Tree of Porphyry, with indication of the characters distinguishing the subordinated groups (e.g., *Folia Latio ribus, Folia angustioribus*, etc.). Hierarchical categories of these groups were not specifically termed; they were designated mono- or biverbally (*Plantago*, *Cornu cervi*, etc.) and might be identified as "proximal genera."

The pre-Linnaean stage in the development of the essentialist taxonomic nomenclature was completed by Joseph Pitton de Tournefort (1656–1708), called "the father of French botany" [Larson 1971; Dayrat 2003] for his influential three-volume *Élémens de botanique* [Tournefort 1694].[3] From the perspective of the dominant trend, his most significant contribution apparently was establishing the first explicitly ranked taxonomic hierarchy with four fixed categories, namely Class, Section, Genus, and Species. His Class corresponded to the "book" used by herbalists to divide their voluminous treatises in parts; it was universally adopted subsequently to mark the major category of the taxonomic hierarchy. His Section (possibly borrowed from Bauhin) corresponded in its rank to the Order of Rivinus (which will be applied by Linnaeus). The main category for Tournefort was the Genus: it corresponded formally to the "proximal genus" of scholastics, but appeared in Tournefort's *Élémens/Institutiones* as a natural unit taking a definite position in the hierarchy of the System of Nature. For this, Tournefort is sometimes called the "father of the genus concept" in systematic botany [Bartlett 1940; Stuessy 2008]. With the Genus now fixed, the Species also ceased to be but a last step in the genus–species scheme, and became the fixed lowest unit in the hierarchy of the System of Nature.

> Focusing on genera, Tournefort had little interest in species; therefore, he was inclined to elevate many morphologically well-delineated species to generic rank; this made his generic classification very fractional due to the recognition of a lot of monotypic genera [Sachs 1906]. This was a noticeable departure from the genus–species scheme of scholastics, in which there could be no such thing as a "proximal genus" not divided into "final species." With this, his species were also quite "narrow" as Tournefort ranked varieties (in their later Linnaean understanding) as full species, in full agreement with scholastic tradition.

Tournefort's attitude towards taxonomic names was almost reverent. He wrote in his *Élémens* that none other than the "Creator of all things Himself has endowed us with the right and ability to name species" [Tournefort 1694: 3] (obviously referring to *Genesis* 2: 19–20) and that "knowing the plants is tantamount to knowing their names, which are given to them in relation with the structure of certain parts of them [therefore] the study of plants should begin with their names" [Tournefort 1694: 1]. He adhered to the strictly essential interpretation of taxonyms: "an idea of character, essentially distinguishing some plants from others, should be invariably associated with the name of each plant" [Tournefort 1694: 2]. Recognizing the particular significance of generic names, Tournefort casually changed those of his predecessors to his own that he deemed more appropriate (Bauhin had done the same before him, and Linnaeus does so after him). Thus, he wrote elsewhere, partly echoing Rivinus:

---

[3] This work is better known in its Latin edition, *Institutiones rei herbariae* [Tournefort 1700].

The Essentialist Route

What is the need, for instance, to follow Morison in calling the bindweed *Convolvulus heteroclitus perenni floribus foliaceis strobili instar*? Wouldn't it be better to establish a separate genus for it and ascribe to it the name *Lupulus vulgaris*, which is widely known?

Tournefort 1694: 38

The practical implementation of Tournefort's ideas of nomenclature was as follows. Classes and sections were numbered, without names, and provided with short diagnoses. All genera were denoted by names, most often monoverbal, rarely biverbal (for example, *Auricula ursi*, *Centarium minus*); these names were not diagnostic, and therefore they might have the meaning of "trivial" ones in the sense of Linnaeus (see Section 5.3). Species nomenclature was strictly binary, but not consistently binomial. According to already established tradition, the complete species taxonym began with the name of its genus, and this rule was often observed for genera with two-word names: for example, in the genus *Primula Veris* there was the species *Primula Veris odorata*. Many plant forms were traditionally considered "species," but designated monoverbally (i.e., they were "generic species" of folk systematics), so they were ranked by Tournefort as monotypic genera, transforming their specific names into generic ones and leaving their sole species unnamed (e.g., *Citreum*, *Aurantia*, *Limonia*). However, in some monotypic genera, species were designated according to the binomial rule: for example, the only species in the genus *Haemanthus* was designated as *Haemanthus Africanus*. In polytypic genera, the name of the first species sometimes duplicated the generic one with no specific epithet: for example, in the genus *Centarium minus*, this name also denoted its first species. Species epithets were usually descriptive (essential); they can be mono-, bi-, and multiverbal, largely depending on the number of species identified in a genus and on the order in which they appeared in synoptic lists. For example, in the genus *Genista*, the species were designated as *Genista Juncea*, *Genista humilior Pannonica*, *Genista Hispanica pumila odoratissima*, etc. Obviously, by modern standards, such "species" were actually varieties, as evidenced by their names with repeated specific epithets (for example, *Glofillaria spinola sativa* and *Glofillaria spinola sativa altera*). Tournefort was quite skeptical about the usual practice of providing long lists of synonyms, believing that they were simply "memory-heavy," and therefore limited them to the most commonly used names.

The Swiss physician and naturalist Karl Nikolaus Lang (1670–1741) was noteworthy for his consistent application of the method of Tournefort to systematic zoology [Bachmann 1896]. In his *Methodus nova*, he used the Tournefortian taxonomic hierarchy consisting of Class, Section, Genus, and Species [Lang 1722]. Taxonyms of the classes were mono- to multiverbal (e.g., *Buccina*, *Cochlea marina longae*, *Testacea marina univalvia non turbinata*); some sections were simply numbered, while others were designated by names (*Strombi ore superius aperto*, *Turbines aperti*). The generic taxonyms were monoverbal (*Patellae*, *Balani*) or multiverbal, in the latter case with the full generic name beginning with the section name (*Strombi sulcati vulgares*, *Turbines aperti lati*). The latter variant was probably an extended interpretation of the rule introduced by Bauhin and his followers for the complete binary names of "species." In *Methodus testacea*, each such name consisted of the monoverbal generic part and the almost universally descriptive multiverbal specific epithet (*Nautilus levis crassus major spina subrotunda, Turbo apertus canaliculatus rectirostrus laevis*). With this, the species nomenclature of Lang could be occasionally binomial: in the genus *Nerita*, the species *Nerita laevis* and *Nerita striata* were distinguished, but there was also *Nerita levis strumosa ore fimbriato*.

Based on the considerations presented in this section, the following important features of the earliest stage of development of taxonomic nomenclature should be highlighted.

Its first advances were largely determined by the adherence of nomenclaturists to the classification genus–species scheme of scholastics. Therefore, the early classifications of plants (and occasionally animals) took the form of an inclusive hierarchy of "primary," "secondary," etc., "genera" (or book chapters, sections, etc.) with non-fixed ranks, while "species" meant only the last step of deductive division. Later, an obvious aspiration for a more rigid structuring of the taxonomic hierarchy became evident. Following this new trend, non-canonical categories, such as Class, Order, Section, Family, began to be used in classifications along with non-ranked "genera." In the most advanced version (Tournefort), a system of several strictly fixed and subordinated ranks/categories was elaborated. A clear rank distinction between the "proximal genus" and the "final species" was especially significant: they became genus and species in an almost modern sense, although still predominantly classificatory rather than biological. The latter was evident from the fact that monotypic genera began to be recognized and designated often as just genera, and not "final species."

The names of taxa of all categories, be they scholastic "genera" of different levels of generality or non-canonical categories, was predominantly descriptive, most often essential, and therefore usually multiverbal. Along with this, monoverbal specific epithets that performed a "label" function (in particular, eponyms) were becoming more widespread. The "double" nomenclature was consistently applied in some treatises, according to which the main genuine (essentialist) and auxiliary trivial ("formal") names, as Linnaeus will call them later, were used simultaneously to designate the same plants. The strict adherence to Latinization of nominative nomenclature, even in books written in native languages, led to the fixation of grammatical rule for the formation of binary taxonyms consisting of "generic" names followed by "specific" epithets.

However, the absence of any generally accepted standard rules for the formation and assignment of taxonomic names was reflected in that each leading systematist followed his own nomenclature system, and this multitude gave rise to a surplus of synonyms. All this became a serious obstacle for classifications to perform an important function of a unified and stable information retrieval system. Therefore, the task of elaboration of a certain fixed system of nomenclature principles and rules to be adopted by the whole (or at least majority) of the taxonomic community became quite urgent. This task was solved in the first approximation by the Linnaean reform of the mid 18th century.

5.3 LINNAEAN REFORM

For the nomenclature reform to happen in systematics of the 18th century, dictated by the logic of the development of natural science, it was necessary that two factors, objective and subjective, coincided.

The first factor was the need for actively developing systematics for a universal, rigorously regulated professional language to describe the System of Nature. Indeed, in the 18th century, the aspiration to regularize scientific language through the development of rules for the formation of the relevant notions and terms, i.e., nomenclature in its general sense, was inherent not only in biology. Thus, this aspiration was sufficiently manifested in chemistry: the conceptual revolution in it, associated mainly with the name of the French chemist Antoine Laurent de Lavoisier (1743–1794), involved reformation of its descriptive language including elaboration of a new nomenclature system [Whewell 1847].

The second factor manifested itself in the person of the Swedish naturalist Carl von Linné (Lat. Carolus Linnaeus; 1707–1778), whose life fell into the "golden age" of classical science under the patronage of the Royal House of Sweden [Skuncke 2008]. The great mission of the botanist-reformer, which Linnaeus took upon himself, was due to his passionarity and high self-esteem [Boerman 1953; Bobrov 1970]. Thus, he characterized himself in one of his letters as "the second Adam," thereby emphasizing the right, as it were, granted to him by the Supreme Being to give new names to the objects of all three "Kingdoms of Nature" [Koerner 1996; Harrison 2009]. And it was Linnaeus himself who called the whole of his enterprise *Linnaean reform* in his autobiography [Bobrov 1970; Müller-Wille 2007].

Linnaeus' views on regulative nomenclature were consistently developed and presented in a series of three books. The series began with *Fundamenta botanica*, about 40 pages long with a list of 365 declaratively (without comments) presented canons (sometimes also called aphorisms or precepts) that were grouped in 12 chapters [Linnaeus 1736]. This was followed by *Critica botanica*, devoted exclusively to taxonomic nomenclature, as evidenced by its subtitle: *in qua Nomina plantarum generica, specifics, & variantia* [Linnaeus 1737a]. In this book, unlike in *Fundamenta botanica*, each canon was commented on in detail: justified and explained, and examples of the application of the stated nomenclature rules were given. *Philosophia botanica*, more than 300 pages long, completed the series [Linnaeus 1751], though its comments on the canons were not as detailed and thorough as in *Critica botanica*.[4] This long-lasting series of books by Linnaeus, that began in the 1830s and finished in the 1750s, resulted in a rather advanced nomenclature system that culminated in his reform and, with it, the whole of essentialist nomenclature. It should be rightly called **Linnaean Canons**.

In the text below, references to particular paragraphs are given mainly after *Philosophia botanica* indicated by numbers; quotations are taken mainly from its last English edition [Linnaeus 2003]; italics in these quotations are in the

---

[4] For English translations of *Critica botanica*, see: [Linnaeus 1938] and partly [Linsley and Usinger 1959], and of *Philosophia botanica*, see [Linné 1775; Linnaeus 2003].

original throughout. In addition, some important provisions of Linnaeus' nomenclature are given with reference to two books devoted to the botanical classification system: *Genera plantarum* [Linnaeus 1737b, section *Ratio operis*] and *Species plantarum* [Linnaeus 1753, section *Lectori aequo*].

The great attention paid by C. Linnaeus to nomenclature was evident from his having singled out in botany (actually in systematics) two equal "foundations [...] *arrangement* and *nomenclature*" [§ 151]. Echoing Tournefort, he believed that "if you do not know the names of things, the knowledge of them is lost too" [§ 210], therefore "nomenclature [is] the second foundation of botany" [§ 210].

The general basis for the nomenclature system of Linnaeus, as a religious person [Petri 2001], was shaped by the natural-philosophical world picture, the core of which was the belief that God "is Nature, for all things are born of Him [...] Nature is irrevocable God" [Linnaeus 1766, Vol. 1: 11], and therefore "Nature is the law of God" [Linnaeus 1788: 113]. Accordingly, Linnaeus called botany a "divine science" and believed that, to describe plants by giving them genuine ("true") names according to their nature (essence) was to make explicit what is hidden in them [Stearn 1955, 1959]. This means that his nomenclature system, just like those of other scholastic systematists, was fundamentally conceptualized. He pointed out that the "theoretical [arrangement] establishing classes, orders, and genera [...] is the foundation of nomenclature" [§ 151, 152]. With this, he emphasized that "The names of plants ought to be definite, and accordingly placed in definite genera" [§ 151], and the "names of classes and orders [...] may be [given] only for natural classes" [§ 254] This is one of the most fundamental provisions of Linnaeus, and it fits the *principle of verity* in both taxonomic and nomenclatural meanings. Obviously, for Linnaeus, definite genera and natural classes (taxa in general) were those that were parts of the Natural System. And although at the time of Linnaeus there were no consistently elaborated taxonomic theories with explicitly formulated criteria of naturalness (they do not appear until the 19th–20th centuries, see [Pavlinov 2018, 2021]), this does not change the general meaning of this theory-burden principle.

The theory dependence of the basics of Linnaeus' nomenclature is clearly evident from his asserting the general rules for the formation of names of taxa of different ranks. For him, as well as for other scholastic systematists, the category of genus was of particular importance: he emphasized that "all genera are natural" as the "work of Nature," while classes and orders are "work of Nature and Art." For this reason, he paid most attention to the names of just the genera, while hardly touching on the higher categories (considered in the section on generic names). According to this natural philosophy, it was important for Linnaeus to make evident from the names of plants to which genera they belonged, so "when the species [were] referred to genera, the species should take the name of the genus" [§ 284]; in contrast, "the names of the class and the order must never be included in the name of the plant" [§ 212].

The main nomenclature provisions of *Linnaean Canons* can be classified as follows. First of all, they are divided into *theory-dependent* and *theory-neutral*: the former encompass canons that presume the essential interpretation of plant names through their connection with plant features, whereas the latter do not. Second, they can be divided into *rank-dependent* and *rank-independent* according to their relation to taxonomic hierarchy: the former encompass specific rules for the formation of names of taxa according to their position in this hierarchy. Rank-dependent canons are theory-dependent to the extent that taxonomic hierarchy is theory-dependent in itself. Finally, these rules can be divided into semantic (stability, priority, uniqueness, etc.), grammatical, and etymological.

Different principles and rules were fixed by Linnaeus in varying degrees of detail, which most probably reflected their importance to him. Some canons were formulated quite clearly as rigorous prescriptions and prohibitions on the formation and choice of names: uniqueness and universality, the number of words in them, which names were genuine (legitimate) and which were erroneous, etc. Others were rather outlined in the form of recommendations to which Linnaeus himself did not always adhere (for example, priority); they receive a clearer formulation in the works to appear after Linnaeus.

> In connection with the essentialist understanding of descriptive names, Linnaeus listed quite scrupulously those anatomical features that, in his opinion, could be considered as the basis for their formation. Following the tradition established by Jung, he devoted much of Sections V and VI to this important issue and divided, first of all, the characters into "the *factitious*, the *essential*, and the *natural*" [§ 186]. This partonomic enterprise involving standardization of anatomical parts of plants and their terminology developed by Linnaeus is sometimes considered to be one of his greatest achievements [Lyubarsky 2018], though this was probably important to him not in itself, but as a basis for the development of genuine taxonomic names [Pavlinov 2013, 2015a].

For a more rigid structuring of the classification basis, as a prerequisite for the consistent application of the nomenclature rules, Linnaeus fixed the hierarchy of taxonomic categories (ranks) and termed them Classis, Order, Genus, Species, and Varieties [§ 155]. These terms were almost entirely borrowed from Tournefort; the difference was in replacing Section by Order (of Rivinus) and in adding Varieties. Linnaeus clearly indicated in *Critica botanica* [§ 251] both their subordination ("Genera consist of Species, Orders of Genera, and Classes of Orders") and scholastic background ("Classis corresponds to *Genus Summum*, and Order to *Genus intermedium*"). In addition, he mentioned Family in the introductory section of the first edition of *Systema Naturae* [Linnaeus 1735] and in *Philosophia botanica* [§ 78], and employed the highest-rank Empire of Nature in the last editions of *Systema Naturae* [Linnaeus 1758–1759, 1766]. The introduction of Varieties was remarkable for its anti-scholastic character, since it clearly indicated that the "species" was no longer the last step in the genus–species scheme, but a fixed category in the hierarchy of the System of Nature that could be subjected to further subdivisions.

However, Linnaeus' taxonomic categories were not definitely settled: for example, the major groups of plants (fungi, algae, etc.) were called "species" in *Fundamenta botanica* [Linnaeus 1736], "families" in *Philosophia botanica*, and "tribes" in the 12th edition of *Systema Naturae* [Linnaeus 1766]. Moreover, in a brief outlining of the "theory of botany" (section *Delineatio Plantae*) in the second volume of the latter edition, Linnaeus indicated only two main categories, "genus" and "species," with "class" and "order" being included in the "genus" in its general, basically scholastic understanding [Linnaeus 1767]. In addition, he sometimes applied auxiliary categories between "class" and "order," as well as between "genus" and "species."

Carl Linnaeus' main concern was that the descriptive names of taxa should adequately express the essential features of the organisms allocated to them: this clearly indicates that *Linnaean Canons* were developed within the framework of the essentialist nomenclature concept. This treatment is evident from the following canons. The "generic names that show the essential character or habit of the plant are the best" [§ 240]. In agreement with this was the claim (ascending to Bauhin) that "generic names that are inconsistent with any species of their genus are unsatisfactory" [§ 232]. This rule was also true for suprageneric categories: "the names of classes and orders should contain a feature that is essential" [§ 253]. Accordingly, in the case of species, "The specific name is the essential definition" (§ 257), and "the specific name must declare its own [particular] plant at first sight, since it contains the definition that is *inscribed on the plant itself*" [§ 258]. C. Linnaeus called the essential species names *genuine* [§ 257].[5]

Linnaeus' adherence to the essentialist treatment of taxonyms evidently explained his adherence to the principle of monosemy dating from Rivinus and Tournefort. In fact, it had a deep semantic meaning for Linnaeus: since the essence of each definite (natural) genus was thought of to be unique, then the genuine name reflecting it should also be unique. Linnaeus wrapped up this principle, as applied to genera, in several somewhat repetitive formulations, with their meaning best expressed by asserting (ascending to Rivinus) that "Where there is a single genus, there will be a single name" [§ 210].

Regarding the species, very significant (in perspective) was Linnaeus' dividing their designators into two main categories: the above-mentioned *genuine* names, which he called the *specific* (*nomina specifica*), and the *trivial* names (*nomina triviale*). This distinction was certainly theoretically loaded: according to Linnaeus, the genuine specific name is associated with the essence, so it "is the essential definition" (see immediately above), while the trivial name is not associated with the essence and is "free of any law" (see below). Thus, the genuine names were a part of the essentialist concept of the nomenclature, while the trivial ones belonged to its nominalist concept.

The genuine (definite) names were of paramount importance for Linnaeus, therefore many of his canons were addressed just to them. He scrupulously prescribed

---

[5] The essentialist species name was called *legitimum* in the original Latin edition of *Philosophia Botanica*. Its newest English edition offers a literal translation of this term as "legitimate" [Linnaeus 2003]. However, in the first English edition, prepared by Linnaean admirer Hugh Ross (1717–1792) and known to C. Linnaeus, this term was translated as *genuine* [Linné 1775]. This translation went back to *Fundamenta botanica*, where the corresponding term is introduced as *Nomen specificum genuinum* [Linnaeus 1736: § 288]; it seems to be more correct and therefore is used throughout this book.

with respective rules how it was obligatory, possible, undesirable, or unacceptable to establish new names or choose them among previously established ones. Linnaeus placed each of these rules in a separate canon, with comments and examples, listing erroneous names and providing those genuine (legitimate) names that appeared most often to be his own, contrary to the principle of priority. Of these rules, the taxonomic community will accept the following: obligatory Latin spelling of all names [§ 247]; the generic name must be substantive and not adjective [§ 235]; the complete designation of species should consist of "a generic name and a specific one" [§ 212, 256]; "The specific name must always follow the generic name" [§ 285] and it must not be part of the latter [§ 287]. The rules limiting the length of taxonyms were important for Linnaeus: the generic name must be monoverbal [§ 221] and contain no more that 12 letters [§ 249], and the genuine specific name must contain no more than 12 words [§ 291]. Linnaeus justified the need for monoverbal generic names somewhat pragmatically: they "should be easy to remember, whereas only a few [botanists] need to remember species names" [Linnaeus 1737b: *Ratio operis*]. Many of these linguistic rules considering essentialist names will not be retained in the nominalist regulative nomenclature.

A very important feature of *Linnaean Canons* was that they recognized the content dependence of genuine specific names. This seemed to be an evident consequence of the scholastic genus–species "formula," evidenced by the well-known aphorism of Linnaeus: "a character does not make a genus, but the genus makes the character" [§ 169]. Accordingly, the genuine specific name, as far as it was designed to express "a character" that was "made by genus," depended on the composition of the genus [Sharp 1873; Svenson 1945; Heller 1964; Pavlinov 2013, 2015a]; this general provision was expressed by several nomenclature rules. First, "The genuine name for a species should distinguish the plant from all those of the same genus" [§ 257], so it "should include no words except those needed to distinguish it from others [species] of the same genus" [§ 292]. Added to this was that "No specific name may be given to a species that is the only one in its genus" [§ 293].

The genuine specific names were heavily loaded semantically by performing simultaneously three basic functions, viz., classification (indication of the genus to which it belongs), diagnostic (indication of the species character), and designation (individuating the species) [Heppel 1981; Pavlinov 2013, 2015a]. The first function was a consequence of the principle of binary nomenclature, which went back to the classification scheme of scholastics and was obliged to designate organisms by genus–species binomens. The second function was provided by identification of these names with the content-dependent diagnoses of species; this identification also provided the designation function by making it possible to recognize particular species in the semantic space by their unique genuine taxonyms.

The lexical structure of the genuine specific names, shaped by Linnaeus in *Systema Naturae* and *Species plantarum* according to these rules, depended in part on the anatomical specifics of the respective name-bearers and in part on the species richness of the genera. They could be three- or four-word in the small genera (e.g., *GNIDIA foliis oppositis lanceolatis*), whereas they were usually longer in the specious ones (e.g., *VERONICA floribus solitariis, foliis digitato-partitis pedunculo longioribus*). It is noteworthy that the monotypic genera were provided with brief diagnoses in the main text, which could serve as the genuine names of their only species, while the latter were denoted by the monoverbal trivial names in the margins (e.g., *HARTOGIA* and *HARTOGIA capensis*).

As Linnaeus put "trivial names aside [being] concerned only with definitions" [§ 257], i.e., with genuine specific names, so there were almost no rules for the former in *Philosophia botanica*. He provided an important (from a subsequent perspective) observation that such name "free from any laws [...] consist of a single word [...] freely taken from any source" [§ 257]. He saw the advantage of trivial names over genuine ones in the fact that "the definition often turns out to be longwinded, so that it cannot conveniently be used in all cases, and is liable to be changed later on, as new species are discovered" [§ 257]. In the introductory section of *Species plantarum*, trivial names were considered in a separate paragraph *TRIVIALIA nomina*, where their subsidiary character was noted and it was recommended that they should be placed in the margins of books next to the descriptions of plants [Linnaeus 1753].[6] With this, the trivial names did not usually duplicate any parts of the multi-word genuine names; this might be exemplified by two of seven species of the plant genus *Myrtus* (as it was defined by Linnaeus in *Species plantarum*): they were designated in the text as *MYRTUS pedunculus multifloris, foliis ovatis obovatis* and *MYRTUS foliis alternis*, while in the margin they were given as *caryophyllata* and *pimenta*, respectively [Linnaeus 1753].

> A separate chapter of *Philosophia botanica* contained the rules for synonyms defined as "Variant names given to the same plant by phytologists" [§ 318]. Linnaeus believed that "A complete system of synonyms is a thing very necessary to botanists" [§ 318]. He suggested that "Vernacular names of regions should either be excluded [from], or else placed at end" of the list of synonyms [§ 324]. A number of rules concerned citation of sources of synonyms, including the names of their authors, dates, and titles of publications [§ 322]. Many of these rules are still in force.

It can be concluded from the above consideration that the nomenclatural system developed by Linnaeus himself, regarding the designations of species, was not so much strictly binomial as *double* [Pavlinov 2013, 2015a]. This system presumed that two kinds of names were used simultaneously that differed in their functions, the main ones being genuine and the accessory ones being trivial. In the works of Linnaeus devoted to the System of Nature, the genuine names were provided in the main text as the headings of species accounts, while the trivial names were placed in the margins as the elements of information retrieval system. For example, in the

---

[6] It is to be remembered that the practice of placing short specific names in the margins of books, thus making them the elements of an information retrieval system, had begun with book printing and was implemented in the works of several of Linnaeus' predecessors of the 16th–17th centuries (e.g., [Wotton 1552; Zalužiansky 1592; Tournefort 1694]).

The Essentialist Route

(a)

48   MAMMALIA FERÆ. Ursus.

Meles.   3. U. cauda concolore, corpore supra cinereo subtus nigro, fascia longitudinali per oculos auresque nigra.
Meles unguibus anticis longissimis. *Faun. suec.* 15.
Syst. nat. 6.
Meles. *Gesn. quadr.* 686.
Taxus. *Aldr. digit.* 264. *Jonst. quadr. t.* 64. *Raj. quadr.* 185.
Habitat in Europa *inter rimas rupium & lapidum.*
Venatur cuniculos, antra fodit inhabitatque; stercus in eodem loco deponunt inhabitantes extra antrum; die latet, corradit Lathyrum in hyememi, folliculum putorium supra anum haurit hybernans.
Anatome. *E. N. C. d.* 2. *a.* 5. *obs.* 32. *& d.* 3. *a.* 3. *obs.* 163.

(b)

142. PLANTAGO. *Cal.* 4 fidus. *Cor.* 4-fida: limbo
163.   reflexo. *Stamina* longissima.
Caps. 2-locularis, circumscissa.

Scapo nudo.

major.   1. P. fol. ovatis glabris, scapo tereti, spica flosculis imbricatis.

asiatica.   2. P. fol. ovatis glabris, scapo angulato, spica flosculis distinctis.

media.   3. P. fol. ovato-lanceolatis pubescentibus, spica cylindrica, scapo tereti.

virginica. 4. P. fol. lanceolato-ovatis pubescentibus subdenticulatis, spicis floribus remotis, scapo tereti. *Flos in America corollas explicat & stamina exserit, in Europa vero vix; Folia semitrinervia & Scapus pubescentia.*

FIGURE 5.2 Examples of typical "double" nomenclature used by Linnaeus with descriptive multiverbal genuine names in the main text and designative biverbal trivial names in the margins: (a) page 48 in vol. 1 of the tenth edition of *Systema Naturae* of 1758; (b) page 142 in vol. 2 of the 12th edition of *Systema Naturae* of 1767. (From the Peter H. Raven Library, Missouri Botanical Garden.)

first volume of the tenth edition of *Systema Naturae*, the European badger (placed in the genus *Ursus*) was designated in the text as *U. cauda concolore, corpore supra cinerco* […] (12 words in total), and next to it in the margin was denoted the trivial epithet *Meles* [Linnaeus 1758]. In the second volume of the 12th edition, the greater plantain (genus *Plantago*) was designated as *P. folis ovatis glabris, nudo scapo tereti* [...] (eight words in total) and *major*, respectively [Linnaeus 1767].

Linnaeus' closest disciples were "Linnaeans" (as they were called in [Stafleu 1971]) in the sense that they, like their teacher, consistently used the "double" nomenclature. The Danish zoologist Johann Christian Fabricius (1745–1808) should be mentioned first of all for his being sometimes called the "Linnaeus of entomology," and his works are considered the greatest in the early history of insect studies [Tuxen 1967]. His fundamental *Philosophia entomologica* firmly reproduced the structure of the *Philosophia botanica* of Linnaeus, including the titles of some chapters (for example, VII. *Nomina*; VIII. *Differentiae*) [Fabricius 1778]. He formulated the fundamental canons in the same manner as Linnaeus had done: the chapter "*Nomina*" began with the canons "Naming is the second foundation of entomology" and "If the names are lost the knowledge also disappears" [p. 101]. Fabricius asserted the monosemy of generic names in the same manner as Rivinus had done: "one genus, one name, so no confusion" [p. 105]. With this, he justified this principle essentialistically by asserting that "generic character is always the same, so is its name" [p. 105]. However, all his rules for denoting species in the section "*Nomina*" he addressed to trivial names, whereas the section "*Differentiae*" was devoted to the "*differentia specifica legitima*" [p. 122]. J. Fabricius implemented "double" nomenclature in his works that reproduced the style of the latest editions of Linnaeus' *Systema Naturae*. Thus, in his multivolume *Entomologia systematica*, as in Linnaeus' *Systema Naturae*, the trivial names were placed in the margins [Fabricius 1798].[7]

The general style of the Linnaean "double" nomenclature was preserved in many posthumous reprints of his works. For instance, it was followed in the *Entomologia fauna Suecica*: multiverbal genuine specific names were provided in the main text, and one-word trivial names appeared in the margins [Linnaeus 1789]. Thus, the species of the beetle genus *Leptura* were listed in the main text as *deaurata antennis nigris, femoribus posticis dentatis*, etc., with corresponding trivial names *aquatica, melanura*, etc., in the margins. In the reprint of the botanical part of *Systema Naturae* prepared by Linnaeus' disciple, the Swedish botanist and physician Johan Andreas Murray (1740–1791), all species were denoted by multiverbal genuine names in the main text, and by monoverbal trivial names in the margins [Murray 1784].

The same "double" nomenclature was also applied in some original monographs published in the second half of the 18th century. A typical example was *Systema regni animalis* by the German naturalist Johann Christian Polycarp Erxleben (1744–1777), in which all species were designated by multiverbal genuine names in the main text and one-word trivial ones in the margins. For instance, in the mammal genus *Simia*, the first species was designated as *S. ferruginea lacetorum piliis* [...] (six words in total) and *satyrus*, respectively [Erxleben 1777]. Several zoological monographs published in England can be mentioned, in which nomenclature was clearly "double"

---

[7] It is noteworthy that Fabricius will later become one of the early convinced binomialists in the treatment and exclusive use of trivial names (see Section 6.1.2).

with multiverbal genuine names in the main text and monoverbal trivial ones in the margins (e.g., [Latham 1790 ; Kirby 1802]). A slightly defferent variant was applied in *Handbuch der Naturgeschichte* (first edition of 1779) by the German physician, naturalist, and anthropologist Johann Friedrich Blumenbach (1752–1840). In it, both variants of taxonyms appeared in the main text: for each species, its description began with the trivial name, which was followed by its German name, and then the multiverbal genuine name was provided to perform the function of Latin-language diagnosis. Thus, in the bird genus *Strutio*, the description of one of its species began as follows: *camelus*, der Straus. *S. pedibus dldactylis digito exteriore parvo mutico spinis alarum binis* [Blumenbach 1782].

As can be seen from the foregoing, the nomenclature system of C. Linnaeus summarized many basic provisions that had appeared in the books of his predecessors, and his reform of taxonomic nomenclature was aimed, in fact, mainly at improving its essentialist concept. Due to this, the main results of his reform appeared to be a culmination of this concept: he completed its development, just as his classification principles completed the scholastic research program in the conceptual history of systematics [Pavlinov 2018, 2021].

Elaboration of the first full-fledged nomenclature rulebook as an array of clearly expressed and properly arranged numerous rules became probably the most important result of Linnaean reform and its most significant contribution to the subsequent development of the whole of regulative nomenclature. In fact, *Linnaean Canons* will provide a powerful impetus to the further rationalization of the latter and will become a kind of model for all future nomenclaturists.

With regard to the more particular points of this reform, as it seemed to be conceived by Linnaeus himself, one of its principal advances was the set of rules for names of genera and species that would reflect the essences of organisms allocated to them. In particular, his rules for species nomenclature accentuated the verbose essentialist designators he called genuine specific names. However, along with the latter, he actively used non-essentialist one-word trivial names as auxiliary elements to navigate his systematic catalogues.

It follows from the preceding that, despite a widespread misconception going back to the mid 19th century [Strickland et al. 1843a, 1843b] and present in a number of contemporary historical reviews (e.g., [Nicolson 1991; Dayrat 2010]), establishing binomial nomenclature was not an inherent part of Linnaean reform [Whewell 1847; Svenson 1945; Ramsbottom 1955; Uzepchuk 1956; Stearn 1959; Larson 1967, 1971; Pavlinov 2015a, 2019]. In fact, as emphasized above, the typical Linnaean nomenclature was "double," according to which species were designated by two kinds of names with different functions, the main one genuine in the text and the auxiliary trivial in the margins. And yet, it was both the terminological fixation of the notion of a trivial name and the consistent application of these names by authoritative Linnaeus that will become an important prerequisite for the subsequent shift to strictly binomial nomenclature within the framework of the nominalist concept.

Contrary to another misconception, the Linnaean nomenclature system did not presume the principle of priority in its modern consistent understanding and application;

this was noted by many reviewers of this system [Gray 1821; de Candolle 1867; Lewis 1872, 1875; Heppel 1981; Melville 1995; Dayrat 2010; Pavlinov 2013, 2015a]. A predecessor of this principle was indirectly expressed in *Philosophia botanica* by the canon, according to which "New generic names should not be contrived, so long as adequate synonyms are readily available" [Linnaeus 1751: § 244]; Linnaeus wrote in a similar manner about names in general in *Genera plantarum* [Linnaeus 1737b: § 23]. However, he easily neglected this rule when it contradicted the essentialist interpretation of names. On this basis, Linnaeus, like Tournefort before him, often replaced previous names with those genuine "contrived" by him as more consistent with numerous rules also "contrived" by him [Heller 1964].[8] It is noteworthy that some tremendously loyal "Linnaeans" considered this manner the merit of his nomenclature system (e.g., [Pulteney 1781]).

As will be shown in the next chapter, the further development of regulative taxonomic nomenclature will proceed in a basically non-Linnaean manner, which is evident from the following. On the one hand, it will be based on nominalist ideas, whereas Linnaeus developed the essentialist one. On the other hand, many important working principles were only alluded to (priority, binomiality) or even not presumed (typification) by Linnaeus, so they will in fact be elaborated and consolidated by both his non-orthodox followers (such as Fabricius) and by his opponents (such as Adanson).

---

[8] It could be noted in parentheses that such practice of replacing "bad" names with "right" ones was very close to the Confucian idea of *zhèngmíng* (rectification of names).

6 The Nominalist Route

The nominalist concept of the professional language of natural sciences began to take shape on a rational basis in the 17th century (J. Locke, R. Descartes, J. Wilkins) [Slaughter 1982; Maat 2004; Scharf 2008], but at that time it did not influence taxonomic nomenclature (see Section 7.1), which continued to develop within the framework of essentialism. The decisive shift from the latter to nominalism in the foundations of the language of systematics was caused by the moving of the whole of biological systematics in the post-scholastic route that began in the second half of the 18th century [Pavliniov 2014, 2015a]. Since scholastic systematics developed the essentialist concept of nomenclature, and Linnaean reform was the latter's culmination, the post-scholastic nominalist nomenclature should be equally rightly designated as *post-Linnaean*.[1]

Clear delimitation of two "foundations" of systematics designated by Linnaeus, viz., classifying and naming of organisms, became an important part of this shift: the former was declared the main goal of systematics, while the latter was assigned an auxiliary role only. Now, contrary to the proclamations of Tournefort and Linnaeus, the comprehension of Natural Systems should begin not with the comprehension of the names of organisms, but with careful exploration of the organisms themselves by analysis of all their features. This novel understanding of the basic goal of systematics initiated the emergence of new research programs in it (see Section 1.1), in which natural systematics began to pay much attention to the elaboration of the appropriate descriptive language. It was aimed at abolishing nomenclature from the dictate of the essentialist interpretation of the taxonomic names arising from an essentialist interpretation of the taxa, and to adopt their nominalist interpretation, in which the naming became just labeling.[2]

The reasons for a successful transition from essentialist to nominalist nomenclature were twofold; some of them philosophical, and others more pragmatic. In the first

---

[1] Mark Ereshefsky applies this term only to the most recent phylogenetic nomenclature. In this, he evidently follows the commonly accepted view that the whole of contemporary traditional nomenclature is "Linnaean" [Ereshefsky 2001a,b]. However, it will be shown in this section that this view is not correct.

[2] Once again this obliges emphasis of the obvious theory-dependent nature of both basic nomenclature concepts and their historical changes.

case, it was rejection of the essentialist view of the System of Nature that initiated a corresponding changeover of the interpretation of taxonomic names. The pragmatic reasons were no less evident: one-word trivial specific epithets were simple and therefore easy both to establish and use; they were potentially more stable because they should not be obligatorily changed following changes in the taxonomic treatments of genera and species; finally, they were not subject to numerous prescriptions and restrictions introduced by Linnaeus for multi-word genuine (legitimate) specific epithets.

The "Philosophies" by two respected naturalists of the late 18th and early 19th centuries provided an illustrative example of the shift in emphasis from the language of description to objects of description. One of them was the German botanist Johann Heinrich Friedrich Link (1767–1851): his *Philosophiae botanicae* was mainly devoted to descriptive anatomy and partly to physiology of plants, whereas nomenclature was very briefly considered after an outline of the natural system of plants [Link 1798]. Another was J.-B. Lamarck: the main content of his *Philosophie zoologique* was the description of the whole of animal organisms and the presentation of a completely new system of the animal kingdom, while nomenclature issues were not considered at all [Lamarck 1809].

The main trend in the post-essentialist (post-Linnaean) development of taxonomic nomenclature was set from the very beginning by the search for an optimal solution to the general problem of ensuring the stability and universality of both nomenclature regulators and the names resulting from their application. This was facilitated by the further rationalization of regulative nomenclature due to a more rigid codification of its standards. Specially designed rulebooks began to appear: some were called Rules, others Laws to emphasize their significance, and lastly, all of them became Codes. They gradually increased in volume due to the increase in number of regulated tasks and regulatory norms, which entailed their structuring: in order to navigate the nomenclature principles and rules, the rulebooks became more systematized due to their partitioning into clearly designated thematic sections and subsections. This general integration trend was superimposed by the diversification one that was caused by different solutions to the same nomenclature tasks dealing with N-definitions of taxa and their names. Accordingly, different nomenclature systems appeared on the subject-area and regional bases [Melville 1995; Malécot 2008; Dayrat 2010; Pavlinov 2014, 2015a].

Most of the works issued at the dawn of post-scholastic systematics and nomenclature, like their scholastic predecessors, were expounded in "learned Latin." But from the early 19th century, books began to be published most often in native languages (mainly French, German, English). This norm became quite common after the mid 19th century, so Latin ceased to be the unified international language of the taxonomic community.[3] This trend was accelerated by the geopolitical factor of the active formation of national states, accompanied by the growth of national self-awareness [Lindbeck 1975; Wimmera and Feinstein 2010]. But Latin taxonyms remained most

---

[3] However, there were exceptions, such as *Philosophia zoologica* by the Danish zoologist Jan Van der Hoeven (1801–1868), written entirely in Latin [Van der Hoeven 1864].

universally used in scientific literature, although there were very noticeable exceptions (e.g., [Lamarck 1778, 1809; Cuvier 1817]).

The periodization of the post-essentialist history of taxonomic nomenclature is poorly developed. The zoologist Benoît Dayrat refrained from recognizing any stages in his survey [Dayrat 2010]. The botanist Valéry Malécot distinguished three main stages: (1) 1800–1870—individual rules and the first rulebooks; (2) 1870–1900—the second (after Linnaeus) wave of debates; and (3) 1904–1907—the first international Codes [Malécot 2008]. The following major stages in the post-Linnaean history of nomenclature are recognized here [Pavlinov 2015a, 2019]:

1. The second half of the 18th century—the first formalized proposal for a fully nominalist nomenclature (*Adanson Rules*) and the establishment of binomial nomenclature.
2. The first half of the 19th century—the emergence of early nomenclature rulebooks (*Règles de la nomenclature* by A.-P. de Candolle, *Code of British Association*), presumed to be unified for both botany and zoology, but subsequently developed separately for each subject area.
3. The second half of the 19th century—the "Great Schism" and fragmentation of Codes on subject-area and regional bases, largely due to controversies about how to ensure the stability of names.
4. The 20th century—in-depth development of international Codes and their further "divergence" on a subject-area basis (appearance of Codes for microorganisms and cultivated plants).
5. The end of the 20th and the beginning of the 21st centuries—the emergence of *PhyloCode* and *Draft BioCode*, each solving the problem of unification of taxonomic nomenclature in its own way. The development of digital technologies that stimulated the development of rational-logical nomenclature.

6.1 DAWN OF NOMINALISM

The purposeful development of the nominalist concept of taxonomic nomenclature began in the second half of the 18th century, in parallel with completion of its essentialist version by Linnaeus. The transition to this fundamentally new nomenclature started in the mid 18th century with M. Adanson and lasted for several decades. The use of essentialist names was largely abandoned at the beginning of the 19th century, and the key idea of nominalism—"a name is just a name"—became dominant in its second half. And yet, some early influential nomenclaturists remained committed to certain provisions of the essentialist idea (A. L. de Jussieu, A.-P. de Candolle): they recommended that descriptive names should be applied and preference given to "appropriate" ones reflecting the important distinctive features of organisms.

> A peculiar mixture of two basic treatments of names could be found in *Preliminary Discourse* by naturalist and mathematician John Frederick William Herschel (1792–1871). He emphasized that "the names shall recall

the differences as well as the resemblances between the individuals of a class, and in which the direct relation between the name and the object shall" be established [Herschel 1830: 136, 137]; and yet "a good, short, *unmeaning name*, which has once obtained a footing in usage, is preferable to almost any other" ([Herschel 1830: 139; italics in the original]).

6.1.1 ADANSONEAN REFORM: A FAILED ATTEMPT

The first important attempt to substantiate and formulate the radically new principles of nominalist nomenclature was made by the French naturalist and systematist Michel Adanson (1727–1806). He practiced initially as a "Linnaean," but in adulthood became an opponent to Linnaeus on most theoretical and practical issues of systematics and nomenclature [Stafleu 1963; Dayrat 2003; Pavlinov 2015a, 2018]. With this, Adanson appeared to be the first representative of the emerging "new systematic" of a post-scholastic kind who was seriously concerned with the issues of taxonomic nomenclature. Among his works, the most significant was *Familles des plantes*, with a lengthy historical and theoretical introduction [Adanson 1763].

Its section "*Noms des Plantes, frases & descriptions*" was especially important for the initial development of the foundations of a nominalist concept of nomenclature. It described the principles of hierarchy of taxonomic categories and rules for the formation and choice of taxonomic names. After reviewing ideas of his predecessors, Adanson summarized his own in a quite original small rulebook that deserved to be called **Adanson Rules** [Pavlinov 2013, 2015a]. Although the latter is rarely mentioned in contemporary reviews of the history of taxonomic nomenclature, it is quite significant as the latter's first clearly nominalist version.

The most fundamental idea of Adanson was to distinguish between names (designations) and diagnoses (brief indications of distinguishing features), and not to base the former on the latter. He declared his nominalistic position as a starting point in the consideration of all nomenclature issues as follows:

> a name is a simple or complex sign, arbitrarily chosen by a person of a given society or a given country [...] In general, names denote objects [...] and do not express their nature or at least their most essential features. This presumes that the names as such have no meaning; such names are referred to as trivial [*populaires*] or simple [*primitifs*].
>
> *Adanson 1763: cxxiii–cxxiv*

M. Adanson insisted categorically on the priority of pre-Linnaean names, including those dating back to antique authors; therefore, tribute should be paid to him for being a "founding father" of the principle of priority. He was inclined to apply any trivial names regardless of their etymological and grammar correctness and argued for the inadmissibility of their substitutions for whatever "essentialist" reasons. In this regard, he questioned the etymological substantiation of the names

The Nominalist Route

of animals and plants, and some new names established by him were quite meaningless;[4] this suggestion was eventually adopted in almost all contemporary Codes. With this, Adanson proposed significantly simplifying Linnaean rules for the Latinization of scientific names: adoption of this norm would rescue systematists of the 19th century from their heated debates over the linguistic norms of Latin names enforced by "purist" linguists (see Section 6.2.4). Adanson rejected descriptive names of families and suggested establishing them based on the name stems of their typical genera; this radically new linguistic rule was subsequently universally adopted. He also proposed a similar rule for generic names, for which the names of typical species should be the source, but this was not be adopted. It was suggested to use the composite one-word species taxonyms formed by combining generic names and specific epithets connected with a hyphen (e.g., *Aparine-Mollugo*, *Aparine-Galion*, etc.); with this, the typical species of the genus would remain without an epithet due to tautonymy of generic and specific parts of such composite names. In fact, this proposal meant rejection of binomial nomenclature, so systematists would not accept it, but this principle is recalled later from time to time by various authors (see Section 7.1).

This first attempt to introduce a consistently nominalist concept of nomenclature undertaken by Adanson appeared to be unsuccessful, for which two main reasons can be indicated. His nomenclature reform turned out to be too radical to be immediately accepted by the conservative community of taxonomists. In addition, the monumental authority of Linnaeus played an important role in this: *Adanson Rules* remained in the "shadow" of *Linnaean Canons*. Because of this, Adansonean reform was not appreciated at first [Stafleu 1963; Dayrat 2003; Pavlinov 2013, 2015a]. It was only a few decades later that A.-P. de Candolle, in offering his version of taxonomic nomenclature [de Candolle 1813, 1819], did him justice. In particular, de Candolle adopted the method of forming family names proposed by Adanson, which was included in all subsequent nomenclature Codes; it is now known as automatic typification.

6.1.2 Affirmation of Binomial Nomenclature

The system of binomial nomenclature, as it was established at the beginning of the 19th century, can by formulated in two basic rules: (a) the complete taxonym of species consists of two separate words, of which the first is the generic name and the second is a specific epithet; and (b) the names of genera and taxa of higher ranks consist of single words. The first experiences of applying this nomenclature system as an auxiliary one occurred in some 16th-century works as part of "double" nomenclature (e.g., by A. Zalužiansky). The latter was improved by C. Linnaeus in the second half of the 18th century with the introduction of notions of a *trivial* name for a non-essentialist specific epithet in the two-word genus–species binomen; it is to be recalled that Linnaeus used trivial names also as auxiliary, in addition to the verbose *genuine specific* names. In the course of the early post-Linnaean development

---

[4] For example, the name of one of the genera of mollusks established in Adanson's *Histoire naturelle du Sénégal* consisted of a meaningless set of letters, *Yetus* [Adanson 1757].

of nomenclature, two-word species names began to be used as the main and eventually sole ones, while verbose names were consigned to oblivion. Thanks to this, the dominance of the essentialist nomenclature concept was put to an end and the active development of its nominalist concept began, in which the working linguistic principle of binomiality was given absolute priority.

An early contribution to the promotion of binomial nomenclature was made by one of Linnaeus' students, Carl Alexander Clerck (1709–1765), who manifested himself as a consistent binomialist in his *Svenska spindlar (Aranei Svecici)* [Clerck 1757]. Clerck adopted the following taxonomic categories: Column (*Agmina*), Class, Genus, and Species. All taxa allocated to them were designated by clearly descriptive one-word names: for instance, *Aëreorum* and *Aquaticorum* for columns; *Retiariorum, Saltatorum*, etc., for classes; *Verticalibus, Irregularibus*, etc., for genera; *castaneus, hamatus, lunatus, domesticus*, etc., for species. No multi-word genuine specific epithets of the Linnaean kind could be found in either Swedish or Latin parallel texts. The generic and specific spider names of Clerck, although established a year prior to those in Linnaeus' tenth edition of *Systema Naturae*, were officially acknowledged as nomenclaturally available [Kronestedt 2010].

One of the notable milestones in the movement towards binomiality was laid by a small essay *Vindiciae nominum trivialium* by already-mentioned J. Murray; this essay was devoted to rules for trivial names [Murray 1782]. They were presented in the form of 28 quite Linnaean aphorisms with some additional explanations. Their list began with statements that confirmed the service character of the binomial nomenclature: "a trivial name does not abolish the species definition and species distinction" [§ 1] (here the genuine name was meant). The following were the rules for the trivial names themselves: "any trivial name consists of one word" [§ 2], "trivial generic name precedes and is separated" from a specific name [§ 3], "trivial generic and specific names must be unique" [§ 4]. In more particular rules, Murray deviated from the Linnaean canon-postulated arbitrariness of trivial names, and made the same requirements for them as for genuine ones. He indicated that "the best of trivial names is the one, which distinguishes the species from its relatives or characterizes it" [§ 5], "that trivial name has priority, which indicates a specific feature" of the plant [§ 6], "trivial name should not contradict an idea of genus" [§ 12]. Thus, one can assume that Murray's *Vindiciae nominum* played the same instructive role with respect to trivial names as does Linnaeus' *Philosophia botanica* to genuine names.

Another important step in establishing binomiality was made by J. Fabricius, also already mentioned above. In the section *"Nomina"* of his *Philosophia entomologica*, he formulated rules for the trivial names of species in no less detail than those for genera. He began their representation by saying that "Linnaeus was the first to introduce trivial names to denote species properly" [Fabricius 1778: 115].[5] He especially emphasized the need for their stability: "Trivial names should never be changed without the most urgent necessity" [p. 121]. Close to the end of his scientific career, Fabricius came to the conclusion, in his *Systema Eleutheratorum*, that among the names, "the best are those that signify nothing. It is impossible to reflect

---

[5] Actually, Linnaeus did not introduce trivial names as such, but applied a term denoting them.

essential generic characters by names, so there is no need for this. [...] The character distinguishes genus and the name denotes it" [Fabricius 1801: viii].

The posthumous editions of Linnaeus' fundamental systeamtic accounts of plants and animals, both in the original Latin and in national languages, played an important role in the final assimilation of binomial nomenclature as a norm of the language of biological systematics. The 13th edition of *Systema Naturae* (1788–1793), prepared by the German zoologist Johann Friedrich Gmelin Jr. (1748–1804), should be mentioned first of all. This enormous work was originally published in Latin in 11 books [Hopkinson 1907], and its fragments began to appear very soon in separate editions in various national languages. For example, in the English edition [Gmelin 1792], all complete species names were genus–species binomens (e.g., *Cervus elaphus*, *Psittacus sinensis*) and no multi-word genuine names could be found. With this, an important difference was that, in the original Latin edition, one-word epithets still remained in the margins next to figures of the respective species, while in the English edition, binomens were provided as headings in the main text. In the six-volume fourth edition of *Species plantarum*, preparation and publication of which took 30 years (1797–1826), the species were also designated binomially in the main text (e.g., *Krameria rixina*), while verbose genuine specific names, which Linnaeus himself had so cared about, were absent [Linnaeus 1824]. Linnaeus' name on the titles of these books will facilitate appearance of the commonly accepted view that it was he who "invented" binomial nomenclature. But a skeptic may wonder if Linnaeus himself would enjoy the denial of his verbose genuine names by his followers.

Binomial nomenclature became popular quite early in the English botanical community: many manuals published there became binomial from the early 1760s [McOuat 1996], along with those that applied "double" nomenclature (the latter is explained in Section 5.3). For example, James Edward Smith (1759–1828), a great admirer of Linnaeus, who bought his collection and library and translated a number of his works into English, published several of his own illustrated plant atlases with strictly binomial nomenclature (e.g., [Smith 1789]). The first accounts of flowering plants published sequentially under the auspices of the *Florae Danicae* project acting from the 1760s provide another example of the early application of predominantly binomial nomenclature [Friis 2019]. The binomial system of species nomenclature pervaded Russia just as early [Sytin 1997], and was exemplified by works of the German-Russian naturalist Peter Simon Pallas (1741–1811) (e.g., [Pallas 1776]).

Some monographs of the second half of the 18th century indicated how trivial species names became fixed grammatically as part of the main text. One of the options could be found in *Ichthyologia* (1738) by Linnaeus' colleague, the Swede Peter Artedi (Lat. Petrus Martini Arctaedius; 1705–1735); its posthumous reprint was prepared by the German physician and naturalist Johann Julius Walbaum (1724–1799). For example, Artedi himself designated a species of the true loach as *COBITIS tota glabra maculosa* [...]; in Linnaeus' *Systema Naturae* it was designated as *C[OBITIS] cirris oris 6, capite inermila, compresso* in the main text and *Barbatus* in the margin; while in Walbaum's comments, the following designation was applied in the text: *Cobitis, Barbatula, cirris 6; capite inermi, compresso* [Walbaum 1792: 10].

With regard to subsequent problems with the interpretation of what might be considered binomial nomenclature in early post-Linnaean works, the manual *Anfangsgründe der Naturgeschichte des Thierreichs* by the German naturalist Nathanael Gottfried Leske (1751–1786) was noteworthy. In it, genera were denoted in one word (e.g., *Didelphys*) and species were denoted in the text by both two-word trivial and multi-word genuine names. With this, trivial names were italicized, while genuine names were provided in the standard font: for example, "*Didelphys marsupialis*, mammis octo, pilis nigrescenti flavis" [Leske 1788: 72]. As one can assume, this case illustrated how the contemporary tradition of highlighting genus–species binomials in the texts was being shaped.

A transitional character between the Linnaean proper and post-Linnaean nomenclature systems at the turn of the 18th–19th centuries was clearly illustrated by *Grundriss der Kräuterkunde* (1792) by the German botanist Carl Ludwig Willdenow (1765–1812); it was reprinted in English as *The Principles of Botany* [Willdenow 1805]. The set of his "botanical aphorisms" that appeared in it deserves to be named **Willdenow Rules**. The latter began with the statement that "The true knowledge of Plants consists in the art of arranging, distinguishing, and naming them; and this art depends on the establishment of fixed rules, drawn from nature herself" [Willdenow 1805: § 145].

In the interpretation of specific names, Willdenow turned out to be a consistent essentialist, following Tournefort and Linnaeus in identifying species name and species diagnosis. Indeed, he believed that "The essential difference, or name (*diagnosis*), of the species is a short description containing only what is essential [so] in forming the specific name we must express only the essential difference" [§ 196]. For the essentialist specific name, understood as a diagnosis, a number of linguistic and semantic norms borrowed from Linnaeus were established, for example: "The specific name must not be too long, and if possible should be contained in twelve words," "When a genus consists but of one species, there is no occasion for a specific difference" [§ 196]. With this, in the section "*Nomenclature of Plants*," his undoubted emphasis was on trivial names: Willdenow stated in the introduction to it that "Linnaeus has performed the most eminent service by establishing a generic name, (*nomen genericum*), and a trivial name (*nomen triviale*), to every plant" [§ 206].

The *Willdenow Rules* were highly appreciated by English botanists; in particular, it was noted in one review that "Had something of the same kind been done earlier, botany would not only have rested on a more stable foundation, but botanical language would have been rendered less harsh than it is at present" [Blackwood et al. 1808: 75]. However, these rules remained hardly noticed on the mainland: they were overshadowed by *Candolle Sr. Règles*, which were to appear soon.

A transition to the binomial system had its own features in France [Williams 2001]. There a great influence was retained, on the one hand, by Tournefort, whose names Linnaeus did not treat too respectfully; and on the other hand, by Georges-Louis Leclerc de Buffon (1707–1788) with his ideas about what Nature was and how it should be studied and described. Of particular importance was the fact that the very popular writings of Buffon and his *Encyclopédie* mates contained descriptions of the natural history of plants and animals rather than their systematic cataloging in a Linnaean style. It was probably for this reason that naturalists belonging to the French-speaking ecumene, who saw no special meaning in such catalogues, played an important role in establishing taxonomic nomenclature in its post-Linnaean version.

The naturalist and first evolutionist Jean-Baptiste Pierre Antoine de Monet de Lamarck (1744–1829) was a consistent supporter of the strict delimitation of the two Linnaean "foundations" of systematics, classification and naming: he emphasized this position in the "*Discours Preliminaire*" of the first volume of his *Flore Française* [Lamarck 1778]. Lamarck rejects the essential interpretation of nomenclature and, in a short commentary following Adanson, emphasizes that "Names [...] only signify our ideas; and these signs, completely arbitrary at their first institution, acquire real and solid value only through their constant use which determines their acceptance" [Lamarck 1778: lxxxiii]. On this basis, Lamarck objected to the substitution of some names for others because of their inappropriateness. With this, he emphasized "the usefulness of *trivial* names, which we should rather call specific names" [Lamarck 1778: lxxxiv]; italics in the original). Thus, Lamarck was probably the first to designate just the trivial names as specific rather than the verbose genuine (as Linnaeus did); this treatment became common in the 19th century. Lamarck reiterated this position in his article "*Nomenclature*" in *Encyclopédie méthodique* [Lamarck 1798]: he argued in favor of fixed rules for the formation of scientific names and of the use of trivial names, noting that "One cannot ignore here the obligation which we owe to M. Linné for his having established these simple denominations which supplement, with so much advantage, the long descriptive phrases by which it was once necessary to burden the memory" [Lamarck 1798: 498]. In his multivolume book on invertebrates, he provided the title species names in a strictly binomial form [Lamarck 1815].

The position of Antoine-Laurent de Jussieu (1748–1836), one of the "founding fathers" of natural systematics of the late 18th and early 19th centuries, was very characteristic with respect to the initial formation of binomial (and nominalist in general) nomenclature. In his early work, *Examen de la famille des renoncules*, he unambiguously expressed his attitude towards nomenclature: names "should not be disregarded, but the study of characters is a much more important part of botany" [Jussieu 1773: 218]. Jussieu clarified this position in a small section "*Nomenclatio & Descriptio*" of the introductory chapter in his main work *Genera plantarum* as follows: "this part of science, which is neither paramount nor scientific in the true sense, should nevertheless be promoted and therefore by no means neglected, but, on the contrary, developed more rigorously" [Jussieu 1789: xxiii]. Then he briefly and with obvious approval outlined Linnaeus' nomenclature system, highlighting that he had applied both significant specific and insignificant trivial names. Jussieu himself believed that genera and species should be designated by significant names, as

"The name of species is to be equally simple and easy to understand, but be in addition significant by expressing the proper characters, to distinguish it from the congeners and thus to be really *specific*, while those indicating soil, region, weather, duration, color, taste, smell, use, etc., are worse, these are *trivial*" names.

Jussieu 1789: xxv; italics in the original

Thus, the nomenclature system of Jussieu was eclectic, combining elements of both essentialism and nominalism, and his position considerably influenced the subsequent development of such "mixed" taxonomic nomenclature in the first half of the 19th century, especially in botany.

6.2 THE 19TH CENTURY: BASIC ISSUES

A search for how to solve the general conjoint task of ensuring both stability and universality of nomenclature set the main trend in its development in the 19th century. However, different suggestions of preferable solutions at first led away from the sought goal and caused multiplication of particular nomenclature systems. This diversification trend involved the following basic issues [Pavlinov 2015a].

The system of ranks/categories fixed by Tournefort and Linnaeus appeared to be just "one of many" because of widespread recognition of many others. Different authors applied different terms to designate the ranks/categories they invented, while others easily changed their subordination. All this led to fragmentation and, therefore, erosion of the whole of the initial "Linnaean" ranked hierarchy.

The taxa began to be treated in a narrower sense and underwent splitting. With this, when old genera and higher-level taxa were divided, the old names were often discarded and all new groups received new names. In this way, different authors established their names following different rules: as descriptive or through typification, sometimes they were denoted by certain symbols.

The criteria for choosing valid (correct) taxonomic names were highly controversial. Initially, a noticeable essentialist tradition continued to be retained that gave preference to the most appropriate (genuine) names. Later, other reasons developed within the framework of the nominalistic tradition came to the fore, namely, priority and usage.

Finally, "linguistic purism" should be noted: some authors specialized in replacing names that violated numerous Linnaean linguistic canons with the "right" ones, thus producing extensive synonymy.

All this entailed a lot of new terms and taxonomic designations at the beginning of the 19th century; these terms appeared and changed quite erratically at the whim of various authors guided by their own preferences.

Such disagreements and contradictions created an obvious problematic situation for nomenclatural activity. An active search for ways to eliminate them became one of the most significant drivers of the early development of post-Linnaean nominalist nomenclature. The principal ones are considered in the following sections.

6.2.1 Rank Fragmentation and Rank Dependence

One of the notable features in the early development of descriptive language inherent to post-Linnaean systematics became fragmentation of the ranked hierarchy accompanied by the strengthening of rank dependence of both regulative and nominative nomenclature.

For this reason, Linnaean and post-Linnaean rank-dependent nomenclatures, from a conceptualist perspective, may be considered to be different systems [Needham 1911; de Queiroz 2005, 2012; Pavlinov 2015a, 2019]. This circumstance should be emphasized, because the whole of the contemporary ranked hierarchy in systematics is usually called "Linnaean" (e.g., [Griffiths 1973, 1976; de Queiroz and Gauthier 1990, 1994; Ereshefsky 1994, 1997, 2001a, 2002, 2007b; de Queiroz 1997; Schuh 2003; Vasilyeva and Stephenson 2012], etc.).

Development in this vein was accompanied by the introduction of many non-canonical rank/categories, the names of which were often borrowed from military and social terminology: among them, for example, were cohort, phalanx, tribe, etc. [Gill 1896; Oberholser 1920; Ereshefsky 1997, 2001a; Pavlinov 2014, 2015a]. All these innovations led to the variety of both rank/categories and their designations. In botany of the first half of the 19th century, differences in the taxonomic hierarchies also involved the position of identically termed ranks in particular hierarchies: for instance, sequential subordination of section, order, family, and tribe had almost a dozen variants. This instability of ranked hierarchy could be illustrated by its variation in the arthropod classification of the French zoologist Pierre André Latreille (1762–1833). In one of his early works, its total hierarchy was defined by the subordination: Subclass > Division > Subdivision > Order [Latreille 1801]; later it was changed to the following: Legion > Centuria > Cohort > Order [Latreille 1806].

For certain pragmatic reasons, the ranks/categories of Tournefort–Linnaeus (usually called "Linnaean") were then denoted as *basic* (main, primary), while those introduced for a more fractional presentation of the taxonomic hierarchy were denoted as *additional* (subsidiary, secondary). Among the latter, *auxiliary* ranks/categories were recognized; their position in the ranked hierarchy was fixed terminologically by denoting them either *subordinate* or *superordinate* with respect to the respective basic ranks (for example, sub- and superfamily, etc.). As a result, the total number of taxonomic ranks/categories could reach more than two dozen in some handbooks and manuals on plants and animals. In addition, unnamed *intercalary* ranks/categories were continued to be used in many books of the 19th century, and were designated frequently by different symbols; this old tradition was fixed in one of the earliest rulebooks [de Candolle 1813]. Such unnamed ranking allowed an optimal ratio to be maintained between the number of subtaxa in a taxon at each step of the ranked hierarchy, and at the same time to get rid of the redundant special terminology.

It is worth noting that there was a certain shift in emphasis in assessing the status of taxonomic ranks, which began to manifest itself in nominalist nomenclature. In scholastic systematics, attention was paid mainly to genera, whereas in natural systematics, the species began to come gradually to the fore, corresponding to the transition from a predominantly deductive to an inductive classification scheme. Some nomenclaturists, when considering taxonomic hierarchy, characteristically distinguished between "species" and "groups," the latter encompassing supraspecific categories [Swainson 1836; Strickland 1837].

By the mid 19th century, a certain agreement was reached about the number, subordination, and designation of ranks and respective categories. In the most advanced Codes, ranked hierarchy was prescribed quite rigidly: all ranks were denoted by fixed terms, their sequence was codified, and changes were prohibited [de Candolle 1867; Dall 1877]. This approach ensured the universality and stability of the ranked hierarchy within the taxonomic community that adopted the respective nomenclature system. In particular, in many systematic manuals of that time, a detailed outline of standardized ranked hierarchy could be found (e.g., [Gray 1879]).

In the works of late essentialists (including Linnaeus), who operated with a small number of ranks, rank dependence of nomenclature, as it was expressed lexically, involved only a difference in the number of words in the names of species (multi-word) and taxa of higher ranks (one-word). In the works of nominalists with their many more ranks, rank dependence became significantly strengthened by the rank-specific character of the names of genera and suprageneric taxa caused by a new manner of the formation of the latters' names based on their nominotypes (see below).

6.2.2 PRIORITY *vs.* USAGE

For the choice of valid (correct) taxonomic names to ensure their stability and universality, two main criteria were animatedly discussed throughout the 19th century: the more theoretical was based on the *principle of priority*, whereas the more pragmatic was based on the *principle of usage*. The basic arguments *pro* and *contra* of each principle are summarized in Chapter 3 on nomenclature principles (see Section 3.4).

In most early nomenclature systems, the main emphasis (with certain reservations) was on priority by publication date. The Dutch zoologist Coenraad Jacob Temminck (1778–1858) expressed this position as follows:

> it is preferable to keep an old name [...] rather than to replace it with another name, which is possibly more appropriate or grammatically more correct, [...] since there is nothing worse for the development of natural sciences [...] than all these different views on the names of genera and species.
>
> *Temminck 1815: xi*

But at the same time, the problem arose of establishing additional conditions for the application of this principle, among which the question of a starting point for considering priority of names became most disputable.

Advocates of *absolute priority* suggested a preference for the earliest known names, including those of antiques. This norm went back to the classics of the 17th–18th centuries; it was defended by some English botanists of the early 19th century (e.g., [Gray 1821]), and will be recognized later by *Douvillé Rules* [Douvillé 1882a,b] (see Section 6.3.3). However, most nomenclaturists were inclined towards *fixed priority*, which implied adoption of a certain date, prior to which no names should be considered to be nomenclaturally available (legitimate). They generally agreed to mark such points by the publication dates of the works in which binomial nomenclature was first consistently applied, but they disagreed upon which particular works should be selected. The general list of proposals for the dates and works sought appeared to be rather long, encompassing more than ten dates for botany and slightly fewer for zoology; they were being considered throughout the whole of the 19th century [Pavlinov 2014, 2015a]. Besides, it was proposed to limit application of this principle for the combinations of specific epithets with the names of "true" genera [Bentham 1879; Jackson 1887] (see Section 6.3.2.1).

The principle of usage presumes that, among several competing synonyms, the preferable one is that which is most frequently used in the current literature; it was unreservedly adopted, for instance, by *Berlin Rules* of botanists [Engler et al. 1897] (see Section 6.3.2.1). However, different nomenclaturists following this principle suggested different time intervals within which the frequency of the name usage had to be calculated; these were 30, 50, and 100 years. The most radical was the proposal of the English lawyer and amateur entomologist A. Lewis that no name published before 1842 should ever be used again unless it had "been kept alive by quotation as the true name in some work since 1842." This date was chosen by him arbitrarily as the date of the appointment of the Nomenclature Committee at the British Association [Lewis 1875].

6.2.3 CIRCUMSCRIPTION VS. CHARACTERS VS. TYPES

The stable and commonly acknowledged association of a taxon with its name implies that the former must be N-defined firmly enough to make it easily recognizable. There are three ways of such definition: extensional, intensional, and ostensive (see Section 3.5); they imply indication for a taxon, whether circumscription, diagnostic characters, or nominotype, respectively. So, there are three particular corresponding nomenclature systems: circumscription-based, diagnosis-based, and type-based.

An indication of taxon circumscription, or composition (the *principle of circumscription*), was considered sufficient to establish both a taxon and its name in the very early stages of the development of nomenclature. This was evidenced by the appearance of lists of taxa with subtaxa (e.g., genera with species) but without their characters in many books of the 16th–18th centuries, including Linnaeus' *Species plantarum* [Linnaeus 1753]. According to this, any changes in taxon circumscription would entail the need to establish a new name for it. This norm was treated as a subsidiary in the first Codes/Rules of the 19th century, but it was revived as the basic one, as opposed to the principle of typification, by some nomenclaturists in the second half of the 19th century.

An indication of the diagnostic characters of taxon (the *principle of diagnosing*) as the basic method of its N-definition is most compatible with the essentialist interpretation of both taxa and their names. This interpretation presumes that the change in the essential characters of a taxon entails establishing a new descriptive name for it. This principle is also compatible with nominalist nomenclature, but in it, it does not presume such a rigid link between diagnosis and a non-descriptive name. The need for diagnosing was stated in the first Codes/Rules, according to which, for the name of a taxon to be recognized as available (legitimate), the latter's characters must be indicated in the original description.

The requirement of indication of the types of a taxon as part of its *N*-definition (*method of type, principle of typification*) constitutes an important distinctive feature of post-Linnaean nomenclature. This method was initially used basically not so much in the "typological" as in the nomenclatural sense, according to which types were considered "examples" of taxa [Whewell 1847; Mequignon 1932; Williams 1939; Dupuis 1974; van der Hammen 1981; Petersen 1993; Daston 2004; Pavlinov 2015a; Witteveen 2016, 2020]. For families and orders, typification was first introduced indirectly (as "automatic" typification) by formation of their names based on generic names in works of the second half of the 18th century (Adanson, Jussieu); it became almost universally recognized after the first Codes of the 19th century. For genera, their type species began to be indicated at the beginning of the 19th century [Latreille 1801, 1806], and this practice became widely recognized by both zoologists and botanists (e.g., [Montfort 1810; de Candolle 1813; Swainson 1820–1821; Westwood 1836, 1837; Strickland 1838a, 1845]).

> In the case of "old" genera, for which no type species had originally been fixed, serious disagreements arose regarding the methods of fixing their nominotypes. The following options were considered in the course of a hot discussion on drafts of the first International Codes at the turn of the 19th–20th centuries: (a) the first species in the author's list; (b) the best known or most typical; (c) remaining in the genus after all others have been excluded; (d) chosen by the first reviser; (e) by tautonymy of generic and species names [Gill 1896; Cook 1898, 1901, 1902b; Jordan 1900, 1905, 1907; Shear 1902; Stiles and Hassal 1905; Allen 1906, 1907; Stiles 1906, 1907; Stone 1906, 1907; Coquillett 1907]. A kind of "protocol" of the sequential consideration of these options was proposed based on two "axioms": (a) to follow the "historical method" (i.e., taking priority by date into account); and (b) to proceed from a supposed intention of the author in his original description of the genus [Stiles and Hassal 1905].

For species-group taxa, distribution of the exicates (the authorized herbarium sheets) that was widely practiced in the 17th–18th centuries in botany can be considered a predecessor of "typification"; it was officially recognized as admissible up to the end of the 19th century [de Candolle 1867; Engler et al. 1897]. The type specimens used when establishing species and their subdivisions were mentioned in the mid 19th century (e.g., [Strickland 1845]), and their typification by deliberately designated specimens was proposed in its second half by American nomenclaturists [Coues et al. 1886; Britton et al. 1892; Bather 1897; Schuchert 1897; Thomas 1897; Marsh 1898; Cook 1900]. While consolidating this nomenclature norm, special attention was paid to the need for permanent preservation of the type specimens in the respective repositories [Strickland 1845; Marsh 1898; Swingle 1913].

Quite serious disagreements between the supporters of circumscription or diagnosing or typification of genera and species occurred in the second half of the 19th century. In the case of genera, it was pointed out that typification was settled rather late in the practice of descriptions of new taxa, so its adoption in a retroactive form would entail suppression of generic names that appeared in early works, including canonical ones (Linnaeus, Jussieu, Fabricius, de Candolle, etc.). In the case of species-group taxa, it was indicated that the original specimens of early authors were rarely preserved, and they were often subsequently replaced by others, therefore the characters indicated in original descriptions were generally more reliable [Schaum 1862; Gray 1864; Puton 1880; Kirby 1892]. On the other hand, the proponents of typification emphasized that analysis of preserved specimens was more reliable than the characters indicated in the original descriptions, which were usually not accurately specified [Strickland 1845; Waterhouse 1862; Cook 1900]. Ultimately, the acceptance or rejection of the typification of species became one of the main causes of the "Great Schism" in taxonomic nomenclature in the second half of the 19th century [Hitchcock 1922; Rickett 1959; Pavlinov 2014, 2015a].

6.2.4 CLASSICALITY *VS.* ARBITRARINESS OF NAMES

To the extent that taxonomic names were agreed to be Latin or Latinized, their classicality was frequently considered an important attribute of nominative nomenclature in scholastic and early post-scholastic systematics. From this point of view, a taxon should be designated only by a linguistically appropriate name that meets the criteria of classical Latin. Inappropriate names that did not meet these criteria had to be changed according to linguistic standards or replaced by appropriate ones. This *principle of classicality* (the name must obey the laws of the language) was considered as important as the *principle of genuineness* (the name must indicate the essential features).

Strict adherence to such a "linguistic purism," coupled with the essentialist interpretation of names, served as a source of nomenclature instability. Each author committed to this idea considered it necessary to correct taxonomic names introduced in a grammatically lax form. Thus, in the catalog of generic names of mammals published at the beginning of the 20th century, there were cases when the same genus was allocated to up to ten names, each claiming to be linguistically correct: for example, *Priodon, Priodonta, Priodontes, Prionodon, Prionodus* for a genus of carnivorous mammals [Palmer 1904].

For this reason, with the growing dominance of nominalist nomenclature ("a name is just a name") and emphasizing the stability of names as the principal objective, linguistic requirements began to be criticized and weakened. In many Codes, starting from the earliest, it was presumed that at least non-descriptive names could be arbitrarily composed and their linguistic inaccuracy, with the exception of obvious misspellings, should not serve as the basis for their emendation or replacement.

Nevertheless, the confrontation between "purists" and "arbitrarists" continued throughout the 19th century. In its second half, on the basis of an idea of the "classical" language of systematics, a reform of taxonomic nomenclature was proposed, directed against the "formal" principles of stability and priority. Its author was the French botanist and linguist Jean Baptiste Saint-Lager (1825–1912), who published several articles on this subject. In a small pamphlet concluding this series, he stated that "the scientific language must constantly improve, as the extension of our knowledge" [Saint-Lager 1886: 1] and "the progress of science is too closely linked to the perfection of the formulas used for the expression of ideas" [Saint-Lager 1886: 53]. Therefore, the treatment of names as arbitrary and meaningless labels must be considered wrong. On the contrary, the "spelling and grammar correctness constitute the supreme law of linguistics [...] this law [is] the only rational, the only true and therefore the only one worthy of being admitted" [Saint-Lager 1886: 54].

The *Saint-Lager Reform*, which called for the discharge of taxonomic nomenclature from all "linguistic rubbish," was rejected by the taxonomic community, which was already firmly committed to the nominalist concept. However, the American botanist Frederic Edward Clements (1874–1945), who began as a nomenclaturist and subsequently became a celebrated phytocenologist, returned to consideration of its key idea two decades later. In his work *Greek and Latin in Biological Nomenclature* especially devoted to this issue, he argued that, over time, "biological nomenclature will be in a fair way to become a symmetrical, stable structure, based upon the two cardinal principles, priority and classicity" [Clements 1902: 2]. In anticipation of this, Clements formulated a set of 12 rules for the formation and correction of taxonomic names, most of which he declared retroactive.

This reform, in both its original (Saint-Lager) and later (Clements) versions, failed because of its incompatibility with the nominalist concept of nomenclature. One of the "subjective" causes was that, for most taxonomists of the second half of the 19th century, classical Latin was no longer something very familiar and deserving of unquestioning observance. Therefore, the *principle of arbitrariness* (going back to M. Adanson) prevailed, according to which taxonyms may be any word-like combinations of letters.

However, an idea of the "purity of language" of taxonomic description was recollected by some later authors (e.g., [Sprague 1921]), and its certain manifestations were retained in the descriptive language of contemporary systematics. Indeed, the nomenclature Codes in botany and zoology provide the Appendices with the spelling and grammar rules of classical Latin and Greek to be followed by those systematists describing new taxa.

6.3 THE 19TH CENTURY: CODIFICATION OF NOMENCLATURE

The above-considered diversity of approach to fixing and naming both taxa and ranks/categories became a "headache" for nomenclaturists of that period. As one of them observed, the "existing dissensions upon nomenclature [...] produce the veritable chaos unless we attain to the establishment of fixed laws" [Lewis 1872: 11]. Thus, the situation appeared to be quite similar to the one that had preceded Linnaean reform a century ago. Therefore, an integrative aspiration became strengthening, aimed at the development of a universal and stable system of regulative nomenclature that would ensure the desired universal and stable system of nominative nomenclature.

This aspiration took the form of codification of taxonomic nomenclature that became one of the main trends in its development in the 19th century. It involved the purposeful development of rulebooks bringing together and fixing nomenclature principles and rules, which were sporadically appearing as historical precedents in the works of various authors. With this, however, this general trend was not monolithic due to a variety of opinions on which nomenclature regulators were the most significant and how they should be applied.

> With this, not all agreed with this dominant trend. For example, the English naturalist William Ogilby (1808–1873) believed otherwise:
>
> So long, however, [...] in Zoology, are not less unsettled and fluctuating than their names, it is useless to think of reforming or amending the nomenclature of the science; it is beginning at the wrong end; let us rather set about studying the natural boundaries and relations of the groups, and it will be time enough afterwards to settle their nomenclature.
>
> *Ogilby 1838: 284*

6.3.1 FIRST CODES

According to the opinion of K. Temminck, mentioned above, descriptive systematics of that time was thrown into chaos due to the lack of uniform, universal, and stable rules for handling taxonomic names [Temminck 1815]. The reason was that *Linnaean Canons*, despite the great influence of their author, did not become generally recognized; some of them allowed different interpretations while others appeared redundant under new conditions set by the then emerging nominalist concept of nomenclature. Because of this, systematists at the beginning of the 19th century applied different methods of operational definition of taxa, formation, and substitution of their names (see Section 6.2).

Such a situation made it quite urgent to develop the uniform, strictly formulated standards for taxonomic nomenclature comparable to *Linnaean Canons* in their significance and influence, but more adapted to the new conditions of taxonomic descriptions. This general integrative trend, driven basically by rationalization of the language of systematics, led to the development of the first nomenclature rulebooks in botany and zoology. They were based mainly on the nominalist concept of nomenclature, but still preserved a considerable admixture of descriptive (and partly even essentialist) interpretation of names.

6.3.1.1 Botany

The first rulebook of botanical nomenclature that appeared "after Linnaeus" was the already mentioned *Willdenow Rules* [Willdenow 1805] (see Section 6.1.2). They provided an illustrative example of a transitional nature between "Linnaean" and "post-Linnaean" nomenclature systems at the turn of the 18th–19th centuries, and were briefly discussed above. Though appreciated by British nomenclaturists [Blackwood et al. 1808; Swainson 1836; Strickland 1837], they seemed to have no evident influence on the subsequent development of nomenclature.

The most significant milestone in the early development of the post-Linnaean systematics and nomenclature was the fundamental *Théorie élémentaire de la botanique* by the Swiss phytographer Augustín-Pyramus (Pyrame) de Candolle (1778–1841), or Candolle Sr., one of the key early theoreticians of natural systematics [de Candolle 1813, 1819]. Part I of this book, entitled "*Théorie des classifications, ou Taxonomie végétale*," considered in detail fundamental characteristics of classification for the first time. In particular, Candolle Sr. recognized the following main ranks (categories): Class, Family (equivalent to Order of Linnaeus, which Candolle did not recognize), Genus, Species. In fact, these were rank-groups, within which he also considered subordinate divisions: subclass, tribe, section, variety (race), respectively.

A.-P. de Candolle paid great attention to nomenclature. In the original French editions of his *Théorie élémentaire*, of six sections of the chapter "*Théorie de la Botanique descriptive, ou Phytographie*," two are devoted to taxonomic names. The first and most important of these two sections is entitled "*De la nomenclature*," and is divided into the following Articles: *De la nomenclature en général*; *Des noms de genres*; *Des noms d'espèces*; *Des noms de familles, de variétés, etc.*; *Conclusion*. These sections consist in total of 28 paragraphs in the first edition (1813) and 38 paragraphs in the second edition (1819). In the English edition published in collaboration with the German botanist K. Sprengel [DeCandolle and Sprengel 1821], the chapter "*Phytography, or Descriptive Botany*" was noticeably shorter: it contained Sections "*On the Names of Plants*" with two Articles ("*On the Generic Names*" and "*On Trivial Names*") and a section on "*Synonymy*" (considered together with other sections outlining rules for describing plants); the text is arranged into 21 paragraphs.

This set of nomenclature provisions is known as **Candolle Sr. Règles**, as opposed to *Candolle Jr. Lois* (see Section 6.3.2.1). Their great significance was in that they represented nomenclature in a detailed and ordered manner and provided thorough explanations and justifications of all principles and rules. These *Règles* marked the beginning of the active development of the professional language of systematics in a non-Linnaean way. However, Candolle Sr. called this nomenclature system "Linnean," which was partly true. Its character as such was manifested in the fact that this particular nomenclature system: (a) was predominantly descriptive, although not entirely essentialist; and (b) prescribed in great detail technical norms concerning grammar, etymology, and spelling of taxonomic names, to which Linnaeus had paid great attention.

In the second edition of *Théorie élémentaire* [de Candolle 1819], the section *De la nomenclature* begins with the statement that "all naturalists have agreed on the need for a unique and universal nomenclature; but if they unanimously recognized the general principle, they did not agree so easily on what should determine this universal nomenclature" [§ 210]. After a brief historical overview, de Candolle emphasized that "This method of nomenclature, which was called the *Linnean Nomenclature* [...] entered into force after the publication of the work of Linné in 1753" [§ 213; italics in the original]; he is obviously referring to the first edition of *Species plantarum*. Next, three basic rules are introduced to make this nomenclature system universal: (a) "All names of natural beings and all terms of science are in Latin," (b) "the names are formed according to the rules of general grammar," and (c) "In order the nomenclature can become universal, it must be stable, and its stability [...] is based on this third principle, which [...] is that the first who discovers a natural being or who registers it in the catalog of nature, has the right to give a name to it, and that this name must be obligatory admitted [...], unless it already belongs to another organism or oppose to the essential rules of nomenclature" [§ 218]. As one can see, the last paragraph provides a package of several important principles (rules), viz., stability, priority, and monosemy of names.

The following important rule is proposed for the names of genera: "Generic names are given to the natural beings, [and they cannot be] contradictory to the generic characters. [Even] if the idea expressed by a name is imprecise or improper, this name must be kept, we can only afford to change it if it is in contradiction with the character" [§ 219]. Following this, in an equally essentialist spirit, it is argued that "Generic names are always meant to relate to a collection of species united by a common character; and the best are those who express this character in a precise way." At the same time, it is stipulated (mentioning Adanson) that in some cases it is possible "to give it a name which has no real relationship with [generic character], entirely insignificant name" [§ 220].

Regarding species names, they are

> much less difficult to establish than the names of genera [as] they are not intended to represent a collective being [...] we can say, in general, that any name which does not imply contradiction with the [characters of] plant, and especially which does not belong to any other species [...] The impropriety of a specific name, or the possibility of finding more suitable ones, is not enough reason to change it; this must be substantiated only by a proof that the name expresses an absolutely wrong idea or belongs to another plant.
>
> *§ 226*

As one can see, the principles of verity and suppression of homonymy of specific epithets are implied here. It is also stipulated that, if a species is transferred from one genus to another, its name does not change: this indirectly recognizes the fundamental status of the stability of specific epithets provided by their context-independence.

It is settled that family names are to be regulated by two substantially different rules. One of these, dating back to Linnaeus and even antecedent authors, treats the most appropriate names as those that reflect significant characters of plants. Another rule, going back to Adanson and Jussieu, presumes that the family name should be formed from the stem of the generic name, and the name-bearing genus must be one of the best known in the respective family. These two rules, considered equally suitable by Candolle Sr., will compete throughout the 19th century, creating significant instability of family-group names. For the taxa of subordinate categories, Candolle Sr. suggests the following rule: "significant" ones should be named, while any symbols could be used to designate "insignificant" ones (no criteria of "significance" of subordinate categories were specified). With this, it is proposed that family-group names of different ranks derived from the same generic names are distinguished by different rank-specific suffixes. This very useful proposal of A.-P. de Candolle will be adopted by most subsequent authors.

In the final article [§ 234], Candolle Sr. emphasizes two provisions that will later become key. One of them meant "the need to admit the name given by the inventor to a plant, whenever this name matches the rules." It will later be called the principle ("law") of priority designed to provide stability of plant names. However, it is introduced with numerous reservations, including quite essentialist ones, allowing names to be changed in some cases; for example, "If the name given by the first inventor is false, that is, if it implies direct contradiction with one of the characteristics of the plant." By another rule, Candolle Sr. first introduces the requirement "That the [new] name is accompanied by at least one characterizing sentence sufficient to make [the plant] recognizable": in fact, this was the first formulation of the principle of diagnosing.

The English edition [DeCandolle and Sprengel 1821] includes a small section on trivial names. It is noted there that "Linnaeus earned for himself immortal honor, by inventing what he called Trivial Names" [§ 212]. However, contrary to the original Linnaean idea of the arbitrariness of the latter and following Murray, it is said that they "should be expressive; they must either express the specific character, or denote some striking property [...] of the plant" [§ 225].

A.-P. de Candolle called for the collecting and storing of herbaria, emphasizing their important function in regard to nomenclature. He said that:

> One of the main uses that science extracts from the herbaria is the fixity they give to nomenclature; one can always find with certainty, with their help, what was the plant which used as a type [...] by the original authors, and thus avoid the errors which may result, either from the accumulation of erroneous synonyms or defects or omissions in the descriptions.
> *de Candolle 1819: § 268*

Apparently, this was one of the first indications of the need to preserve the type specimens for species-group taxa.

The English botanist John Lindley (1799–1865) made another significant contribution to the early development of the theory and language of taxonomic science. He adhered to the concept of natural systematics and, therefore, was "anti-Linnaean" in this respect [Pavlinov 2018, 2021]. With this, the issues of taxonomic nomenclature did not bother him too greatly: indeed, his fundamental monograph *An Outline of the First Principles of Botany* lacked any consideration of taxonomic names [Lindley 1830]. However, he discussed some principal ideas about nomenclature in the chapter "*Of nomenclature and terminology*" in his *An Introduction to Botany* [Lindley 1832]. J. Lindley praised a number of Linnaean rules as "undoubtedly excellent in many respects [and] constantly appealed to, by the school of Linnaeus, as a standard of language," but noted that some of them "have, indeed, fallen wholly into disuse" [Lindley 1832: 456]; he meant primarily an essentialist connotation of the rules. His own position was expressed by "an admitted principle, that it is of little real importance what name an object bears, provided it serves to distinguish that object from every thing else. This is the material point, to which all other considerations are secondary" [Lindley 1832: 456]. However, according to generally accepted tradition, Lindley clearly tended to ascribe generic names with "positive meaning, which cannot be misunderstood" [Lindley 1832: 458]. And yet, he considered "superficial or insignificant" many of the Linnaean canons expressing the essential interpretation of generic names, such as the suppression of "barbaric" names, the agreement of the name of the genus with the characteristics of its species, etc.

In developing important nomenclature precepts, Lindley offered the following basic rules: "In *constructing* a generic name, take care that it is harmonious, and as unlike all other generic names as it can be. In *adopting* generic names, always take the most ancient, whether better or worse than those that have succeeded it" ([Lindley 1832: 457; italics in the original]). As one can see, Lindley adopted priority in its absolute meaning and protested against arbitrary replacement of the old names with the new "more appropriate" ones, and therefore implicitly was against the retroactivity of many Linnaean (and also Candollean) restrictive linguistic norms. In doing so, he did not call for rejection of "the names of other persons, because they do not think fit to acknowledge arbitrary rules which you are disposed to obey" [Lindley 1832: 457]. By all this, Lindley's regulative nomenclature appeared to be closer conceptually to the contemporary nominalist one than that of Candolle Sr.

> It is noteworthy that J. Lindley distinguished two basic categories of taxonomic names: "those of the class and order, which are *understood*; and of the genus and species, which are *expressed*" ([Lindley 1832: 454; italics in the original]). This categorization seemed to correspond roughly to the division of taxonyms into designators and descriptors (see Section 2.2); however, it was not discussed in any way by Lindley.

Another English botanist, Samuel Frederick Gray (1766–1828), who also acknowledged priority in an absolute sense, is to be mentioned for having introduced a certain group-specific standardization of suffixes of family names (which correspond to orders and far higher categories in contemporary plant classifications) in his two-volume *Natural Arrangement of British Plants* [Gray 1821]. In order to distinguish those from different classes, he supplied their names with different endings: *-phytae* for algae, *-lameae* for lichens, *-myceae* and *-thecae* for different fungi; however, no such group-specific endings were applied to tracheophytes. This was the first instance of ascribing the classification function to higher-rank taxonomic names by lexical means, within traditional nomenclature in the same manner as in the rational-logical one (see Section 7.1).

6.3.1.2 Zoology

In zoology, the first half of the 19th century was marked by a rather high activity of nomenclaturists; the most notable contribution to the development of nomenclature was made by the British.

The system of naming generic and subgeneric divisions was one of the dominant themes of the discussion in the *Magazine of Natural History* throughout the 1830s–1840s. The reason for this was that, in the practice of generic classifications of the first third of the 19th century, when large genera were divided into smaller ones or subgenera/sections were singled out in them, all taxa often received new names, while the old ones were cancelled. Contrary to this practice, W. J. Swainson based the names of genera and subgenera he recognized on the names of the respective "typical" species [Swainson 1820–1821], thus making these names tautonymous. In order to bring this practice to an end, it was suggested that "Where an old genus is divided into several new ones, new appellations must, of course, be found for them; but, even then, the original name should be retained for that group which is the most typical of the whole" [Strickland 1835: 39]. Following this line, the zoologist John Obadiah Westwood (1805–1893) examined the significance of the type species and how they should be fixed [Westwood 1836, 1837]; he proposed a set of rules for generic nomenclature that was given the informal name *Codex Westwoodianus* [Ogilby 1838]. According to one of these rules, "where an author does not state the particular species which he regards as the type of his genus, we are bound to suppose that he would place it at the head of his genus" [Westwood 1837: 170]. This *Codex* seemed to provide the first explicit formulation, fixing the new tradition ascending to Latreille and others (see Section 6.2.3), of what will later become known as the method of type in its nominalist sense [Pavlinov 2014, 2015a; Witteveen 2014].

The first British rulebook of nomenclature for animals was elaborated by the above-mentioned William John Swainson (1789–1855), a naturalist and opponent of Linnaean "artificial" systematics. His set of rules for the formation and change of zoological names was exposed in the section "*On the Nomenclature and Description of Birds*" in his *On the Natural History and Classification of Birds* [Swainson 1836]. He believed that "Nomenclature [...] is not strictly a part of the science of natural history; yet it is not only a convenient but an essential instrument for making that science more readily understood" [§ 183]. The main provisions of **Swainson Rules**, which he himself called *Laws of Nomenclature* and which H. Strickland subsequently refers to, were as follows. "Scientific names are not only given to every object or species in nature, but also to the different ranks or divisions under which species are comprehended" [§ 185]. "Every group or species for which a new name is proposed, must be properly defined; otherwise it cannot be adopted or noticed" [§ 187]. "Every new group or genus must have a new name [...] When repetitions [of names] are discovered, the name as first imposed or employed, is to be retained [in its initial use], and a new one given to the other group" [§ 189]. "Names of genera must be framed according to resemblances or properties, which, however, must be found not in one species of the genus only, but in the majority of those which are known" [§ 192]. "Names of groups higher than genera should always be derived from the pre-eminent type of the group, or, if of a tribe or order, from the character which is most universal"

[§ 195]. As one can see, *Swainson Rules* represented a mixture of both descriptive (though not precisely essentialist) and nominalist interpretations of the taxonomic names. The priority by date was expressly stated in them, and automatic typification of suprageneric taxa was applied (though not consistently). So they provided a very illustrative example of the prevailing state of the nomenclature systems of that time.

One of the key figures in the history of zoological and partly botanical nomenclature of the second third of the 19th century was the English naturalist Hugh Edwin Strickland (1811–1853) [Melville 1995; Rookmaaker 2011; Pavlinov 2014, 2015a]. He adhered, like A.-P. de Candolle, to the theory of natural systematics [Pavlinov 2018, 2021]; with regard to nomenclature, his main concern was the stability and universality of taxonomic names. In the paper that opened a series of publications on this matter in *The Magazine of Natural History*, he proceeded from the premise that "A complete parallel seems to exist between the proper names of species and of men" [Strickland 1835: 39]. Therefore,

> in order that the object of the specific name may be duly performed, it is essential that it be universally adopted, and, therefore, never, or very rarely, altered. But it is not, I think, essential that the meaning of the name should precisely designate the species; or, indeed, that it should have any *meaning* at all [...] The peculiar names of species ought, I think, to be in the same way regarded as proper names, the end of which is defeated by repeated alterations. And, although viewed in the light of a *memoria technica*, the recognition of species, by means of their names, is certainly facilitated, if the meaning of those names has a reference to the objects which they represent, yet this is so far from being essential, that it is, in some cases, prejudicial, by blinding persons to the distinction between the specific name and the specific character, and causing them to regard the objects of the latter as belonging to the former.
> *Strickland 1835: 38; italics in the original*

He asserted (with reference to Westwood) that "priority seems to be the universal law for the adoption of specific names [...] to the exception, that the same name be not repeated twice in a genus" [Strickland 1835: 40]. With this, however, Strickland understood taxonomic nomenclature as predominantly descriptive and therefore, following the essentialists, believed "that if, as rarely happens, the specific name have a meaning contradictory to the species which it represents, that name should be changed for one that is not contradictory" [Strickland 1835: 40].

In the next paper, H. Strickland published a draft of what he called *Rules for Zoological Nomenclature* [Strickland 1837], summarizing the earlier experiences of his colleagues and referring to them in the proposed formulations of the rules (in the quotations below, italics were in the original throughout). This draft consists of 22 articles with minimal comments on them. They are divided into two main sections; some rules are applied to the already "*established* Nomenclature" (their acceptance, rejection, change), and others are the "Rules in naming *new* species or groups." Strickland emphasized a fundamental status of the "law of priority" by stating that "The *first* name given to a group or species should be perpetually retained" [Art. 4].[6] With this, he considers admissible certain exceptions to this "law" in the case of previously established names: one is homonymy [Art. 5], and another is the case of "a name [...] whose meaning is false, as applied to the object or group which it represents" [Art. 6]: such names must be "expunged" in both cases. As to the new names, "It is *desirable*, but not *essential*, that a name should have an etymological meaning" [Art. 8] and "The *meaning* of a name must imply some proposition which is true as applied to the object which it represents" [Art. 11], while "The name of a species or group should be taken from those characters which are most essential and distinctive" [Art. 15]. By this, *Strickland Rules*, just like *Candolle Sr. Règles* before them, preserve certain evident elements of an almost essentialist treatment of taxonomic names by giving preference to those properly derived from the characters and expunging "false" names (i.e., non-genuine in the sense of Linnaeus). Following the already established tradition, he adopts the automatic (indirect) typification of families and subfamilies by suggesting that "The names of families and subfamilies should be derived from the most typical genus in them" [Art. 18], and they "should each have a distinctive termination" [Art. 16]. In contrast to this, typification of genera is not even mentioned in the project, although Strickland suggested in one of his previous papers that, when the genus was divided into several ones, the original generic name should be preserved for the most typical group [Strickland 1835].

The publication of Strickland's draft was followed by a brief discussion of the typification issue. One of its participants, W. Ogilby, insisted that "the word type be merely synonymous with *example*" ([Ogilby 1838: 282; italics in the original]). H. Strickland agreed with this interpretation and explained that he meant "by the type of a genus [...] that species which is usually selected as an example of the genus," and "by the most typical genus of a family, [...] that genus which seems to afford the best sample of the characters on which the family is based" [Strickland 1838a: 330]. In this regard, he reformulated Art. 18 of his draft as follows: "The names of families and subfamilies should be derived from that genus in them which affords the best example of their characters" [Strickland 1838a: 331; also Strickland 1845]. And yet, it is clear from the latter phrase that a certain "hidden essentialism" occurred in Strickland's consideration of the basis for selection of nominotypical genera.

During discussion of his project, H. Strickland expressed an idea of the need "to form a congress, or in humbler phrase, a committee of naturalists from all parts of the scientific world, to draw up a code of zoological laws" [Strickland 1838b: 199]. In response to this initiative, the British Association for Advancement of Science created, in 1842, a Committee "appointed to consider of the rules by which the Nomenclature of Zoology may be established on a uniform and permanent basis" [Strickland et al. 1843a: 105]. Members of the Committee considered a new version of *Strickland Rules*, prepared by H. Strickland on the basis of the comments on them, at the next annual session of the Section of Zoology and Botany of the British

---

[6] An absence of any fixed starting date for the application of the principle of priority drew massive criticism from proponents of the principle of usage, who accused H. Strickland of "letting the genie out of the bottle" by encouraging the revival of old and already forgotten priority names [Lewis 1872: 72].

Association (Manchester, 1842). The project did not receive unequivocal approval [Jardine 1858; Sclater 1878; Rookmaaker 2011] and was published as part of the report of the Committee [Strickland et al. 1843a, 1843b], along with its brief review [Gould 1843]. This report is later called **Code of the British Association** (*BA Code*) or *Manchester Code* or still *Strickland* (= *Stricklandian*) *Code*. Thus, this Code appeared to be the first to be introduced not by the personal authority of whatever prominent naturalist (C. Linnaeus, A.-P. de Candolle, etc.), but by a certain scientific collegial authority after its public discussion. For this reason, it should be authorized by reference just to the British Association (as is done here), and not personally to H. Strickland (as is often done). This general principle of authorization was stated in the preamble of the *BA Code* and is implemented in all subsequent nomenclature Codes and Rules.

The *BA Code* is divided into two parts entitled "*Rules* for the rectification of the present zoological nomenclature" and "*Recommendations* for the improvement of zoological nomenclature in future" (italics in the original). By contemporary standards (and in comparison, for example, with *Candolle Sr. Règles*), it is not too large: its first and second parts comprise 14 and seven main articles, respectively, each with numerous subparagraphs. Each section is also provided with a more or less extensive rationale, preceded by something like a "slogan" revealing its idea. Thus, the main sections of Part I are opened by the following declarations: "Limitation of the Plan to Systematic Nomenclature" and "Law of Priority the only effectual and just one."

The *BA Code* begins with a paragraph stating that "The name originally given by the founder of a group or the describer of a species should be permanently retained, to the exclusion of all subsequent synonyms (with the exceptions about to be noticed)" [Strickland et al. 1843a: § 1]. It is emphasized that the Code "is strictly confined to the *binomial system of nomenclature* […] originated solely with Linnaeus," so the principle of priority [should not extend] beyond the date of the 12th edition of the 'Systema Naturae'" [§ 1; italics in the original]. The latter thesis is approved by the assertion that "The binomial nomenclature having originated with Linnaeus, the law of priority, in respect of that nomenclature, is not to extend to the writings of antecedent authors" [§ 2]. Thus, this states not only the principle ("law") of priority as the main means of ensuring the stability and universality of nominative nomenclature but also the starting point for its application for the first time: it is the 12th edition of *Systema Naturae* by Linnaeus of 1766. The latter will become the subject of hot discussion and criticism, since this date excluded the earlier widespread names of many recognized naturalists (Artedi, Scopoli, Brisson, Blumenbach, etc.). Also for the first time, it is prescribed that the original generic name should be preserved when dividing the genus; for this, it is specified that,

> When the evidence as to the original type of a genus is not perfectly clear and indisputable, then the person who first subdivides the genus may affix the original name to any portion of it at his discretion, and no later author has a right to transfer that name to any other part of the original genus.
>
> § 5

As can be seen, this rule indirectly presumes typification of genera (though without mentioning species in such capacity), and also is a forerunner of the principle of first reviser.

Several cases of replacement or change of generic and specific names are reserved: homonymy of names; if a name "implies a false proposition" [§ 11]; tautonymy in the genus–species binomen, which requests that "Specific names, when adopted as generic, must be changed" [§ 12].[7] Besides, in Recommendation A there is a long list of the kinds of names which may or should be rejected or changed because of their inappropriateness (geographical, barbarous, wrong, nonsense, etc.). All this indicates an evident departure from the basic conditions of nominalist nomenclature.

An important criterion of the availability (legitimacy) of names asserts that "new genera and species be *amply* defined and *extensively* circulated in the first instance" [Recommendation E] (italics in the original). By this, the principles of diagnosing and publication are set (following A.-P. de Candolle and W. J. Swainson), and the new names featuring on the labels attached to specimens and exicates, in correspondence, in oral reports, etc., appeared to be outlawed. Other recommendations specify the following points: formation of names of families and subfamilies from generic names with the corresponding endings *-idae* and *-inae*; lowercase spelling of any specific epithet; both these recommendations will become universally adopted. However, the lack of a fixed hierarchy of taxonomic categories/ranks seems to be one of its fairly significant shortcomings.

The *BA Code* undoubtedly became a turning point in the development of taxonomic nomenclature. With this, although entitled "*The Nomenclature of Zoology*," it expressly emphasized "that almost the whole of the propositions contained in it may be applied with equal correctness to the sister science of botany" [Strickland et al. 1843a: 121; 1843b: 274]. Thus, it was initially presumed to serve uniformly for both zoology and botany, but this intention was soon broken down by the "Great Schism."

Although *BA Code* is sometimes referred to as regional [Melville 1995], it gained international significance immediately after its publication [McOuat 1996; Dayrat 2010; Pavlinov 2015a]. It was translated and published in France in the same year [Guérin-Méneville 1843], and was recommended a few years later, with minimal changes, by the Association of American Geologists and Naturalists that published a special *Report of Scientific Nomenclature* outlining the main provisions of the *BA Code* [Dana 1846]. This *Report* added a noteworthy recommendation, missing in the latter, regarding name authorship: "It is recommended that the original authority of a species always follow the name *in brackets*; and if the name be subsequently altered, the authority for the same as altered, be added *without* brackets" ([Dana 1846: 9; italics in the original]). This lexical rule generally becomes accepted, though with the opposite meaning.

An attempt to elaborate a nomenclature system based on the *BA Code* was undertaken by the Italian naturalists Charles Lucien Jules Laurent Bonaparte (1803–1857), known for his zoological research, one of the nephews of the famous Napoleon Bonaparte. His draft was placed before the 4th Congress of Italian Scientists in Padua

---

[7] This rule will entail a multiplication of synonyms by inventions of a lot of new species names.

(1842); in it Bonaparte insisted on the importance of an integrated Code for both "branches of natural history" and emphasized the great significance of the principle of priority [Minelli 2008]. This project was not implemented and therefore did not affect the development of nomenclature, though it is sometimes referred to as the *Padua Code*; it is considered the "first BioCode" by the Italian zoologist Alessandro Minelli [Minelli 2008].[8] The reason for its failure was that the naturalists who discussed it at Padua could not agree on a number of important points. In particular, they unanimously objected to considering homonymy of plant and animal generic names on a unified basis, thus making their contribution to the coming "Great Schism."

One of the "non-canonical" versions of post-Linnaean nomenclature was proposed for insects by the French jurist and amateur zoologist Charles Jean-Baptiste Amyot (1799–1866). It was based on quite a different ranked hierarchy, in which the notion of "genus" was rejected for certain etymological reasons, and the notion of "species" was used to replace it [Amyot 1848]. The taxa considered as such "species" were not further divided, and Amyot applied the so-called "*mononymic method*" to denote them. They were assigned one-word taxonyms, so the "specific" names of Amyot coincided with the traditional generic ones. Thus, this nomenclature system seemed to violate formally the principle of binomiality, but as far as Amyot's "species" were actually genera, no contradiction is actually observed.

6.3.2 THE "GREAT SCHISM": MULTIPLICATION OF CODES

The diversification trend became one of the dominant characteristics of the development of taxonomic nomenclature in the second half of the 19th and part of the very beginning of the 20th century. It spawned a number of subject-area and regional Codes in Europe and North America. Two main factors aggravated this "Great Schism": the growing isolation of subject areas of classical descriptive biology (botany and zoology) and different interpretations of precepts to ensure the stability of names (priority or usage, different starting points for priority, etc.); geopolitical contradictions were also in effect [Nicolson 1991; Melville 1995; Dayrat 2010; Pavlinov 2014, 2015a].

6.3.2.1 Botany

The new rulebook of botanical nomenclature emerged half a century after *Candolle Sr. Règles* following the suggestion of the Botanical Congress held in London in 1866. Its draft was prepared by the Swiss botanists Alphonse Louis Pierre Pyramus (Pyrame) de Candolle, or Candolle Jr. (1806–1893), the son of Candolle Sr. This project was considered by the next Botanical Congress (Paris, 1867), which was organized by the Société botanique de France especially for this purpose. The Congress approved the rules as "the best guide for nomenclature to follow in the

---

[8] It should be remembered that the "first BioCode" was actually elaborated by C. Linnaeus.

plant kingdom." Their text was first published in French as *Lois de la nomenclature botanique* and then translated into English [de Candolle 1867, 1868]. It is convenient to call this new rulebook **Paris Laws** to distinguish it from both *Paris Code* of botanists of 1956 and *Paris Code* of zoologists of the 1880s; it is also known as *Candolle Jr. Laws/Lois*.

These *Paris Laws* are very clearly structured: they are divided into chapters, sections, and paragraphs, the titles of which proclaim the principal meaning of the articles they contain. The latter are discussed by Candolle Jr. in the extensive comments section, which is almost twice the volume of the articles themselves, as well as in a separate subsequent publication [de Candolle 1869]. His comments substantiate interpretations of the provisions adopted in *Paris Laws* and also analyze other opinions, sometimes very numerous and diverse. The most important provisions of these *Laws* are as follows (cited after the English edition).

Chapter I begins with the assertion that "Natural History can make no real progress without a regular system of nomenclature, acknowledged and used by a large majority of naturalists of all countries"; moreover, "the principles and forms of the nomenclature should be as similar as possible in botany and zoology" [Art. 1]. An important provision asserts that "Nomenclature comprises two categories of names: 1. Names, or rather terms, expressing the nature of the groups comprehended one within another. 2. Names particular to each of the groups of plants or animals" [Art. 7]. In contemporary terms, entry (1) corresponds to the designations of taxonomic categories (ranks), and entry (2) corresponds to the designations of taxa. Chapter II contains the complete taxonomic hierarchy from Kingdom to subvarieties, 20 categories in total. It asserts that "their relative rank, sanctioned by custom, must not be inverted" [Art. 11]. Of them, five categories are recognized within species; this will become a characteristic feature of all future botanical nomenclature, differentiating it from that in zoology.

Regarding establishing new taxa, it is indicated that "When a group is moved, without alteration of name, to a higher or lower rank than that which it held before, the change is considered equivalent to the creation of an entirely new group" [Art. 51]. In contrast, "An alteration of characters, or a revision carrying with it the exclusion of certain elements of a group or the addition of fresh ones, does not warrant a change in the name or names of a group" [Art. 53].

Chapter III considers particular principles and rules for the names, and begins by stating that "Each natural group of plants can bear in science but one valid designation, namely, the most ancient, whether adopted or given by Linnaeus, or since Linnaeus, provided it be consistent with the essential rules of nomenclature" [Art. 15]. Thus, three fundamental nomenclature principles are set by this article, viz., monosemy, priority, and availability (legitimacy). It is suggested that ormation of the names for the different rank-groups follows different rules. The names of Divisions and Subdivisions, of Classes and Subclasses "are drawn from their principal characters" [Art. 18]. Cohorts and Subcohorts "are designated preferably by the name of one of their principal Orders, and as far as possible with a uniform termination" [Art. 20]. Orders and Suborders, of Tribes and Subtribes, "are designated by the name of one of their genera" [Art. 21]; their names differ in rank-specific suffixes. "Genera, subgenera, and sections, receive names [that] may be derived from any source whatsoever, and may even be arbitrarily imposed" [Art. 25]; "subsections, as well as to inferior generic subdivisions […] may simply be indicated by a number, or by a letter" [Art. 26]. "All species, even those that singly constitute a genus, are designated by the name of the genus to which they belong, followed by a name termed specific" [Art. 31]; "The specific name ought, in general, to indicate something of the appearance, the characters, the origin, the history, or the properties of the species" [Art. 32]. "No two species of the same genus can bear the same specific name, but the same specific name may be given in several genera" [Art. 35]; "Avoid specific names having, etymologically, the same meaning as the generic name" [Art. 36]. "Hybrids whose origin has been experimentally demonstrated are designated by the generic name, to which is added a combination of the specific names of the two species from which they are derived" [Art. 37].

Section 3 describes criteria for publications to be valid, including indication of characters and use of Latin names. With this, the distribution of exicates is equated to publication, while reports at conferences are not [Art. 42].

Additional rules for the formation of taxonomic names are summarized in Sections 4 and 5. It is stipulated that "For the indication of the name or names of any group to be accurate and complete, it is necessary to quote the author who first published the name or combination of names in question" [Art. 48]. When dividing genus, section, or species, the original name is retained for the subdivision which is the "type or origin of the group" (in the case of genus and section) or "distinguished earlier than the others" (in the case of species) [Art. 54–56]. When moving a section or a species to another genus, or a subspecies or a variety to another species, the name of the group being moved does not change unless there is homonymy [Art. 58]. Section 6 begins with the statement that "Nobody is authorized to change a name because it is badly chosen or disagreeable, or another is preferable or better known, or for any other motive" [Art. 59]. But this is followed by a list of cases (five of them) when "Every one is bound to reject a name" [Art. 60]; in particular, the name is rejected, "When it expresses a character or an attribute that is positively wanting in the whole of the group in question, or at least in the greater part of the elements it is composed of." The latter passage clearly indicates the noticeable traces of essentialism in this nomenclature system.

As can be seen, *Paris Laws* represented, in general, a very advanced version of the nomenclature system of that time. With this, one cannot fail to note its somewhat eclectic character. Some rules were proposed *de novo*, often contrary to the established tradition in botany (for example, authorization), some with reference to zoological rules; most of them will be inherited by the subsequent Codes. On the other hand, some rules fixed the provisions of "descriptive" nomenclature going back to *Candolle Sr. Règles* (first of all, those for rejecting and changing descriptive names); most of them will be excluded from botanical nomenclature as it will moves toward its more nominalistic character.

Among the fundamental flaws of *Paris Laws* was the lack of detailed consideration of the principles of priority and typification. In fact, Candolle Jr. repeatedly emphasized the need to preserve the oldest names, but did not indicate any specific starting date or work, although they had already been mentioned in *Candolle Sr. Règles*. Nothing specific about this important rule was said also in his *Commentaries*: Candolle Jr. confined himself to a recommendation not to accept the names of Tournefort and earlier authors, while those of Linnaeus "must remain [as] legitimized by habit and by general assent" [de Candolle 1868: 44]. The notion of type was only mentioned in the case of genera, but in the sense of their "typical subdivisions," and not type species.

The almost simultaneous publication of *Paris Laws* in French and English marked its international status. However, they did not receive wide recognition among botanists. In particular, they were ignored by the English adherents to *Kew Rule* (see below), who even refused to participate in their discussion at the Paris Congress. Nevertheless, they became an important landmark in the development of internationally recognized principles of nomenclature in botany and partly in zoology. They will be referred to in the *AA Code* (1877, see Section 6.3.3), in the preamble of *Paris Code* of zoologists (1890s), and they will form the basis of *Vienna Rules* of botanists (1905), officially recognized as the first international one (see Section 6.4.2).

> It is to be noted that A. de Candolle was aware of the transitory nature of his *Laws*, just like that of any other nomenclature system. He concluded an introduction to them with the surmise that, over time,
>
> "it will become necessary to effect some great revolution in the formulae of science [...] Perhaps there will then come to light something very different from the Linnaean nomenclature,—something will have been devised for giving definite names to definite groups. This is the secret of futurity, of a yet very distant period. In the meantime, let us improve the system of binominal nomenclature introduced by Linnaeus.
>
> *de Candolle 1868: 16*

In the 1860s, the above-mentioned **Kew Rule** (also known as *Bentham Rule*; see [Kuntze 1891]) suggested a very peculiar interpretation of the relationship between the priority of the names of species and the "truth" status of the genera that include them. Its authors were English botanists at Kew Botanical Garden, especially George Bentham (1800–1884). G. Bentham argued that the genus had greater taxonomic significance than the species, therefore placing a species in a "true" genus, with the result that it took its "true" place in the System of Nature, had greater taxonomic significance than the original description of that species. Accordingly, it was not the specific epithet as such, but the genus–species binomen, that had a real nomenclature meaning, providing that its generic part designated the "true" genus.[9] It was concluded from this reasoning that it made sense to apply the principle of priority

---

[9] This proposition was evidently based on one of the important provisions of *Linnaean Canons* asserting that a species should be named not by itself, but as a member of a "definite" genus (see Section 5.3 on the latter). So, *Kew Rule* was actually the most "Linnaean" one in this respect.

only to those specific names that were associated with the "true" generic names [Bentham 1858, 1879, 1880; Hiern 1878; Jackson 1887]. In particular, the latter author argued that "Our practice is to take the name under which any given plant is placed in its true genus as the name to be kept up" [Jackson 1887: 69]. It was presumed that the stability of "true" genera warranted greater stability of the "true" genus–species binomens than the re-establishment of old species names according to the principle of priority in its canonical formulation [Gray 1883; Hallier 1900]. It was agreed in addition that the nomenclature applied in major publications by recognized monographers should be given the highest significance when considering the issue of "true" genus–species binomens [Bentham 1879; Jackson 1887].[10]

The *Kew Rule* gained certain popularity due to its being applied in the authoritative *Genera plantarum* by G. Bentham and Joseph Hooker (1862–1883) and the botanical nomenclator *Index Kewensis* (which began publication in 1893). Yet it remained a very particular, albeit remarkable, episode in the history of botanical nomenclature [Nicolson 1991; Stevens 1991; Bonneuil 2002; Pavlinov 2014, 2015a]. *Kew Rule* is later rejected by botanists when adopting the *Vienna Rules* of 1905.

In the 1880s, the revision of *Paris Laws* became the most remarkable event in the history of botanical nomenclature. The first important step in this direction was made by the author of its first version, A. de Candolle, in his extensive *Nouvelles remarques* [de Candolle 1883]. He examined the development of rules for naming plants and animals after the publication of the first version of *Paris Laws*, and commented in detail on criticisms of their main provisions by various taxonomists. In the introductory section of this paper, Candolle Jr. emphasized that the main trend in the development of nomenclature over the past several years was the noticeable convergence of the rules in botany and zoology on a number of fundamental points, including the recognition of the principle of priority as one of the basic means for ensuring the stability of names. But this important comment did not drive botanical and zoological branches of nomenclature towards their integration.

The first part of this paper concerns (with reference to zoologists) reasons for choosing the particular works as a starting point for the application of the principle of priority. The following criteria are considered most important by Candolle Jr. in this regard: a standard ranked hierarchy is to be consistently applied; the ranks of the newly established groups must be indicated definitely (as family, genus, species, etc.); these groups must be characterized pertinently by their diagnostic features. The second part of the article considers the issues omitted in *Paris Laws*, including anatomical nomenclature with reference to the recently published A. de Candolle's *La phytographie* [de Candolle 1880], the nomenclature of fossil organisms with reference to *Douvillé Rules*, and the significance of lower taxonomic categories in the light of the theories of Alexis Jordan and Charles Darwin. In addition, the standard names of the main divisions of the plant kingdom, the suggestions of *Kew Rule*, and some other particular provisions are considered. The third part of the paper contains an updated revision of *Paris Laws*

---

[10] As some contemporary historians of botany suspect, the authors of *Kew Rule*, by doing so, freed themselves from the need to trace older synonyms in "peripheral" and "non-significant" literature [Bowker 1999; Bonneuil 2002; Endersby 2008].

(called again *Lois de la nomenclature botanique*), and it is indicated that "the rules of nomenclature apply to all classes of plants, both fossil and living" [Art. 7]. Although Candolle Jr. notes in the introductory section that "The law of priority is increasingly recognized as the fundamental principle of good nomenclature" [de Candolle 1883: 4], this revision of *Laws*, like the previous one, does not specify any particular work or date as a starting point for the application of this "law."

The next step in the development of *Paris Laws* was undertaken at the Geneva Botanical Congress (1892). Its participants approved their officially recognized status and complemented them with the following important provisions: 1753 (*Species plantarum* by Linnaeus) as a starting point for the priority of generic names; rejection of names based only on images and exicates; avoiding names that were out of use for either 50 or 100 years; preservation of generic names differing only in their endings; the standard treatment of the principle of priority was confirmed (*Kew Rule* rejected). In addition, an official list of plant generic names protecting against further replacements and corrections (*nomina conservanda*) was approved at the Geneva Congress and published for the first time.

Another "hot spot" in botanical nomenclature, after *Kew Rule*, emerged in the early 1890s with **Kuntze Code** authorized by the German botanist Karl Ernst Otto Kuntze (1843–1907). In the introductory section of the first volume of his *Revisio generum plantarum* (1891–1898), he emphasized the importance of fixed priority and proposed 1735 as a starting point for applying this principle to generic names, and 1753 to species names. K. Kuntze justified his decision as follows: "*Systema naturae editio princeps prima* of 1735 is accepted as the first consistently implemented Linnaean nomenclature and systematics of genera, and *Species plantarum* of 1753, as the first consistently implemented Linnaean nomenclature and systematics of species" [Kuntze 1891: lxxvi]. In addition, the year 1763 (Adanson's *Familles des plantes*) was proposed as the starting point for considering the priority of suprageneric names. Kuntze exposed his full-length trilingual (in German, French, and English) *Codex nomenclaturae botanica emendatos* in the first part of the third volume of his *Revisio* [Kuntze 1893]. With this, he emphasized in the afterword that he did not annul but rather supplemented and specified the provisions of *Paris Laws*; this was in general true, as most of the latter's articles were repeated almost literally by Kuntze.

The following significant provisions of *Codex emendatos* should be highlighted. It is suggested that the laws of nomenclature "can be altered only by competent authors at an international congress" [Art. 70]. It is specified that "The annulments and alterations of the existing laws shall have no retroactive force and shall be applicable only to new or subsequently renewed denominations" [Art. 71].[11] It is stipulated that "Publications are only admissible for competition for valid nomenclature as long and as far as they [...] appear in the Latin, English, French or German languages" [Art. 69]; Kuntze added Italian later. An equation of the distributions of the authorized herbarium sheets to publications is suppressed. Contrary to the principle of priority, it

---

[11] Following this rule would reduce the undesirable significant retroactive effect of many articles of *Paris Laws*.

is agreed (with reference to the *Paris Laws* of 1892) to suppress the "forgotten" names of genera, species, and varieties that had not been in usage within the last 100 years.

Contrary to the latter clause and based on his decision on starting dates, Kuntze revived old long-forgotten plant names and replaced those newer ones that turned out to be junior homonyms of those he re-established. According to calculations by his colleagues, the "Kuntzean reform" entailed the appearance of about a thousand generic and at least 30,000 genus–species binomens, which became "new" in the context of the then well-established nominative nomenclature in botany. This renovation was comparable in scale with that by Linnaeus in the mid 18th century. But since Kuntze was not Linnaeus, his proposals were strongly criticized and categorically rejected by his contemporaries who were not inclined towards such radical alterations [Greene 1891a; Ascherson and Engler 1895].

Berlin botanists led by Paul Friedrich August Ascherson (1834–1913) and Adolf Engler (1844–1930), in opposition to *Kuntze Code*, prepared their own short nomenclature rulebook. Their suggestions were considered at the Versammlung deutscher Naturforscher und Aerzte (Congress of German Naturalists and Physicians) (Vienna, 1894), and the results were published within a year [Ascherson and Engler 1895; Smith 1895]. It contained a review of contemporary ideas supplemented by a massive critique of *Kuntze Code*; because of this, it was informally called *Berlin Protest* [Greene 1891b; Britten 1898; Nicolson 1991]. A more developed and formalized set of rules was published as *Nomenclaturregeln für die Beamten* in several years [Engler et al. 1897], and immediately reprinted in French and English [Briquet 1897; Robinson 1897]. It was followed soon by a not very significant upgrade [Engler et al. 1902] known as **Berlin Rules** (or *Berlin Code*). In the preface to them, the authors emphasized the critical situation with botanical nomenclature and the undesirability of its drastic reform (meaning *Kuntze Code*).

The original *Berlin Rules* contain only six articles, and their updated version comprises 14 articles; they appear in an order corresponding to the weight of the criticism addressing *Kuntze Code*. The names once synonymized should not be used again in future to denote recognized taxa; this is the rule "once a synonym always a synonym," evidently borrowed from the zoological *American Ornithologists' Union Code* of 1886 (AOU Code, see Section 6.3.2.2). The year 1753 (date of the first edition of *Species plantarum* by Linnaeus) should be taken as the beginning of the application of the principle of priority for both genera and species. However, a generic name, even the oldest, should be discarded if it had not been in circulation for the past 50 years (this corresponds to the principle of usage). It is acknowledged that binomial nomenclature is considered not applicable to hybrids. The description of a new species requires indication of its diagnostic characters, with such indication, whether on the exicate label or in the publication, being equally admissible. The author has no right to change, at his will, a generic or specific name once established by him, unless for a very serious reason.

The Swiss botanist John Isaac Briquet (1870–1931), a pupil of Candolle Jr., in his review of *Berlin Rules*, praised most of them and urged colleagues to follow them [Briquet 1897]. However, in fact, these *Rules* are not so much a full-fledged rulebook as a brief list of "anti-Kuntzean" recommendations [Britten 1898]. For this reason, they lack many important provisions that formed the basis of almost all fairly advanced nomenclature systems of that time. For instance, the principle of binomiality is not mentioned in them (probably because it was considered self-evident). They miss the rules for preserving and changing taxonomic names, their typification, etc. They lack clear criteria as to which competing synonym should be considered the most used.

American botanists remained somewhat aloof from the development of nomenclature standards by Europeans in the mid 19th century, but their noticeable activity in this regard began to be seen towards its end [Britten 1895; Parkinson 1975; Pavlinov 2014, 2015a]. One of the initial stimuli toward the rise of their activity became the aspiration to fix the principle of priority as an indispensable condition for the stability of botanical nomenclature. To this end, Emerson Ellick Sterns (1846–1926) published an appeal to American botanists to sign an *Agreement of the Botanical Nomenclature League of North America*, in which he called:

> to use [...] those names which [...] conform most closely to that law of priority which requires: I. That the first published specific or varietal name of a plant, given to it in accordance with the binomial system of nomenclature, whether appropriately or not, shall be perpetually and strictly maintained [...] II. [...] That when two or more generic names have been regularly applied to the same genus, the earliest shall be maintained, to the entire exclusion of any of later date.
>
> *Sterns 1888: 232*

This appeal received a certain amount of support [Britten 1888], and it was emphasized referring to the above-mentioned zoological *AOU Code* that adoption of this principle would establish a certain unity of principles of taxonomic nomenclature among American systematists [Merriam 1895].

In response to this appeal, the Committee on Botanical Nomenclature was appointed at the Botanical Club of the American Association for the Advancement of Science by an initiative of Nathaniel Lord Britton (1859–1934). It was approved at the first meeting of the Club in Rochester (1892) that *Paris Laws* of 1867 be adopted, except where they conflict some other important provisions [Britton et al. 1892]. A brief summation of the latter was published as **Rochester Rules** (or *Rochester Code*); their principal content is as follows [Bessey 1892; Britton et al. 1892; Kuntze 1893; Nicolson 1991; Pavlinov 2014, 2015a]. The priority of publication "is to be regarded as the fundamental principle of botanical nomenclature" [Rule I]. "The botanical nomenclature of both genera and species is to begin with the publication of the first edition of Linnaeus' *Species plantarum* in 1753" [Rule II]. When a species is transferred to another genus, "the original specific name is to be retained, "unless it is identical with the generic name, or with a specific name previously used in that genus" [Rule III]. With this, the original author of such name must be provided in brackets, followed by the name of the author of the new binomial [Rule VII]. For the name of a new genus to be legitimate (available), it must be published with a description, and "the citation of one or more previously published species as examples or types" is needed [Rule V]. "Similar generic names are not be rejected on account of slight differences in spelling, except in the spelling of the same word" [Rule VI].

At the next meeting of the Botanical Club (Madison, 1893), its members considered amendments to *Rochester Rules* taking into account the nomenclature novelties adopted at the above-mentioned Geneva Botanical Congress [Britton 1893]. The most significant allows tautonymy of the genus–species binomen prohibited by other botanical rulebooks. In addition, when considering the priority of generic names published in the same work, it is suggested that consideration be given to "precedence by place" (by page and by line). A group of supporters of these *Rules*, that appeared to be substantially different from those developed by European botanists, was called "the neo-American school of nomenclature" [Rickett 1953].

The *Rochester Rules* were heavily criticized by adherents of the principle of usage [Bessey 1895; Britten 1895; Robinson 1895, 1898; Ward 1895; Fernald 1901a,b]. One of their leaders, Charles Edwin Bessey (1845–1915), on behalf of many dozens of American botanists, published "A protest against 'Rochester Rules' " with a critical review of their basic provisions [Bessey 1895]. Another leader of this "anti-Rochester" group, Benjamin Lincoln Robinson (1864–1935), distributed a circular with *"Recommendations regarding the nomenclature of systematic botany"* [Robinson 1895], subsequently called **Harvard Rules** [Nicolson 1991; Pavlinov 2014, 2015a]. The latter are very brief and contain only the following five concisely annotated rules: "1. Ordinal names, having been established by long usage, should not be subjected to revision upon theoretical grounds." "2. Long-established and generally known generic names should be retained." "3. In specific nomenclature the first correct combination is to be preferred.—For these reasons it seems best to adopt the principle of priority under the genus."[12] "4. The varietal name is to be regarded as inferior in rank to the specific [and] No specific name should be altered, because of preexisting varietal names for the same plant."[13] "5. The principle of 'once a synonym always a synonym,' while recommended as an excellent working rule for present and future, may not justly be made retroactive" [Robinson 1895: 263].

The American botanist John Hendley Barnhart (1871–1949) contributed to this nomenclatural activity by proposing a concise set of rules for the names of orders and families [Barnhart 1895], which were not formulated clearly enough in *Rochester Rules* and were absent from *Harvard Rules*. He wrote that the family name should consist of the stem of the legitimate (accepted) name of a recognized genus belonging to it (dating back to Adanson and Candolle Sr.). It is also suggested that the correct (valid) family name must be the oldest, and its authorship must be indicated in the same manner as for the generic name. Family names, similarly to those of genera and species, cannot be taken from works prior to the first edition of *Species plantarum* by Linnaeus.

In connection with preparations for the International Botanical Congress to be held in Vienna (1905), at which the new international rules for botanical nomenclature were supposed to be adopted (see Section 6.4.2), the Committee on Nomenclature of

---

[12] Probably taken from *Kew Rule*.
[13] This recommendation is later included in botanical nomenclature.

the Botanical Club of the American Association elaborated its draft rules. The latter was based on *AOU Code* and *Rochester Rules*, and it was preliminarily considered at the next meeting of the Club in Washington (1903) and approved at its meeting in Philadelphia (1904), and after its publication in a trilingual version it became known as the **Philadelphia Code** [Anonymous 1904; Nicolson 1991; Pavlinov 2014, 2015a]. It contains Part I with four basic principles and Part II with 19 canons. Its most important innovation is detailed prescription of the rules ("canons") for typification of the names of species-group taxa: they are largely borrowed from the zoological *AOU Code* and appear here for the first time in botanical nomenclature [Hitchcock 1905; Petersen 1993; Pavlinov 2014, 2015a]. In this regard, an important notion of nomenclatorial type is introduced for the first time in the list of basic principles. It is noteworthy that, although 1753 is recognized as the starting point for the principle of priority, it is suggested that pre-Linnaean authors of generic names, if there were any, should be indicated (in brackets).

However, participants of the Vienna Congress rejected the *Philadelphia Code*. The latter's authors declared their strong disagreement with this decision and adherence to the main provisions of their nomenclature rulebook [Hitchcock 1921, 1922]. So, they published it shortly with some amendments as the **American Code of Botanical Nomenclature** [Anonymous 1907]. Since N. Britton still served as the main driving force behind this Code, it is sometimes referred to as the *Brittonian Code*, by analogy with *Stricklandian (= Strickland) Code* [Nicolson 1991]. The difference between the *American Code* and its Rochester predecessor is mainly in clarification of the rules for typification of genus-group taxa. In particular, it is recognized that "The publication of a new generic name as an avowed substitute for an earlier invalid one does not change the type of a genus" [Anonymous 1907: Canon 15]. The recommendation by *Vienna Rules* to write diagnoses of newly described taxa in Latin is not accepted. Given the substantial difference in the way genus-group taxa are defined, the *American Code* was called "type-basis" in contrast to *Vienna Rules* [Hitchcock 1922], which (in this terminology) were "circumscription-basis."

In botany, along with canonical nomenclature regulating naming the taxa of wild flora, plant breeders began to develop their own rules for cultivated plants [Stearn 1952a; Malécot 2008]. Unlike the case of wild flora, the first and most active in this respect appeared to be North American botanists. The *Rules for American Pomology* were agreed by several regional Horticultural Societies in 1847, which was the first move towards a unified nomenclature of cultivated plants. As stated in the introduction, "These rules [...] are calculated to stamp a character of scientific precision and accuracy on the nomenclature and description of fruits, which will make Pomology rank, as it should, among other branches of Natural History" [Anonymous 1847: 273–274]. The main emphasis of these *Rules* with 13 articles is on the quality of fruits when considering establishing new names for them [Art. I]; to the latter consideration is added the precept that "Before giving a name to a new fruit, its qualities should be decided by at least two seasons' experience" [Art. X]. An evident non-nominalist character of this nomenclature system is evident from the provision that "Characteristic names, or those in some way descriptive of the qualities, origin, or habit of fruit or tree, shall be preferred" [Art. VIII], to which was added that, "if the name proposed is inappropriate, or does not come within the rules, then the describer shall be at liberty to give a name" [Art. II]. The principle of publication is established in very strong form: "The name of the new variety shall not be considered as established, until the description shall have been published in at least one Horticultural or one Agricultural Journal" [Art. V]. It is prescribed that "No new names shall be given, which consist of more than two words, excepting only when the originator's name is added" [Art. VII]; however, this provision does not seem to presume anything like canonical binomiality, as it does not refer to the generic–species binomen. Unlike taxonomic nomenclature, these *Rules* consider traditional pomological names and do not presume their Latinization. An upgraded version of these *Rules* appeared within a year in the same journal [Anonymous 1848].

6.3.2.2 Zoology

The multiplication of the specialized Codes of zoological nomenclature began in the mid 19th century with the *Gesetze der entomologischen Nomenclatur* that were discussed and approved by the International Congress of Entomology in Dresden (1858). Its draft was prepared by Ernst Hellmuth von Kiesenwetter (1820–1880), and it is called **Dresden Code** or sometimes *Kiesenwetter Code*. Although no obvious agreement was reached among Congress participants on major issues (such as the relation between priority and usage of names), they decided to publish the "rules, the adherence to which is deemed desirable in the allocation of new names" [Wessel 2007: 162]. This Code was published in the same year in German [Kiesenwetter 1858] and soon after in French [Mulsant 1859–1860], i.e., it became international to a degree. It addressed entomologists, with Fabricius cited most frequently as the main authority. So, its authors did not seem to care much about the fact that they thus set a dangerous precedent in zoology for the simultaneous functioning of several nomenclatural rulebooks [Dayrat 2010; Pavlinov 2014, 2015a].

In general, *Dresden Code* is noticeably more primitive than *BA Code* in both its content (22 paragraphs) and structure [Sclater 1896; Pavlinov 2015a]. Its most important provisions are as follows [Kiesenwetter 1858; Mulsant 1859–1860]: the use of the oldest names (priority) without indicating any starting date; the importance of typifying generic names; a sufficiently clear indication of the conditions of availability of names (publication, diagnosing, or figuring). This Code deals with the names of genera and species [§ 2, 9] (other rank-groups were not even mentioned), for which standard grammatical rules are considered [§ 3–8]. A new name must be published in a scientific book or in a journal; its publication should contain either a diagnosis or an appropriate figure to make the object recognizable [§ 10, 12]. The earliest name should be used to denote a taxon, unless it is a homonym of an older name [§ 14]; with this, if it were impossible to select an appropriate name on the priority basis, "the choice […] may be more or less free according to considerations of convenience" [§ 15]. A rather detailed description of typification is commendable; it should be applied whenever a species or genus is split or united [§ 17–19]. According to the established "descriptive" tradition, the new "names should, as far as possible, be characteristic, i.e., express a special quality of the denoted object" [Section for establishing new names, § 3]. It is to be highlighted that the homonymy of generic names is suggested to be considered only for zoological taxa [§ 14]; by this, the independence of zoological and botanical nomenclatures is implicitly acknowledged, and this contributed to the initial "divorce" of zoological and botanical nomenclature.

An updated version of *BA Code* was prepared by the Committee on Nomenclature at the British Association and approved at its Birmingham Meeting in 1865. This version was then called *Birmingham Code*, and its text was published in a report of the Association with the following basic provisions [Jardine 1866].

It is stated at the very beginning "That Botany should not be introduced into the Strickland rules and recommendations"; "That the permanence of names and convenience of practical application being the two chief requisites in any code of rules for scientific nomenclature"; and "that the XIIth edition of the *'Systema Naturae'* is that to which the limit of time should apply, viz. 1766." However, a special reservation is made that "as the works of Artedi and Scopoli have already been extensively used by ichthyologists and entomologists, it is recommended that the names contained in or used from these authors should not be affected by this provision" [Jardine 1866: 28]. It is also stipulated that Brisson's generic names of birds should be retained, but this was not extended to his species names, as he "still adhered to the old mode of designing species by a sentence instead of a word" [Jardine 1866: 31]. Some restrictions and recommendations contained in the first version of this Code are softened in the new edition; they apply to the species names established as toponyms or eponyms. With this, suppression of the genus–species tautonymy is approved, and it is prescribed that, contrary to the respective condition in the first version of *BA Code*, if a species becomes a sole member of a new genus and its name becomes a generic name, "it is the generic name which must be thrown aside, not the old specific name" [Jardine 1866: 28].[14]

As a matter of fact, this renewed edition of *BA Code* was largely a reproduction of the 30-year-old nomenclature supplemented by only a few changes, which, for instance, the authors of *AOU Code* considered unsuccessful [Coues et al. 1886]. Due to this, this version of *BA Code* failed to play any significant and noticeable role as a driver of nomenclature in zoology, though, with the exception of *AOU Code* (see below). With this, it became a kind of trigger that stimulated a new round of active discussions of important issues in nomenclature. Among them were, first of all, the principle ("law") of priority and the initial date of its application [Dayrat 2010; Pavlinov 2014, 2015a]. The above-mentioned A. Lewis commented sarcastically that "Those respected gentlemen could not realise in 1842 the condition of entomological nomenclature in 1872" [Lewis 1872: 2]. The American zoologist William Henry Edwards (1822–1909) believed that "Strickland's code [...] does not work at

---

[14] This condition yielded a great number of newly established generic synonyms and is subsequently rejected.

all satisfactorily and never received the approval of naturalists from various fields of activity" [Edwards 1873: 22]. As noted by one of the discussants, "the recognized rules of nomenclature are in no way satisfactory [and] the attempts made by the British Association [...] have not been successful" [Agassiz 1871: 355]. He was echoed by a comment that, "although the Zoological Nomenclature Rules were authorized by the British Association, they did not appear to have resulted in a clear improvement in the situation" in zoology [Crotch 1870: 59].

Where the updated version of *BA Code* undoubtedly succeeded, there appeared to be a clear marking of the "Great Schism" between zoological and botanical sections of taxonomic nomenclature. This was created by the above-mentioned declaration of the irrelevance of botanical nomenclature to the zoological one.[15] Due to this, taxonomic nomenclature continued its movement away from *Linnaean Canons* that had been designed as part of a universal and unified descriptive language for the whole of natural history.

One of the most advanced versions of the rules of zoological nomenclature was elaborated by the Committee on the Revision of the Nomenclature And Classification of North American Birds, appointed at the first annual meeting of The American Ornithologists' Union in New York (1883). The draft of the new Code was presented by the naturalist and historian Elliot Coues (1842–1899) at its next meeting, where it was approved and then published as an official document of the Union [Coues et al. 1886] to become known as **American Ornithologists' Union Code** (*AOU Code*). It was based largely on *BA Code*, as stated in the Committee resolution and in the preamble of the Code, where it was referred to as *Stricklandian Code*. The *AOU Code* consisted of three main parts: "General Principles" (five of them), "Canons of Zoölogical Nomenclature" (51 of them), and "Recommendations for Zoölogical Nomenclature in the Future" (nine of them). Most of the Canons (hereafter abbreviated as Ca.) were commented on more or less extensively; these comments, similar to *BA Code* and unlike *Paris Laws* of botanists, were placed in the main text rather than in an appendix. The main precepts of *AOU Code* are briefly characterized below.

First of all, the importance of the "principle of an inflexible law of priority" [Coues et al. 1886: 11] is emphasized in the Introduction as one of the basics for zoological nomenclature. And yet, *AOU Code* does not absolutize this principle, allowing some restrictions to its application. Their aim is "to guard, as far as may be possible, against needless or undue rejection of names in current usage in favor of obscure earlier ones which rest upon descriptions so vague or imperfect" [Coues et al. 1886: 11]. Among other principles, the main provision of the nominalist nomenclature is declared that "A name is only a name, having no meaning" [Pr. V]. It is asserted that "Zoölogical nomenclature has no necessary connection with botanical nomenclature, and names given in one of these two systems cannot conflict with those of the other system" [Pr. IV]: this is the ultimate approval of the independence of the zoological and botanical systems of nomenclature.

---

[15] It should be noted that this was done at the insistence of the botanists present at the British Association meeting [Jardine 1866].

The desirability of unification of the starting date for the application of the fixed priority for the names of all categories in zoological taxa is emphasized, and this date is set as follows: "Zoological nomenclature begins at 1758, the date of the Xth edition of the 'Systema Naturae' of Linnaeus" [Ca. XII]. This decision is substantiated in a lengthy commentary, which considers arguments *pro* and *contra* either earlier (works of Tournefort, Lang, Klein, earlier editions of Linnaeus' *Systema Naturae*) or later (the latter's 12th edition) dates. Thus this date appeared for the first time in zoological nomenclature, and will be adopted subsequently as the universal starting point in zoology.

Of particular interest is the section "Of the Binomial System as a Phase of Zoölogical Nomenclature." Its preamble reads:

> that the rigidity and inelasticity of [a binomial] system, which has been followed for more than a century, unfits it for the adequate expression of modern conceptions in Zoölogy, and that therefore a strict adherence to it is a hindrance rather than a help to the progress of science. It believes that strict binomialism in nomenclature has had its day of greatest usefulness and necessary existence; and that at present it can only be allowed equal place in nomenclature by the side of that more flexible, elastic, and adequate system of trinomials.
>
> *Coues et al. 1886: 29*

This general idea is developed in the section "Of the Trinomial System as a Phase of Zoölogical Nomenclature," in which provisions for the formation of three-word subspecies names are formulated.

> Trinomial nomenclature consists in applying to every individual organism [...] three names, one of which expresses the subspecific distinctness of the organism from all other organisms, and the other two of which express respectively its specific indistinctness from, or generic identity with, certain other organisms.
>
> *Ca. XI*

By this, the *principle of trinomiality* appears in *AOU Code* to become an important innovation comparable to canonical binomiality in its potential significance.

> Like the similar provision of botanical *Paris Laws* of Candolle Jr., this principle assigned a certain nomenclatural meaning to the basic idea of classification Darwinism implying the particular importance of infraspecies taxa to be recognized and named [Pavlinov 2018, 2021]. It is not officially acknowledged by future Codes, but special reservation will be made in them that the use of three-word subspecies names would not violate the binomial principle applied to species. This clarification appeared in response to those authors who believed that the use of tri- and quadrinomials (for subspecies and infrasubspecies taxa and hybrids) made nomenclature "non-Linnaean."

Rather strict precepts for the principle of typification are of importance. In the case of genus-group taxa, it is stated that "Type of a genus to be determined by the 'process of elimination,' if no type is originally mentioned" [Ca. XXIV]. All issues of allocation of old names entailed either by splitting or uniting genera or recognition of subgenera in them should be resolved taking into account the belonging of the type species. *AOU Code* formulated for the first time a requirement to fix type specimens for newly established species-group taxa. According to this provision, "The basis of a specific or subspecific name is [...] the original type specimen or specimens, absolutely identified as the type or types of the species or subspecies in question" [Ca. XLIII]. This extension of the principle of typification to the species-group taxa caused another serious controversy between the nomenclaturist communities of North America and Europe involving both zoologists and botanists, but it subsequently receives universal recognition.

Two important novelties appeared in *AOU Code* that were widely recognized first, but were rejected in the 20th century. One of them was the rule *"once a synonym always a synonym"* (junior synonyms were evidently meant). Another presumed that a family-group name should be rejected if its nominotypical generic name proved to be a junior synonym: it entails considerable confusion of names in this rank-group.

The *AOU Code* quickly became popular in North America not only among ornithologists but also among other vertebrate zoologists, as well as some entomologists. Moreover, as noted above, American botanists elaborated their *Rochester Rules* taking its certain provisions into account.

The next significant step in the history of zoological nomenclature occurred at the turn of the 1880s–1890s; this was an attempt to integrate it by developing a new Code with initially claimed international status. Its draft was prepared by the French zoologist Raphaël Anatole Émile Blanchard (1857–1919), so it is sometimes called *Blanchard Rules*. The latter were based on *Chaper Rules* elaborated a few years before (see Section 6.3.3), and in fact represented their substantially revised version. In fact, Blanchard's introductory review of current zoological nomenclature represented the consideration of *Chaper Rules* (with references to their articles), and the main nomenclature provisions were borrowed from them [Blanchard 1889]; M. Chaper published his arguments in favor of his *Rules* in the same issue [Chaper 1889].

The following comments of Blanchard on his *Rules* are noteworthy [Blanchard 1889]. These *Rules* are intended to apply to all plants and animals, both extant and extinct, so it is called *De la Nomenclature des êtres Organisés* (just like *Chaper Rules*).[16] The binomial nomenclature (incorrectly called binary by Blanchard) is supplemented by the trinomial (trinary) one, already quite popular at that time. In the section *"Loi de Priorité,"* several versions of the starting point for application of the principle of priority (from 1700 to 1766, the same as in *Chaper Rules*) are considered, emphasizing the contribution of Tournefort to development of binary (binomial) nomenclature. Blanchard believes that the starting dates for generic and specific

---

[16] For this, it can be considered as another attempt to develop a kind of *BioCode*.

names should be 1700 in botany and 1722 in zoology; these dates are indicated in his draft [Art. 75–76].

The draft *Blanchard Rules* were considered and a significantly revised version of them was provisionally approved at the First International Zoological Congress (Paris, 1889), with recommendations for their consideration at the next congress [Blanchard 1889]. Its final text differed from Blanchard's draft in great brevity due to many clarifications and detailing having been omitted: there were 100 articles in the draft and only 53 in the Congress version; the latter can be called the *zoological* **Paris Code** (to distinguish it from *Paris Laws* of 1867 and *Paris Code* of 1956 of botanists). When considering the principle of priority, it is stated that "The binary nomenclature was founded by Tournefort in 1700, Lang was the first to apply it to zoology in 1722" [§ 43], so "The year 1722 is therefore the date on which zoologists must go up to find the oldest generic or specific names" [§ 44]. Both texts of Blanchard's draft (as a *Rapport présenté au Congrès International*) and that approved by Congress (still designated as *De la Nomenclature des êtres Organisés*) were published in the *Bulletin de la Société zoologique de France* [Blanchard 1889]. The English translation of this Code, as *Rules of Nomenclature*, in a significantly abbreviated version was published shortly after in the *American Naturalist* [Fischer 1892].

Based on the results of the preceding discussions, R. Blanchard elaborated a new draft version (again as *Nomenclature des êtres Organisés* of 100 paragraphs and with the same references as *Chaper Code*), and presented it for consideration at the Second International Zoological Congress (Moscow, 1892). The results were first published in the materials of the Congress as a brief *Deuxième Rapport*, and then in the *Mémoires de la Société Zoologique de France* in a more extensive version [Blanchard 1892, 1893]. The latter consisted of the following parts: (a) additional clarification on the articles of the first version of *Paris Code*; (b) Blanchard's detailed comments on most of the articles in his second draft as compared to the first one; (c) a review of the articles considered at the Moscow Congress to be included in the revised version of *Paris Code*, with comments on them; (d) the full text of the revised version of *Paris Code* with 63 Articles, as it was approved at the Moscow Congress. One significant provision of this new version is the approval of 1758 (tenth edition of Linnaeus' *Systema naturae*) as the starting date for the application of the principle of priority [§ 45]. It was first set in *AOU Code* (see above); from then on, it is finally established in zoological nomenclature. The section "*Du nom de Famille*" states that a family name should be devalidated if the generic name, on which it was based, proves a junior synonym of another generic name [§ 43]. This article is also borrowed from *AOU Code*, and is later canceled. Due to these important novelties, this version of zoological nomenclature might deserve being called **Moscow Code**.

Notwithstanding its adoption, *Paris (Moscow) Code*, which claimed to be international, could not put a stop to the diversification trend in the development of zoological nomenclature in the late 19th and early 20th centuries.

German zoologists made a noticeable contribution to this trend. The Deutsche Zoologische Gesellschaft, on the initiative of Carl Gustav Carus (1789–1869), prepared and published its *Regeln für die zoologische Nomenclatur* known as

Deutsche Zoologische Gesellschaft Code in 1894. Its appearance was mainly due to an excessive—in the opinion of its authors—"liberality" of other zoological rulebooks in their treatments of the principle of priority and the rules for the replacement and correction of the scientific names of animals [Carus et al. 1894]. This Code contains several provisions that will influence the further development of zoological nomenclature. Following *AOU Code* and zoological *Paris Code* (in its final version), it unequivocally sets 1758 as the starting date for application of the principle of priority, without allowing any exceptions. Also, like *AOU Code*, name changes are only allowed in cases of their evident incorrect spelling or grammar; this refusal to consider the "appropriateness" of descriptive names is fully consistent with the nominalist interpretation of nomenclature. Replacement of a specific epithet due to its tautonymy with a generic name in the genus–species binomen is suppressed; this rule is soon universally adopted. Finally, this Code does not consider taxa higher than family-group rank (orders, classes, etc.), which corresponds to the established tradition in zoological (unlike botanical) nomenclature not to regulate the names of megataxa.

The **Nomenclature in Ichthyology** could become another "narrow subject" rulebook; a draft was prepared by a group of American ichthyologists led by David Starr Jordan (1851–1931) at the very beginning of the 20th century. It remained only as "a provisional version based on the Code of the American Ornithologists' Union," and yet, its published reviews showed that it had some remarkable clauses [Allen 1905; Fisher 1905]. One of the main objectives of the authors of this Code is to simplify nomenclature rules: they believe that "The value of a code depends not on the authority behind it, but solely on its simplicity, usefulness, and naturalness" [Fisher 1905: 28]. For this reason, in particular, it is recommended to solve an important issue of fixing types for the genus-group taxa guided by the "the first species rule" since "All processes of fixing types by elimination or by any other means resting on subsequent literature, lead only to confusion and to the frittering of time on irrelevant questions" [Ca. XI]. A forthrightly nominalist position is expressly asserted (following *AOU Code*) in that "a name is a word without necessary meaning [therefore] Questions of etymology are not pertinent in case of adoption or rejection of names" [Ca. XVII]. Accordingly, it is suggested that non-homonymous generic names differing minimally in their spelling and regardless of their etymology should be considered. Called "*one-letter rule*" [Allen 1905], it is then universally adopted in zoological nomenclature.

> When discussing the latter rule, the American zoologist Joel Asaph Allen (1838–1921) pointed out an important issue regarding homonymy of generic names. "The different branches of zoology have now become so extended and specialized that the […] rule of divorce might well be extended to the different branches of zoology" [Allen 1905: 432–433].

Several draft rulebooks appeared in the late 19th and early 20th centuries due to "breaking" nomenclatural activity of entomologists.

One of them was prepared by the English politician and amateur entomologist Thomas de Gray, Sixth Baron of Walsingham (1843–1919). It was published in 1896 and became known as **Merton Rules** after the Walsingham's family estate [Coues and Allen 1897]. The *Rules* were addressed primarily to entomologists, but eventually to all zoologists, and their main objective was set by their authors as follows: "to insure absolute obedience to the Law of Priority [and] to define a method by which the recognition of antecedent work can be consistently secured" [Walsingham and Durrant 1896: 1]. *Merton Rules* provide very detailed regulation of the formation, preservation, and change of generic and species names. In this regard, unlike "continental" rulebooks, the rules of typification are considered in detail. In general, the value judgment of the reviewers of these *Rules* fitted well into the conservative formula "what is good is not new; what is new is not good" [Coues and Allen 1897].

A group of American entomologists prepared and published their own detailed **Entomological Code** in the early years of the 20th century [Banks and Caudell 1912]. This contains 119 articles and is most remarkable for its democracy: its introductory section declares that "No person nor committee has authority to interpret, limit, or extend the precepts here laid down; every user interpreting the rules for himself" [Art. 1]. This provision appears as a protest against the ruling powers ascribed to the newly appointed International Commission on Zoological Nomenclature (see Section 6.4.1). Important innovations are as follows: a *quadrinomial nomenclature* was introduced for infra-subspecies forms [Art. 53], specific homonyms are divided into primary and secondary for the first time [Art. 58], typification methods for the genera and species are described scrupulously, and a detailed categorization of type specimens is suggested [Art. 70].[17] Some of these provisions will be incorporated in subsequent zoological nomenclature.

The last narrow subject-area rulebook in zoology appeared to be **Rules of Entomological Nomenclature** proposed by the British National Committee on Entomological Nomenclature and brought before the Third International Entomological Congress (Zurich, 1925). These brief *Rules* contain 28 articles; they are similar in many respects (e.g., in typification) to those in the American *Entomological Code*. The members of the Committee considered it "as a basis for future deliberations regarding a revision of the rules of Zoological Nomenclature" deserving to be forwarded to the International Commission on Zoological Nomenclature [Anonymous 1928: 1R]. But this goal seemed not to be reached.

6.3.3 Prototypes of *BioCode*

Several nomenclature projects appeared in the last third of the 19th century that claimed their general biological rather than particular subject-area status. By this, they tried (unsuccessfully) to revive an integrative "Linnaean" tradition by overcoming the "Great Schism," and thus could be considered the prototypes, to a degree, of the contemporary

---

[17] More than curious was a special prohibitive rule, according to which "A specific name of an insect based wholly on characters of internal anatomy, on habits, or on anything other than external characters, or invaginated parts of the exoskeleton, is invalid" [Art. 49].

BioCode. Two most significant nomenclature systems of this kind, American (of Dall) and European (of Chaper and Douvillé), are considered in this section.

The preparation of the American rulebook began when, towards the end of the 1870s, the Committee on Zoological Nomenclature at the American Association for the Advancement of Science organized a survey of leading systematic biologists in North America on the most essential principles and rules of nomenclature in botany and zoology; 27 questions were posed [Dall 1877]. The main objectives of the Committee were set as follows: (a) to clarify the points of agreement and disagreement between zoologists and botanists with regard to these principles and rules; and (b) to bring them together into a single Code to the extent that it might be feasible. Based on the materials from this survey (47 responses were received), the combined nomenclature provisions were published as *Nomenclature in Zoology and Botany*, later known as **American Association Code** (*AA Code*), or *Dall Code*. This was an extensive document of 84 paragraphs with nomenclature provisions and their justifications [Dall 1877]. Many of them repeated those of *Paris Laws* of botanists and partly of *BA Code* of zoologists; in fact, it was a kind of their compilation with a pretension to synthesis. In a brief review of this *Code*, the American zoologist Edward Cope called it *biological* [Cope 1878]: this was perhaps the first case of such designation of taxonomic nomenclature, which went on to become quite popular in the 20th century.

The introductory section of *AA Code* emphasizes that "The principles and forms of nomenclature should be as similar as possible in Botany and Zoology" [§ V]. A table with parallel botanical and zoological hierarchies is presented in a separate section, and it is emphasized that "No classification containing inversions […] can be admitted" [§ X]. The standard formula for the principle of monosemy is provided [§ XII]. It is emphasized that the names "should only be changed for the most important reasons" [§ XIII]. Of importance is the explicit recognition of rank dependence of the formation of names [§ XIV]. According to established tradition, the names of taxa of the highest categories "are taken from some one of the principal characters" [§ XV], while the names of botanical families/orders, subfamilies/suborders, tribes and subtribes, zoological families and subfamilies "are designated by the name of one of their principal genera" [§ XVI], with the addition of the appropriate standardized endings. It is specified that the "Names of higher rank than genera, with the fluctuation of their limits […], are not rigidly subject to the lex proritatis" [§ XIX]. Finally, for the names of genera and their subdivisions, it is stated in a nominalist manner that they "may be taken from any source whatever, or may be framed in an absolutely arbitrary manner" [§ XX].

Regarding the designations of species-group taxa, the principle of binomiality is emphasized [§ XXVI], and the preference for a descriptive nature of specific epithets is highlighted [§ XXVII]. It is indicated that the "Names of subspecies or varieties are formed like specific names […] preceded by the abbreviation subsp. or var." [§ XXXI], so that the full name of the intraspecific form would not look like a violation of the principle of binomiality.

The list of main criteria of a valid publication contains the following provisions. Appearance in replicated books and journals goes first, the distribution of exicates in botany is feasible, while names on museum labels or reported at meetings are not considered to have been published. It is repeated following *Paris Laws* that "A species is not to be considered as named unless both generic and specific names are simultaneously applied to it" [§ XXXIX]. However, "A generic name accompanied by a description, but without a reference to any definite species, may be considered as published when the organisms to which it is intended to refer, are unmistakable," and something like "reciprocal validation" of the generic and specific original descriptions is permissible [§ XLI]. A change of authorship is required only when transferring a taxon from one rank-group to another, but not when changing its rank within the same rank-group; nor is it changed because of "change of the diagnostic characters or the circumscription of a group" [§ XLVI]. When dividing a genus, its original name is retained for the subdivision that contains the type species (if fixed) or the most well-known species; or the one most "typical" for the original genus [§ LIV].

The section "Of a starting point for Binomial Nomenclature" is quite remarkable. It begins with a rather lengthy discussion of the disagreements on this important point, and arguments *pro* and *contra* different works are considered. It is suggested that the crucial point is to make clear in which of them binomial nomenclature using one-word species epithets was applied with sufficient certainty. This discussion is summed up by the following rather vague recommendation:

> The scientific study of different groups, having a value greater than or equal to that of a class (classis), having been begun at different epochs, and the inception of that study in each group respectively being usually due to some "epoch-making" work, the students of each of the respective groups as above limited may properly unite in adopting the date of such work as the starting point in nomenclature for the particular class to which it refers.
>
> § LVIII

Consideration of the issue concerning rejection or changes of names is quite typical for the Codes of the 19th century. On the one hand, it asserts the rather general rule that a "name cannot be changed under the pretext" of any substantive motives (not "agreeable," "poor Latin", etc.) [§ LXV]. On the other hand, there is a rather long list of cases when the name should be replaced: some of them are quite obvious (homonymy, violation of the rules of publication or priority), while others are redundant and bear the traces of essentialism or linguistic purism ("positively false", composed of multilingual parts, etc.).

As can be seen, *AA Code* was indeed quite an advanced compendium of the nomenclature rules. It just remains to be regretted that it did not receive common recognition, since botanists and zoologists have already diverged quite widely. But some of its provisions were adopted in the above-considered highly influential *AOU Code*.

A predecessor of *BioCode* that was not as profoundly developed as *AA Code* was prepared by the French mining engineer fascinated by paleontology and zoology, Maurice Armand Chaper (1834–1896), on behalf of the Société Zoologique de France. Its draft was considered at the Meeting of the Société Zoologique (Paris) in 1881 and published under the pretentious title *A la nomenclature des êtres organisès*

[Chaper 1881]. These **Chaper Rules** comprised a short list of provisions and extensive commentaries on them. Their author emphasized the importance of the unity of zoological and botanical nomenclature, the pragmatic nature of the proposed rules, the significance of the principle of priority, the importance of applying Latin not only to the names but also diagnoses, and the need to designate a type species for each new genus. Although Chaper intended his rules to apply to both "Linnaean" kingdoms of living nature, it was suggested that homonymy of generic names should only be considered within each kingdom. In *Chaper Rules*, the notions of binary and binomial nomenclature were treated identically, probably for the first time. This incorrect (see Section 4.2.2) interpretation will be repeated in some subsequent codes and preserved in contemporary botanical nomenclature.

When considering the principle of priority, Chaper indicated the following works as possible starting points for its application: *Institutiones rei herbariae* [Tournefort 1700] for the whole of botany, whereas several works were indicated for animals, viz., a manuscript on mollusks by Tournefort of 1708 (published posthumously in 1742); *Methodus nova et facilix testacea* [Lang 1722]; the first edition of *Systema Naturae* [Linnaeus 1735]; *Tentamen methodi ostracologicae* [Klein 1753]; and the tenth edition of *Systema Naturae* [Linnaeus 1758]. However, *Chaper Rules* did not decide on a particular starting date for zoology, which seemed to be its undoubted disadvantage.

The *Chaper Rules*, despite being addressed to both plants and animals, remained largely unnoticed by botanists, as their author was not especially known among them. However, they left a significant trace in zoology by framing a basis for the zoological *Paris Code* that is later transformed into the first *International Rules of Zoological Nomenclature* (see Section 6.3.2.2).

A compatriot and colleague of M. Chaper, Joseph Henri Ferdinand Douvillé (1846–1937), acting on behalf of the International Geological Congress (Paris, 1778), prepared another set of nomenclature rules simultaneously with *Chaper Rules* and with significant congruence with them. The draft was presented at the International Geological Congress (Bologna, 1881) and published under the title *Règles à suivre pour ètablir la nomenclature des espèces* along with comments on them [Douvillé 1882a,b]. These **Douvillé Rules** (also known as *Bologna Code*) contained 11 articles; among their significant advantages were an introduction on the predecessor to the principle of first reviser and the direct indication of a nominative genus as the type of corresponding family. Another fundamentally important novelty became the assignment of the exclusive right to consider the treatments of nomenclature at the Geological Congress; this condition is later codified by botanists and zoologists with reference to their respective international authorities. This Code refused to fix any particular start date for the principle of priority, but it considered mandatory that the name, even if published before Linnaeus, should conform to the principle (law) of binominal nomenclature.

6.4 THE 20TH CENTURY: TRADITIONS AND INNOVATIONS

Taxonomic nomenclature continued to develop rather dynamically throughout the 20th century, moving away from its origins to become far less "Linnaean." One of the active nomenclaturists at the turn of the 19th–20th centuries emphasized that taxonomic nomenclature:

> is not studied merely to preserve the Linnaean or the DeCandollean traditions. The taxonomic problems of to-day are very different from anything contemplated by Linnaeus, and if the system of nomenclature popularized by him could not be modified to serve practical purposes it would undoubtedly be discarded, as occasionally threatened.
>
> *Cook 1902b: 311*

The dominant route of the development of taxonomic nomenclature in the 20th century remained the nominalist one, and the main driving force regulating its development remained its rationalization. The latter manifested itself in an increasingly strong codification of both regulative nomenclature itself by further development of the rulebooks and the forms of the latter's regulation.

Along with this, the emergence of new taxonomic theories gave rise to several new theory-based nomenclature concepts [Pavlinov 2015a]. In particular, adherents of phenetic and phylogenetic theories rejected traditional nomenclature and suggested its new versions (considered in Chapter 7). According to original intentions, each was designed to consolidate taxonomic nomenclature on a new conceptual basis, but in fact they contributed to its further split since they did not receive unanimous recognition.

6.4.1 Major Trends

Further rationalization of nomenclatural activity became most noticeable in the 20th century, involving the development of new important organizational forms. First of all, officially recognized international Commissions/Committees for the ruling of all nomenclature issues were established in each subject area of systematic biology; these were *International Commission on Zoological Nomenclature* (from 1895) in zoology, *International Interim Committee* (in 1926), subsequently changed to *Executive Committee of Nomenclature* (in 1935) and now *General Committee on Botanical Nomenclature* (from 1950) in botany, *International Committee on Systematic Bacteriology* (from 1930, *International Committee on Systematics of Prokaryotes* from 2000), *International Committee on Nomenclature of Viruses* (from 1966), and *International Commission on the Taxonomy of Fungi* (from 1982) (see subsequent sections for more detail). They were assigned an exclusive right to change and/or interpret the fundamental provisions of the relevant nomenclature systems and to submit the latter's new editions for consideration at the respective international congresses. The official status of these authorities and the documents they consider was increased with the establishment of the *International Union of Biological Sciences* (*IUBS*) with respective Divisions for the main branches of systematic biology. The principal issues relevant to the development of the subject-area

Codes, including their changes, approvals, etc., began to be considered by these Divisions at *IUBS* General Assemblies, and this provision is indicated in the complete official titles of contemporary Codes. The appearance of specialized international journals as forums for the discussion of various nomenclature issues began to play a very important role in shaping the trend towards integration: these are *Bulletin of Zoological Nomenclature* (published from 1943) in zoology, *Taxon* (from 1951) and *Regnum Vegetabile* (from 1953) in botany, *International Bulletin of Bacteriological Nomenclature and Taxonomy* (from 1950, *International Journal of Systematic Bacteriology* from 2000) in microbiology, *Mycotaxon* (from 1974) and *IMA Fungus* (from 2010) in mycology, and *Bionomina* (from 2010) for the whole of taxonomic (biological) nomenclature.

An aspiration to overcome the "breaking" trend in the development of nomenclature was evidently reinforced at the turn of the 19th–20th centuries and came to dominate in a few decades [Pavlinov 2015a, 2015b]. However, this integration trend could not overcome the basic "subject-area schism" and thus proceeded within each of these areas. With this, botanical and zoological systems of nomenclature were developed in parallel and even synchronously throughout the 20th century aimed at resolving the same problems. Thus, the first international rules/codes of nomenclature in both botany and zoology were adopted practically simultaneously at the corresponding international congresses in 1905. Each overcame the regional "schisms" between the European and American nomenclature systems in the 1930s. They both introduced important novelties in a similar manner: the "systematic group" or "category" received the unified official designation "taxon"; species-group taxa were ruled to be typified by the specimens; the basic structure of the Codes was similarly changed (for example, vocabularies/glossaries were added); the criteria for valid publication of nomenclatural acts were specified to incorporate the development of new information technologies.

Along with this, the diversification trend appeared to be no less clearly manifested. The development of the main subject-area Codes (or *special* Codes according to *Draft BioCode*) in zoology and botany, aimed at a stronger formulation and detailing of their principles and rules, led to their growing discrepancy [Ride 1988; Ochsmann 2003; Kraus 2008; Pavlinov 2015a, 2015b]. In addition to these "old" subject-area Codes and nomenclature of cultivated plants that had split in the mid 19th century, the separate nomenclature Codes for prokaryotes and viruses appeared in the second half of the 20th century [Lapage et al. 1992; Mayo and Horzinek 1998].[18] The specific issues of the nomenclature of fungi and lichens are actively discussed in recent literature with respect to unification of the naming systems for different ontogenetic phases [Ciferri and Tomaselli 1955; Culberson 1961; Hawksworth 1974, 2001; Weresub and Pirozynski 1979; Sigler and Hawksworth 1987; Jørgensen 1991; Hennebert and Gams 2003; Hawksworth 2011; Lendemer 2011; Taylor 2011; Hong et al. 2012; Schoch et al. 2014; Hawksworth et al. 2017], and the possibility and even desirability of the development of a separate *MycoCode* for them was considered by the

---

[18] For the main contemporary subject-area Codes, it is proposed to designate the onymology sections developing them as *botanonymology*, *zoonymology*, and *bacterionymology* [Dubois 2000; Aescht 2018].

Amsterdam Declaration [Hawksworth et al. 2011]. In addition, two new versions of traditional nomenclature were proposed in botany and zoology, viz., *Reformed Code of Botanical Nomenclature* [Parkinson 1990] and *Linz ZooCode Proposal* [Dubois et al. 2019]. This diversification trend manifested itself most prominently in the development of rational-logical and phylogenetic nomenclature concepts (considered in Chapter 7).

In connection with the separation of botanical and zoological Codes, which differ in how they technically resolve several nomenclatural tasks and with no real hope of integration, the problem with the nomenclature of so-called "ambiregnal" protists emerged [Corliss 1962, 1995; Patterson and Larsen 1992; Anonymous 2001; Frolov and Kostygov 2013]. Many are considered either plants or animals in various traditional classifications, so they formally fall under either botanical or zoological nomenclature. In this regard, it was suggested that their macrotaxa, differing in the fundamental properties of biological organization, "could be addressed by an agreed nomenclatural allocation of taxonomic groups to particular Codes" [Anonymous 2001: 7]. On the other hand, it is proposed that a certain unity for them should be preserved (including a common name, *Protozoa*), including application of the uniform nomenclature norms to them, which seems to be important for many practical purposes [Frolov and Kostygov 2013].

Both integration and diversification trends in the most recent development of taxonomic nomenclature culminate paradoxically enough in the appearance of two conceptually quite different nomenclature rulebooks. One of them is *Draft BioCode*, which claims to overcome finally the "Great Schism" of the subject-area Codes [Greuter et al. 1996, 1998; Anonymous 2001; Greuter 2004] in a traditional manner (see Section 6.5). Another is *PhyloCode*, that was also meant to become a unified nomenclature system for all organisms, but on a very specific non-traditional conceptual basis [de Queiroz and Gauthier 1994; de Queiroz and Cantino 2001; Cantino and de Queiroz 2010, 2020; de Queiroz 2012 (see Section 7.2)]. The *BioCode* still remains a draft and seems to have no evident impact on contemporary nomenclature, while recent publication of *PhyloCode* [Cantino and de Queiroz 2020] promises to cause another "Great Schism."

Within the framework of traditional nomenclature, typification of the species-group taxa became one of the most hotly discussed issues in the first third of the 20th century. This rule was suggested at the end of the 19th century by American nomenclaturists, contributing to another "Great Schism" in taxonomic nomenclature by separating the American "type-based" and European "diagnosis-based" systems. This contradiction was overcome in the 1930s in botany and zoology, but several other problematic issues were highlighted; one concerned the number of type specimens to be fixed. Initially, typification implied indication (fixation) of a single type specimen, but it became complicated in the first half of the 20th century (see Section 3.5). When describing species-group taxa, several type specimens suggested fixing as the elements of the type series in order to reflect within-species/subspecies variability. In this regard, different categories of types began to be recognized, and their number reached several dozens by the mid 20th century. Current Codes recognize only a few of these categories not associated with the particular manifestations of variability; exceptions are hapantotypes in zoology and epitypes in botany. However, a question of the admissibility of fixing non-traditional kinds of type specimens (such as phototype, genetype) is now hotly debated from other perspectives (see Section 3.5).

The nominalist interpretation of the type specimens as just "examples" of the species and subspecies led at first to a disrespectful attitude towards them. Due to this, if the original specimens kept in collections ceased to meet certain nomenclature requirements for some reasons, systematists of the 19th century replaced them quite often with newer, more "appropriate" ones. With the recognition of the principle of typification for species-group taxa, the requirements for preserving the original specimens were tightened [Marsh 1898; Schuchert and Buckman 1905; Swingle 1913; Mayr 1942, 1969]. The need for long-term preservation of type specimens in (preferably officially recognized) repositories, as well as indication of the latter in the published descriptions of new taxa, is mentioned in all contemporary Codes. They also impose on the collection/herbaria curators the obligation to make the type specimens easy to access and to publish their catalogues.

Active paleontological research in the second half of the 20th century brought specificity to the development of nomenclature applied to extinct organisms. This is caused by an impossibility in many cases to allocate isolated fossil items, such as plant pollen and leaves or animal traces and bones, to the same species. To resolve this particular problem, *parataxonomy* and *ichnotaxonomy* are separated from *orthotaxonomy*: the latter classifies the whole organisms, parataxonomy deals with their fossil disarticulated remains, and ichnotaxonomy considers fossil traces of organisms. Accordingly, *paranomenclature* and *ichnonomenclature* systems are introduced separately from *orthonomeclature* in order to regulate naming of *parataxa* (or *morphotaxa*) and *ichnotaxa*, respectively [Vialov 1972; Sarjeant and Kennedy 1973; Simpson 1975; Hahn 1981; Melville 1981; Bengtson 1985; Rasnitsyn 1986, 2002; Mikhailov 1991; Mikhailov et al. 1996; Eriksson et al. 2000; Palii 2013]. By establishing these two nomenclature systems, it is presumed that para- (= morpho-) and ichnotaxa are classified and named independently of orthotaxa, regardless of respective isolated fossils possibly belonging to the same species [Seilacher 1964; Simpson 1975; Rindsberg 2018].[19] The *Palaeontological Data-Handling Code* was proposed as a consistent set of rules for regulating and handling information on fossils, including their names [Hughes 1989].

Currently, the rules for naming such "non-orthotaxa" are officially considered in traditional botanical and zoological Codes, which put restrictions on applying certain standard rules for them. Suppression of synonymy of the names of para-, morpho-, ichno-, and orthotaxa is not mandatory due to the above-mentioned problem of reliable

---

[19] Although ichnotaxa and their nomenclature are discussed basically with respect to fossil organisms, they seem to hold true for objects of cryptozoology, such as traces of the legendary Bigfoot, which was recently described as a new taxon [Kim et al. 2008; Ketchum et al. 2012].

allocation of isolated anatomical remains and traces of fossil organisms to orthotaxa. To avoid homonymy of their names, certain pre- and suffixes are used to designate specifically morphotaxa (such as oolithotaxa of fossil egg shells) and ichnotaxa. This makes the nomenclature systems for fossils partially independent both from each other and orthonomenclature, which may be considered as another manifestation of the contemporary diversification of specific subject-area nomenclature systems.

Since the mid-20th century, descriptive systematics began to face new challenges influencing the development of taxonomic nomenclature and giving rise to the opinion that its traditional ("Linnaean") format is largely obsolete (e.g., [Ehrlich 1961; Michener 1963; Sokal and Sneath 1963; Oldroyd 1966; Cellinese et al. 2021], etc.). These most recent challenges are threefold.

First of all, it is necessary to highlight the global impact of digitalization of the whole of research activities in contemporary systematics, which affects the semiotic structure of its descriptive language. This led to a revival of interest in rational-logical nomenclature, which entailed the appearance of its most advanced form, numericlature. This is supplemented by the development of electronic databases and Internet resources: they require the elaboration and recognition of new standards for both designations of taxa and publication of their descriptions [Kennedy et al. 2005; Patterson et al. 2006, 2010; Fransson 2008; International Commission 2008, 2014; Welter-Schultes et al. 2009; Tuominen et al. 2011; Dubois et al. 2013; Minelli 2013; Remsen 2016; Winston 2018; Vlachos 2019; Williams 2021, etc.]. These new standards include the emergence of certain digital networks such as *Global Names Architecture* (GNA) (http://globalnames.org), *Life Science Identifiers* (LSID) (www.lsid.info; [Fransson 2008]), *Global Biodiversity Information Facility* (GBIF) (www.cbd.int/cooperation/csp/gbif.shtml), etc.

The particular character of describing taxonomic diversity based on analysis of molecular genetic data also put its noticeable mark on the ways of delineating and naming taxa. This entails certain specifics of N-definitions of taxa by referring to diagnostic features that do not agree completely with traditional ones. With this, new "technical" forms of taxonomic N-designators emerge that differ significantly from traditional ones by using some elements of numericlature (e.g., [Ratnasingham and Hebert 2007; Morard et al. 2016; Minelli 2017, 2019]. Widening of the usage of "non-morphological" kinds of data in the descriptions of new species-group taxa may yield the need for significant modification of the traditional principle of typification (see Section 3.5).

Another challenge that faces this principle relates to the practical needs of nature conservation. In this regard, the need to soften the current requirements of traditional typification is discussed in order to minimize the impact of collecting activity on populations of rare species and subspecies (e.g., [Loftin 1992; Norton et al. 1994; Collar 2000; Donegan 2008]; see also Section 3.5). Ensuring the stability of their scientific names by softening the requirements of the principle of priority is also considered [Heywood 1991; Dubois 2010b].

Thus, both nomenclature systems and the nomenclature concepts underlying them continue to evolve. The same two main groups of causes, just as one or two centuries ago, are responsible for their inevitable contemporary dynamics.

One of them are "external"; they are associated with the development of theoretical foundations and methodology of biological systematics. This development leads to the complication of ideas about the structure of "taxonomic reality" and its various manifestations, including the elaboration of new theoretical concepts (such as phenetic and cladistic), which justify their own versions of the language of taxonomic descriptions. On the other hand, methods for identifying and diagnosing taxa are being improved, and now involve characters that do not fully comply with some provisions of the traditional Codes. All this leads to awareness of the need to "adjust" taxonomic nomenclature, as part of the professional language of descriptive systematics, to the new challenges.

Other reasons are "internal"; they are consequences of the imperfection of this language, which is inherent in any linguistic system. The latter is intrinsically fuzzy, so the definitions and formulations in it admit different interpretations and exclusions, which inevitably generate uncertainties in possible resolutions of particular nomenclatural tasks. This fundamental feature of the descriptive language of systematics contradicts the key provision of the basic principles of rationality and designative certainty (see Sections 3.1 and 3.2). Accordingly, the current developments of the Codes are largely driven by the same needs for eliminating these uncertainties, but the removal of some of them by tightening the rules inevitably reveals or gives rise to others.

6.4.2 BOTANY

In the history of botanical nomenclature at the turn of the 19th–20th centuries, the first and most important event was the elaboration of a new rulebook intended for international recognition and application. Work on it began at the Botanical Congress in Geneva (1892), which confirmed the credentials of the *Paris Laws* of 1867 and expressed the desirability of their further enhancement. The need for a new Code that would incorporate the latest changes in botanical nomenclature was recognized at the First International Botanical Congress (Paris, 1900). For this, the *Commission internationale* was appointed, with J. Briquet as its head; he prepared a draft of the new rulebook based on *Paris Laws* and presented it at the Second International Botanical Congress (Vienna, 1905). American botanists also submitted their draft *Philadelphia Code* for consideration, but its main provisions that differed from those of *Paris Laws* were refused. As a result, the appended draft of the Commission was approved and became the trilingual **Règles internationales de la nomenclature botanique** [Briquet 1905, 1906, 1912; Maiden 1906, 1907; Cheeseman 1908; Oshanin 1911]; it was designated the contracted name of *Vienna Code*, but later becomes known as *Vienna Rules* to distinguish it from *Vienna Code*, adopted a century later [Flann et al. 2015] (see below). The adoption of *Vienna Rules* reduced the diversity of the regional botanical Codes, officially recognized at the beginning of the 20th century, from four or five to two—International (rather, European) and American [Weatherby 1949; Smith 1957; Nicolson 1991; Pavlinov 2014, 2015a]. The main disagreement between the last two Codes was the denial or recognition of typification of species-group taxa.

These *Vienna Rules* were in fact a revised version of *Paris Laws* in its Geneva version, complemented by some important proposals from the *Berlin Rules*, including adoption of 1753 as the starting date for the application of the principle of priority for both generic and specific names in botany. An important proposal for the regulation of amendments of the international *Rules* is borrowed from *Kuntze Code*, while *Kew Rule* was outlawed. The main drawback of *Vienna Rules* (from a contemporary perspective) was the absence of the principle of typification: the term "type" was mentioned only regarding the split of the genus-group taxa. French was recognized as the basic language of *Vienna Rules*, reflecting their "French" roots (which was also true for the zoological *Règles*; see below), later this function is taken over by English.

It was specificied that *Vienna Rules* applied only to recent vascular plants. There was a particular reason for this: specialists on lower plants (algologists, mycologists) considered it absurd to refer to the works of Linnaeus, who had little knowledge of these plants, and suggested special starting dates and works for them [Shear 1902; Earle 1904; Durand 1909]. This important issue was considered at the next International Botanical Congress (Brussels, 1910), where specific starting points for algae and certain groups of fungi were recognized, viz., *Species Muscorum* [Hedwig 1801], *Synopsis methodica Fungorum* [Persoon 1801], *Systema mycologicum* [Fries 1821–1829 (1832)]; a reservation was also made for several groups of algae [Shear 1910; Oshanin 1911; Briquet 1912; Korf 1983]. Subsequent discussion of fungal nomenclature showed a serious collision between Hedwig's and Persoon's names [Singer 1960; Demoulin et al. 1981]. Such a solution obviously contradicted the requirement of the principle of universality, but it was justified from a practical standpoint.

The next significant event in the unification of botanical nomenclature was the establishment of an *International Interim Committee* at the Ithaca Congress (1926) to harmonize *Vienna Rules* and *American Code* [Nickolson 1991]. Their integration happened at the Cambridge Congress (1930) as the importance of typification of species-group taxa was recognized and unified criteria for effectiveness of publications of their names were adopted [Anonymous 1934]. This was approved by *Cambridge Code* (published in 1935), which also recognized typification of the names of the order-group taxa by family names. This was followed by *Amsterdam Code* (approved in 1935, officially published in 1950), which established a permanent *Executive Committee of Nomenclature*, and very soon by *Stockholm Code* (approved in 1950, published in 1952). The International Botanical Congress at Stockholm (1950) was significant for its "institutional reform." First, the *International Association for Plant Taxonomy* (IAPT) was appointed with the *General Committee on Botanical Nomenclature* designated to regulate nomenclatural activities in systematic botany. Second, the journal *Taxon* was established. In addition, the botanical nomenclature rulebooks that were officially (titularly) designated as *Rules* before 1952 from that year on became *Codes*.[20]

---

[20] In botany, each Rule/Code was traditionally named after the city in which was held the International Botanical Congress that approved it [Parkinson 1975; Malécot 2008; Pavlinov 2015a,b].

The next *Paris Code* (approved in 1954, published in 1956) set out an introductory Preamble with the guiding principles of independence of Codes, typification, priority, publication, monosemy, Romanization, and retroactivity. It was followed by the *Montreal Code* (approved and published in 1959) and the *Edinburgh Code* (1964), which added specific provisions concerning fossil plants and limited the principle of preservation of names to genus and family rank-groups only. Several subsequent updates were mostly technical, and nevertheless they traditionally received specific names: *Seattle Code* (1969), *Leningrad Code* (1975), *Tokyo Code* (1993), and *St. Louis Code* (1999). The latter considered the status of separate *Permanent Nomenclature Committees* (for Algae, Bryophyta, Fossil Plants, Fungi, and Vascular Plants) established under the auspices of IAPT. The *Vienna Code* (2006) was significant in that it: (a) officially recognized 1793 (*Genera plantarum* by Jussieu) as the starting date for the priority of family-group names; (b) recognized the valid status of publications only if they received ISBN/ISSN; and (c) provided an official Glossary for the first time.

The *Melbourne Code*, adopted in 2011 [McNeill and Turland, 2011; Anonymous 2012], changed the official name of the botanical nomenclature to reflect most recent ideas about the level of divergence of the major groups of the traditional "plant kingdom": from then on it became **International Code of Nomenclature for algae, fungi, and plants**. It introduced special standards for the names in some groups of fungi, allowed Internet publications (subject to a number of provisions), allowed diagnoses in English to replace Latin ones, and changed some standard nomenclature terms. This *Code* was followed by a number of publications concerning fungal nomenclature (see above).

The above-mentioned *Reformed Code of Botanical Nomenclature* [Parkinson 1990] seemed to have no significant impact on the most recent development of nomenclature in botany. It was reported that the changes it suggested, "as compared to the official botanical Code, are not substantial but are rather semantic and structural, reflecting its author's idiosyncratic convictions" [Greuter 2004: 23].

As noted above, a possibility of the adoption of a separate *MycoCode* for the nomenclature of fungi is considered [Hawksworth et al. 2011]. This seems to be especially relevant for microscopic fungi, for which systematics and nomenclature face the same problems as in prokaryotes and protists [Taylor 2011]. It also concerns the nomenclature of lichens.

The current version of botanical nomenclature is *Shenzhen Code*, adopted by the International Botanical Congress held in Shenzhen (China, 2017). As with previous Codes, it took effect as soon as it was ratified by Congress; its final documentation was published in 2018 (www.iapt-taxon.org/historic/2018.htm). For mycologists, the Code is important for the ruling to transfer decision making on all matters related to the naming of fungi from International Botanical to International Mycological Congresses [Hawksworth et al. 2017; Parra et al. 2018]. Following this, *Fungal Nomenclature Session* of the 11th International Mycological Congress (San Juan, Puerto Rico, 2018) considered some proposals concerning fungal nomenclature and its regulation [May et al. 2018; Aime et al. 2021].

6.4.3 ZOOLOGY

The most important event in the history of zoological nomenclature on the eve of the 20th century became the appointment of the *International Commission on Zoological Nomenclature* by the Third International Zoological Congress (Leiden, 1895) [Melville 1995]. French, English, and German became the official languages of the documents developed by the Commission; from 1953, this status was retained for the first two with the priority of English. To strengthen the authority and powers of the Commission, it was determined at the Fourth International Zoological Congress (Cambridge, 1898) that: (a) it operates on a permanent basis; (b) its membership is expanded to make it really international; and (c) any initiatives to amend provisions of the officially recognized *International Rules of Zoological Nomenclature* are submitted to the International Zoological Congresses solely by the Commission. A statement of the latter's status and powers was included in these Rules.

The establishment of two official institutions was also of great importance for the future development of nomenclatural activity in zoology. Firstly, it was the *Bulletin of Zoological Nomenclature* issued from 1943 as the official organ of the Commission for the publication of: (a) reports on the activities of the Commission; (b) the main results of this activity in the form of certain resolutions of the Commission; and (c) public discussions of issues concerning zoological nomenclature. Secondly, it was the *International Trust for Zoological Nomenclature*, established in 1947 to provide organizational and financial support for the ongoing activity of the Commission and publication of the Bulletin [Melville 1995].

Besides preparation of the new versions of Code, the key points in the Commission's activities are as follows. It is empowered to make certain particular decisions that do not affect the basics of nomenclature principles and rules. They are published in the form of: (a) "Declarations," including some particular corrections of the acting Code; and (b) "Opinions" and "Directives" concerning particular nomenclature inquires; the first of them were published in 1907 in the materials of the Boston Congress [Melville 1995]. The Commission publishes the "*Official Lists and Indexes of Names and Works in Zoology*" that are conserved or rejected for nomenclature purposes by its plenary powers [Melville and Smith 1987; Smith 2001] (www.iczn.org/other-publications/the-5th-code/). The updated versions of these *Lists* are now regularly published online (e.g., http://iczn.org/sites/iczn.org/files/officialists.pdf).

The first major outcome of the activities of the International Commission became the preparation of the draft of a new nomenclature rulebook in systematic zoology based on the zoological *Paris Code* of the 1880s (see Section 6.3.2.2). It was considered and preliminarily approved at the International Zoological Congress in Leiden (1895). At the next congress in Berlin (1901), the Commission was entrusted with preparation of the new rulebook. Its text was finally agreed and approved at the working meetings of the Commission during the Berne Congress (1904) and published in 1905 in three languages (with French recognized for arbitration) as the official **Règles international de la nomenclature zoologique** [R. L. 1905; Oshanin 1911].

These *Règles* were rather laconic: their main part contained 36 articles grouped into nine sections. In the introductory section, trinomial nomenclature for the subspecies was approved. Articles dealing with the names of taxa of different rank-groups were considered as follows: first for families/subfamilies, then for genera/subgenera, and lastly for species/subspecies. The "botanical" rule was preserved, according to which, when a species was transferred to another genus or when it was divided, authorship was indicated not only for the specific epithet, but also for the change made. The principle ("law") of priority was adopted in a rather rigorous form: no exceptions stipulated that might agree with the principle of usage. The debate on the principle of typification was reflected in the detailed regulation of the choice of the type species for the genus-group taxa. The main part of the *Règles* did not consider the type specimens of the species-group taxa, but it was recommended in the Appendix that the types should be fixed in the descriptions of new taxa and stored in repositories. Two previously adopted rules were excluded: suppression of tautonymy in the genus–species binomen and "once a synonym always a synonym."

In the first half of the 20th century, through the efforts of the International Commission, the particular amendments to the *Règles international* were summarized and issued several times, each authorized by ongoing International Congresses. Thus, restrictions on the application of the principle of priority were suggested in Monaco (1913) and approved in Copenhagen (1953). Mandatory indication of diagnostic characters in the descriptions of new genera and species or reference to their previously published indication was adopted at the Budapest Congress (1927). Fixation of types for species-group taxa was recommended at the Padua Congress (1930), which eliminated one of the important disagreements between the International (European) and American (of ornithologists and entomologists) Codes and created a prerequisite for their integration. All these amendments made the adoption of a new rulebook quite relevant: the need for this was recognized at the Paris Congress (1948) and confirmed by *Copenhagen Decisions on Zoological Nomenclature* (1953) [Schenk and McMasters 1956; Hemming 1957, 1958]. The draft of a new set of rules was published in the *Bulletin of Zoological Nomenclature* in 1958 [Bradley 1958]; it was considered and generally approved at the London Congress (1958), and its final version was officially published as **International Code of Zoological Nomenclature** (ICZN) in 1961 [Anonymous 1961]. It is notable for Article 13, in which the names of taxa of phylum-group, class-group, and order-group are considered on a general basis; it is subsequently excluded.

The second revised edition of ICZN was adopted at the Washington Congress in 1963 and published in 1964 [Anonymous 1964]. Among its most important innovations were the following: (a) introduction of the principle of preservation (conservation) of names to restrict the application of the principle of priority, taking into account their usage within the 50-year time interval; (b) official refusal to regulate the names of orders and other higher-rank taxa; (c) suppression of infra-subspecies taxa; and (d) suppression of naming hybrid specimens as such.

Preparation of the third, substantially revised, edition of ICZN began in 1973, its draft was published for public discussion in 1977 and adopted in principle by the Division of Zoology of IUBS at its XX General Assembly (Helsinki, 1979); it was published in 1985 [Anonymous 1985]. The most noticeable differences from the previous edition were: (a) expanding the list of conditions for the valid publication of names; (b) changing the conditions for the availability of intercalary names of intrageneric and intraspecific taxa; (c) clarification of the rules for names of the family-group taxa; and (d) clarification of the rules for typification of the species-group taxa.

Finally, the fourth (current) edition of ICZN was adopted by the International Commission on Zoological Nomenclature, ratified by the Executive Committee of IUBS, and published in 1999 [Anonymous 1999]. It introduced provisions concerning ichnotaxa and partly parataxa, specified criteria for the publication and availability of newly established names, criteria for rejection of "forgotten" names, and some rules of typification including a significant tightening of the notion of nomenclatural type. Electronic publication was made available with amendment to the Code in 2011 [Editorial 2012; Zhang 2012]. A package of proposals for the fifth edition of ICZN is now being prepared [Editorial 2014; Francisco Welter-Schultes personal communication, February 2021].

The scientific names of zoological taxa of higher categories (orders, classes, etc.) were and still are not officially regulated since the time of *BA Code*. However, certain unified rules (standardization of endings, priority, typification, etc.) began to be introduced by some authors long ago (e.g., [Gadow 1893; Berg 1932]), and they were being quite actively discussed up to now [Stenzel 1950; Rodendorf 1977; Rasnitsyn 1982, 2002; Starobogatov 1991; Dubois 2000, 2005, 2006b, 2011a, 2015; Alonso-Zarazaga 2005; Dubois et al. 2019]. It is proposed that the so-called *duplostensional nomenclature* system will provide an unambiguous methodology and explicit criteria for the nomenclatural consistency of taxonyms of zoological taxa above the rank of superfamily [Dubois 2015].

The international *Linz ZooCode Committee* was created on the initiative of the French zoologist Alain Dubois in 2014 to develop a package of proposals aimed at sufficient renovation of zoological nomenclature in general [Dubois et al. 2016, 2019]. General concepts and particular provisions and suggestions, including very extensive new terminology, were discussed at several sessions of this Committee; they are summarized as *Linz ZooCode Proposal* (LZP) [Dubois et al. 2019]. Its main differences from ICZN are as follows: (a) complete coverage of the whole of the taxonomic hierarchy (as in *BioCode*); (b) a substantially updated list of nomenclature principles and rules; (c) a very detailed regulation of both the formulation and resolution of nomenclatural tasks and the nomenclatural objects involved in them; and (d) an extensively renewed terminology. A significant novelty of the latter seems to be a serious disadvantage of LZP: it will probably hinder its recognition by the taxonomic community (indeed, any system is conservative enough to be "afraid" of excessive novelty).

No specific nomenclature was developed for domestic animals, unlike for cultivated plants (see Section 6.4.5). Their names are officially regulated by the ICZN, but experts interested in their nomenclature consider particular options and suggest specific identifiers for them [Uerpmann 1993; Odening 1979; Wyrwoll 2003; Parés-Casanova 2015].

6.4.4 MICROBIOLOGY

The taxonomic names of bacteria (and prokaryotes in general) were initially regulated by the botanical Code. The first step towards developing a specific nomenclature for microorganisms was taken at the First International Microbiological Congress (Paris, 1930): the International Microbiological Society was created with a permanent *International Committee on Systematic Bacteriology*. A respective nomenclature system was first presumed to be a modification of the botanical Code, but it soon became clear that a separate Code was needed to take into account the specifics of identifying and describing bacterial taxa. Its draft was reviewed and approved at the third (New York, 1939) and then with some changes at the fourth (Copenhagen, 1947) International Microbiological Congresses; after approval it was published as the **International Code of Nomenclature of Bacteria** (ICNB) (= *Bacteriological Code*) [Buchanan et al. 1948]. It was quite similar in structure and content to the botanical one; it proclaimed interdependence with the latter with respect to homonymy, recognized the fundamental importance of the principles of monosemy, typification, priority (with certain reservations), and binomiality. At the next congress (Rio de Janeiro, 1950), microbiologists decided to prepare a new version of their nomenclature, now specified as the **International Code of Nomenclature of Bacteria and Viruses**. It was published in 1958 after approval at the Stockholm Congress [Buchanan 1959].

The nomenclature systems of bacteria and viruses split in the 1960s, and subsequent editions of *Bacteriological Code* in its original interpretation (i.e., without viruses) were published in 1975, 1980, and 1990 with some modifications [Lapage et al. 1992]. Because of the fundamental reconsideration of the megataxonomic system of prokaryotes, *International Committee on Systematic Bacteriology* (ICSB) was renamed as the *International Committee on Systematics of Prokaryotes* (ICSP) in 2000, and its nomenclature rulebook was renamed as **International Code of Nomenclature of Prokaryotes** (ICNP) [Tindall et al. 2006; Oren 2008; Parker et al. 2019]. The *Judical Commission* at the ICSP is responsible for considering, between Congresses, any issues concerning the interpretations of ICNP and recommendations for its changes [De Vos and Truper 2000].

It is stated in the chapter on "Principles" of ICNP that "The nomenclature of prokaryotes is not independent of botanical and zoological nomenclature," which means recognition of the homonymy of bacterial and other taxonomic names. Therefore, "When naming new taxa in the rank of genus or higher, due consideration is to be given to avoiding names which are regulated by the International Code of Zoological Nomenclature and the International Code of Nomenclature for algae, fungi and plants." The year 1980 is adopted as the initial date for application of the principle of priority: this was because of development of substantially new methods of classifying bacterial taxa. Typification of taxa of all major ranks from species to class is required; typification rules are identical to those in botany [Rule 15]. In connection with recognition of a certain interdependence of the nomenclature of prokaryotes and plants (with algae and fungi; see above), some nomenclature rules for cyanobacteria (classified earlier as archaic algae) are discussed [Stanier et al. 1978; Friedmann and Borowitzka 1982; Komárek and Golubić 2004; Oren 2004, 2005; Pinevich 2015].

As a result of the split of the bacteriological and virological nomenclature systems, the *Provisional Committee on Nomenclature of Viruses* was appointed [Lwoff 1964]. It received the status of *International Committee on Nomenclature of Viruses* in 1966 [Gibbs et al. 1966] and transformed into the *International Committee on Taxonomy of Viruses* in 1971 [Matthews 1983; Adams et al. 2017]. Its activity resulted in a set of principles and rules that were called originally *Rules of Nomenclature of Viruses* and later **International Code of Virus Classification and Nomenclature** (ICVCN). The latter underwent several revised editions (1980, 1998, 2002, 2012) prepared and published by the International Committee [Anonymous 1982; Francki et al. 1991; Murphy et al. 1995; Mayo and Horzinek 1998; King et al. 2011]. Consideration of its successive versions involved active discussion on rules for the formation of names of viral species (e.g., [Flores et al. 1998; Bos 2000, 2002; Gibbs 2000, 2003; Van Regenmortel 2001]); this was largely caused by the ambiguity of their substantive interpretation [Van Regenmortel 1997; Gibbs 2003; Andino and Domingo 2015; Siddell et al. 2020].

The ICVCN edition adopted in 2012 is currently in effect [Anonymous 2013]. It is notable for the fact that it officially regulates the rules of both systematization and naming of viruses, ensuring a close relationship between these two "Linnaean foundations" of systematics. The particular features of this Code are as follows: the principle of usage is declared as the basis for stability of viral nomenclature; species names are not necessarily binomial; the names of all superspecific taxa have standard rank- and group-specific endings. However, the Executive Committee of ICNV recently recommended adopting the standard binomial system to name viral species [Siddell et al. 2020].

6.4.5 CULTIVATED PLANTS

Updated variants of the nomenclature of cultivated plants appeared during the second half of the 19th century [Plumb et al. 1884; Miller 1899; Cogniaux 1911; Stearn 1986; Hetterscheid et al. 1996]. This nomenclature was partly regulated by the botanical Code until the mid 20th century [Stearn 1952; McNeill 2004; Malécot 2008]. In 1952, at the Congress of the International Society for Horticultural Science (London), the *International Commission for the Nomenclature of Cultivated Plants* was appointed and the independent **International Code of Nomenclature for Cultivated Plants** (*Cultivated Plant Code, Cultivated Code*) was officially adopted based on the draft prepared by the English botanist William Thomas Stearn (1911–2001) [Stearn 1952, 1953]. This Code has been updated several times since then under the auspice of the Congresses of the International Society (1958, 1961, 1969, 1980, 1995, 2004, 2009) [Spooner et al. 2003; Brickell et al. 2004a, 2004b, 2009; Trehane 2004; Malécot 2008]; its current ninth edition was approved by IUBS in 2016 [Brickell et al. 2016]. *International Cultivar Registration Authority* was appointed to be responsible for registering cultivar and cultivar group names.

The nomenclature system developed for cultivated plants is referred to as *cultonomy* [Hetterscheid et al. 1996]. One of the "hot points" in its most recent development became a more careful consideration of its relation to traditional botanical nomenclature [Van Raamsdonk 1986; Jeffrey 2003; McNeill 2004, 2008; Spencer and Cross 2007]. The nomenclature in question is fundamentally pragmatic and therefore somewhat simpler than the scientific one [Stearn 1986; Vrugtman 1986; Ochsmann 2003; Spooner et al. 2003; Brickell et al. 2004b, 2009; McNeill 2004; Spencer et al. 2007]. It adopts a simpler ranked hierarchy of *cultons*, *cultivars*, and *cultivar-groups*; in general, any human-altered plants can be termed *cultigens* [Hetterscheid and Brandenburg 1995; Brickell et al. 2004b; McNeill 2004; Spencer and Cross 2007; Spencer et al. 2007]. The following basic provisions are recognized: the primacy of priority over usage, a softened version of typification, and admissibility of changing old names if they do not fit this Code.

6.5 DRAFT *BioCode*

The "founding fathers" of the now predominant traditional nomenclature system conceived it as universal with respect to its taxonomic scope. However, the "Great Schism" in taxonomic nomenclature, which began in the mid 19th century, continued and was partly strengthened in the 20th century, so it currently consists of particular subject-area codes, viz., botanical, zoological, bacteriological, virological, and cultivated plants, each declaring mutual independence as one of the basic principles. In this context, an attempt was made to elaborate an integrated version of traditional nomenclature at the end of the 20th century.

This project began to develop in the early 1990s; its initiators and most active participants were the German botanist Werner Greuter and the English mycologist David Hawksworth. Through their efforts, the *International Committee on Bionomenclature* (ICB) was appointed at the XXV General Assembly of the IUBS (Paris, 1994), which elaborated an initial draft of **International Rules for the Scientific Names of Organisms** (*Draft BioCode*) [Hawksworth et al. 1994]. Its third edition was published in several printed media and on the Internet in 1996. According to this project, *BioCode* was supposed to be put into action in 2000 [Greuter et al. 1996, 1998]. However, the ICB members, at their meeting during the next General Assembly of the IUBS (Naples, 2001), decided to make major revisions to the *Draft BioCode* [Anonymous 2001; Greuter 2004], which resulted in the preparation of its next draft version, published in 2011 [Anonymous 2011; Greuter et al. 2011].

The structure of *Draft BioCode* is as traditional as its content; it comprises a Preamble and three main sections: Principles, Rules, and Authority. It is rather laconic, as evidenced by a relatively small number of articles—35. In contrast to the Botanical and Zoological Codes, this one does not have any appendices with grammar and spelling rules, nor does it provide a Vocabulary. The latter was published separately, and also includes terms used in phytosociology [Hawksworth 2010].

The Preamble, after the standard declaration on the need for a universal stable nomenclature, indicates the sphere of application of *BioCode*: "all kinds of organisms, whether eukaryotic or prokaryotic, fossil or non-fossil, and of fossil traces of organisms (ichnotaxa)." It does not presume regulation of names falling within the scope of phylogenetic, viral, and cultivated plant nomenclature. The relevance of the current subject-area Codes of bacteriological, botanical, and zoological nomenclature (called "special" in *Draft BioCode*) is acknowledged with respect to the regulation of names established prior to the final adoption of *BioCode*.

The *Draft BioCode* is based on the following standard basic principles: freedom of taxonomic decisions, binomiality (incorrectly called binarity) of species names, typification (with the exception of some taxa of higher categories), priority, monosemy (for the taxa from species for family rank-group), Romanization, stability of names, and retroactivity. The principle of independence of nomenclature for major groups of organisms is not mentioned.

The *Draft BioCode* defines the following main categories of names: *established*, *acceptable*, and *accepted* (this categorization largely corresponds to *recognized* in *PhyloCode*), which must meet certain criteria of publication and some others. The latter include mandatory registration of the names and nomenclature acts in certain *Registration Centers* for particular groups of organisms, and the registration date becomes that date of the establishment of both names and nomenclatural acts. Contrary to omitting the principle of independence, restriction on homonymy of generic names regulated by the subject-area (special) Codes is retained: such names are considered *hemihomonyms*.

The ranked hierarchy recognized in *Draft BioCode* establishes the following rank-groups (categories): order and higher-rank categories, family and infrafamilial categories, genus and infragenus categories, species and infraspecies categories; *primary* and *secondary* ranks/categories are recognized within each rank-group. The ranked hierarchies adopted in the subject-area Codes are preserved in *Draft BioCode*, with specific rules for the formation of names acting within each rank-group. The principle of priority must be followed within each of the primary rank categories, but not within the secondary ones. Special lists of *protected names* are supposed to be compiled, to which the principle of priority does not apply.

The family-group names are typified according to the standards adopted in respective subject-area Codes. For higher categories, both typified names with group- and rank-specific standard suffixes and (as an exception) traditional descriptive names are admissible. In botanical binomials, genus–species tautonymy is not allowed.

The prospects for the ultimate adoption of *BioCode* as a basic unified regulator for most subject-area sections of taxonomic nomenclature are not very clear. In fact, it appeared to be just a "sum" of subject-area Codes, simply compiling specific rules from them without their apparent consistent integration. With this, it was suggested by biocoders that the subject-area Codes retain their preemptive rights to regulate N-definitions of the names of respective taxa that were established before the adoption of *Biocode*. This makes evident the main "technical" problem with the *BioCode*, which hinders its successful promotion: it turns out to be too cumbersome in one respect and surplus in others.

All this will undoubtedly have a braking effect on its development and application. Indeed, the adoption of any amendments affecting different major groups would require reconciliation of the opinions of a great number of groups of interests. And if *BioCode*, to overcome this problem, retains powers for the current subject-area authorities, then its sought integrative destiny would remain but a "good wish."

7 The Conceptualist Route

Taxonomic nomenclature, as part of the professional language of systematics, developed from the very beginning in the conceptualist route because of its certain theory-ladenness. Its main tone was originally set by the essentialist concept, according to which one of the main purposes of taxonomic names was to reflect properly the essences of organisms. The shift from the essentialist to the nominalist route at the turn of the 18th–19th centuries also had a conceptual character: it was associated with a different basic natural philosophy, which made it possible to deprive the language of systematics of its essentialist content and proclaim that "a name is just a name." This made the subsequent development of nominalist nomenclature ostensibly theory-neutral, but elaboration of the principle of trinomiality with reference to the Darwinian evolutionary model in the second half of the 19th century meant that its freedom from a biologically sound theory was only partial. The struggle against the "Linnaean legacy" throughout the 20th century was also basically theory-dependent, since it was generated by certain theory-laden concepts. In the first half of the 20th century, this struggle was headed by biosystematics; in its middle, it was associated with classification phenetics based on a positivist philosophy of science; and at its end, a phylogenetic worldview seized the initiative in this [Pavlinov 2018, 2021].

One can rightfully conclude from the preceding that the whole of the history of taxonomic nomenclature, considered in a general large-scale capacity, followed the conceptualist route widely understood. In this regard, as correct as it might be in general, the theoretical backgrounds of several nomenclature systems do not fit the conceptual frameworks of the essentialist and nominalist ones; they were driven by their own forces and followed their own routes. Therefore, somewhat conditionally and for the sake of convenience, their development can be considered within a more specifically treated *conceptualist route* to highlight their extreme and explicitly stated theory-laden nature. These systems are arranged into two main nomenclature concepts. In one of them, the emphasis is on strengthening the rational-logical nature of taxonyms designed mainly to perform the classification function. Another concept rigidly links taxonomic nomenclature to evolutionary theory. These two concepts are considered in the present chapter.

7.1 RATIONAL-LOGICAL NOMENCLATURE

As repeatedly emphasized above, a certain rationalization of the professional descriptive language of biological systematics, as part of the latter's classification method, was one of the key conditions for its initial scientification. It set the main trend in the development of the whole of taxonomic nomenclature, which ultimately led to its codification. However, traditional nomenclature, in following this route, retained one of the important features of natural language, viz., its verbal structure; and the respective principle of verbality is accepted by default or expressly stated in contemporary Codes. This means that taxonyms must be lexemes borrowed from a certain natural language or imitating them. Thus, the currently dominant traditional taxonomic nomenclature did not succeed in diverging far from the folk one in this respect, so it remains irrational and therefore pre-scientific in a sense.

Along with the main "verbalist" trend in the history of taxonomic nomenclature, another one involved the development of *rational-logical* nomenclature. The latter's core is the requirement for taxonyms to perform primarily or exclusively a classification function by indicating, in one way or another, the position of taxa designated by them in the taxonomic system; the already mentioned philosopher J. Herschel called such nomenclature *systematic* [Herschel 1830]. The taxonyms considered in such a vein can be likened to the museum locator labels that inform about the position of specimens in the repository [Felt et al. 1930; Felt 1934; Pavlinov 2015a]. By their form, those nomenclature systems developed within the framework of this concept look like versions of the nominalist concept dealing with the "non-essentialist" verbal designations of taxa. However, the concept in question presumes a strictly symbolic designation, so it radically breaks with verbal tradition, which makes it more consistent with the designative features of "natural philosophy," as distinguished from descriptive "natural history" by the medieval English philosopher Roger Bacon (1219–1292) [Whewell 1847].

This general concept goes back to the ideas of the rationalist philosophers of the 17th century (René Descartes, Francis Bacon, etc.) on the characteristic relationships between rationally organized Nature, a single rational knowledge about it, and a single unified rationally organized language describing it [Foucault 1970; Slaughter 1982; Gaidenko 2003]. It was best expressed by a natural philosophical metaphor, according to which "The Book of Nature is written in the language of mathematics"; its author is the Italian physicist, astronomer, and mathematician Galileo Galilei (1564–1642).[1] According to this, the language of science, as part of its unified rational method, should not just be universal, like the Latin of scholastics, but maximally formalized and basically free from any traditional semantic motivation. Based on this idea, Descartes developed the general concept of "algebraic" language based on the decimal system; Bacon's intention was to develop a "pictographic" language using letters of a certain natural language [Slaughter 1982; Scharf 2008]. The Baconian

---

[1] This metaphor goes back to an idea of the early Christian theologian Aurelius Augustinus Hipponensis (354–430), who likened Nature to the *Book of Nature*, which can be read in the same way as the *Book of Revelation* [Harrison 2006].

idea gained a certain popularity in early systematics, while the Cartesian one does so in the 20th century, giving rise to numericlature.

The author of the best-known version of the universal "*philosophical language*" (today called *analytical*) of Baconian kind was the English natural philosopher, religious figure, and linguist John Wilkins (1614–1672). For him, "philosophy" was in many ways a "taxonomy," since both concerned the ordering of knowledge about the "nature of things" (in their general understanding); in particular, he was interested in representing the structure of universal knowledge about Nature in the form of an hierarchical classification [Slaughter 1982; Maat 2004]. Accordingly, Wilkins, in his *An essay towards a real character*, attempted to develop a universal "philosophical language" as follows [Wilkins 1668]. First, a strictly ordered general hierarchy of the System of Nature was built, represented in the form of the Tree of Porphyry; next strictly defined places were determined for all "things" at the lowest level of that hierarchy; lastly, these "things" were assigned strictly formalized deductively defined designations with a certain method of coding, which indicated their positions in the System of Nature thus construed.

To implement this idea, Wilkins divided Nature into 40 basic categories called "*genera*," next divided them into "*differentiae*" (usually nine in each "genus"), and the latter were divided into "*species*" (usually 15 in each "differentia"). The whole hierarchy of Nature was shaped by the strictly fixed sequences of divisions finished with particular "things" (in our case, plant and animal species in their nearly contemporary understanding), and the position of each division in each particular sequence is indicated by the corresponding lexical morpheme. According to this system, a unique syllable of two letters was assigned to each "genus," consonants corresponded to "differentiae," and vowels corresponded to "species" in them. On this basis, the fixed sequences of divisions in the Tree of Porphyry appeared to be represented by the respective fixed sequences of the combinations of letters. As a result, each "thing" received a fixed designation in the form of a unique combination of letters corresponding to the place it occupied in the final row of the "tips" of the tree; for example, the elephant should be designated as *zibi*. If necessary, these letter designations could be converted into digital ones by numbering the sequences in a certain way; with this coding method, the elephant could be designated by a combination of numbers *18.1.4.1* (see [Maat 2004] for details). So, an aspiration to make the linguistic systems of Bacon and Descartes mutually compatible was evident in the method of J. Wilkins.[2]

One of the first variants of the formalized system of nomenclature based on the Wilkins' method has been developed for Linnaeus' *Genera Plantarum* by the Swedish mathematician and engineer Christopher Polhem (1661–1751); his text was published only in our days [Scharf 2008]. Like those of Wilkins, Polhem's designators were four-letter uninomials; Polhem believed that his symbolic combinatorics allowed unique designators to be allocated to almost one and a half million plant species.

---

[2] The botanist J. Ray (mentioned in Section 5.2) was involved in the Wilkins' project in providing him with the hierarchical classification of several "higher genera" of plants. However, Ray commented on the whole project quite negatively; the main criticism was that the scheme proposed by Wilkins was too dogmatic and simplified to describe the real diversity of organisms [Slaughter 1982].

The nomenclature system of the German physician and phytographer Nathaniel Matthaeus von Wolf (1724–1784), suggested in his *Genera plantarum* [Wolf 1776], was somewhat different. His designators of plant genera were binary, consisting of two-letter combinations: the first "word" indicated family allocation, the second was generic; no nomenclature was developed for the species [Ross 1966; Scharf 2008].

The general idea of Wilkins will be repeated (without reference to it) in the mid 20th century by two American microbiologists: according to their proposal, "the name of organism shall consist of two words, each of which contains nine letters, comprising three syllables" [Siu and Reese 1955: 400]. These letters corresponded to the respective containing taxa of different ranks, so a certain fixed-rank hierarchy was presumed. This nomenclature system was implemented at the generic level: for example, the mold genus *Aspergillus* should be denoted "Fimmabbat."

Another version of the "philosophical language" was elaborated by the French botanist Jean-Pierre Bergeret (1752–1813) in his three-volume *Phytonomatotechnie universelle* [Bergeret 1783–1785]. His nomenclature system differed essentially from that of Wilkins in that it was justified inductively [Scharf 2008; Pavlinov 2014, 2015a]. Bergeret proceeded not from certain natural philosophical ideas about the universal System of Nature, but from analyses of the diversity of anatomical features of plants. This meant that the resulting basic hierarchy was not so much taxonomic as partonomic (for their differences, see Section 1.1), in which terminal partons were anatomical elements, each designated by a certain symbol. At the next step, the taxa got characterized by the sets of these fixed anatomical elements, and the combinations of the symbols ascribed to them became formalized designators of terminal taxa. Respectively, these designators reflected the place of taxa in the general hierarchy. In practice this approach looked as follows. First, a universal table of fixed traits (15 in total) with their fixed modalities was compiled; these modalities were assigned with unique letters, and each of their combinations was encoded by a standard 15-letter "formula." The latter then was allocated to particular plant genera characterized by the respective combinations of modalities: for instance, for *Plantago major* (broadleaf plantain), it looked like GIQGYABIAHUQZEZ. The "formula" with repeating letters could be given in an abbreviated form: for example, for *Amanita muscaria* (fly amanita), the complete "formula" was AAAAAAAALAAAAYZ, and the abbreviated one was A^8LA^4YZ. So this version of botanical nomenclature applied certain elements of the rational nomenclature used in chemistry. An evident shortage of this "phytonomatotechnie" was that it did not recognize and designate either higher-rank taxa or species.

The general rational-logical concept was implemented several times in a much less formalized manner based on the simple idea that "closely related forms should have close names" [Harting 1871: 28], and their names should "quickly and accurately indicate systematic position of their carriers" [Tornier 1898: 576]. To this end, to traditional verbal taxonyms were added certain markers indicating taxonomic position of the taxa. The above-mentioned botanist S. Gray seemed to be the first to apply this rule within the framework of traditional nomenclature [Gray 1821] (see Section 6.3.1.1). Simultaneously with him, the French botanist Louis-Marie Aubert du Petit-Thouars

(1758–1831) proposed supplying generic names with the suffixes indicating that genera belonged to particular families [Petit-Thouars 1822]. A similar suggestion of the Mexican naturalist Alfonso Luis Herrera (1868–1942) presumed indication of belonging genera to each of three "Kingdoms of Nature" [Herrera 1899; Evenhuis and Pape 2010]. P. Harting and G. Tornier, mentioned above, developed (independently of each other) sample systems of standard codification of the names of phyla and higher-rank taxa of animals and plants similar to that developed by Bergeret. There were also other proposals to insert group-specific alpha-numeric markers in taxonyms [Jordan 1911; Felt et al. 1930; Felt 1934; Kluge 1999a]. A variant of the monoverbal nomenclature supposed that the one-word generic names perform partly descriptive and partly classificatory functions [Reynier 1893; Raspail 1899].

The mastering of positivist philosophy of science by systematics in the 20th century contributed to the development of nomenclature systems with taxonomic designators performing a classification function [Pavlinov 2015a, 2018]. This philosophy presumed the unconditionally nominal (philosophically treated) status of ranks/categories of whatever levels of generality, so the traditional descriptive rank-dependent designations of taxa did not make any sense [Cain 1959; Michener 1963, 1964; Sokal and Sneath 1963]. According to a moderate treatment, the principle of verbality was retained, but traditional species binomials were rejected, so that their uninomial taxonyms would not change with the change in their generic allocation [Cain 1959; Michener 1964]. The American zoologist Dean Amadon (1912–2003) invented a unified supraspecific category "*suneg*" (mirrored from "genus"), defined as "A term identical with the type genus of a family, subfamily, or tribe [...] to be used, except in strictly taxonomic publications, in lieu of generic names for all the species in any given family, subfamily, or tribe" [Amadon 1966: 55].

In the most consistent embodiment of the rational-logical idea, *numericlature* was suggested to designate taxa and/or ranks in a Cartesian manner by a unique combinations of numbers [Jahn 1961; Michener 1963; Little 1964; Hull 1966, 1968; Yochelson 1966; Griffiths 1981; Dayrat 2010; Pavlinov 2015a]. The main goal of such a formalized system is presumed to make numerical taxonomic designators the elements of the multiuser information retrieval systems most compatible with computer-based digital technologies [Kennedy et al. 2005; Page 2006; Patterson et al. 2006, 2010; Schindel and Miller 2010].

The main function of digital designators is to indicate the place of taxa in the hierarchical taxonomic system by respective combinations of numbers assigned to them following certain rules. Accordingly, the development of numericlature begins, just as in the case of Wilkins' "philosophical language," presumes elaboration of a total (now digital) codification of a total unified taxonomic system for all living beings, which would serve as the basis for the formation of an individual *reference number* for each taxon in it. For example, in one version, the genus *Plasmodium* is assigned the number "1310101D," where "1" refers to the phylum Protozoa, "3" refers to the class Sporozoa, "101" refers to the family Plasmodiidae, etc. [Jahn 1961]. According to another approach, the reference number for *Musca domestica* is "10-7-26-081-052-0325" [Michener 1963; Little 1964]. D. Hull proposed *phylogenetic numericlature* as a "system of identification, positional, and phyletic numbers for taxa that makes possible a significant relationship between numerical classification and phylogeny" [Hull 1966: 14]; this idea was partly implemented in a cladistic classification of Insecta [Hennig 1969].

Thus, the system of reference numbers seems to be analogous in its structure and function to the system of DOI, which is the Internet standard for the unique numeric designations of published items [Garrity and Lyons 2003]. Reference numbers are used for registration of zoological taxa in ZooBank, not instead of but in addition to traditional verbal naming [Pyle and Michel 2008]. The creation of International Species Information System (ISIS) provides another example of the successful implementation of numericlature [Flesness 2003].

Various systems of fully formalized taxonomic designators were not adopted as the basis for reforming taxonomic nomenclature for a number of quite obvious reasons. First, their use weakly complies with the mnemonic function of nomenclature, which is of great importance in practical systematics. Secondly, any system of reference numbers, as far as it depends on a standard codification of the overall structure of classifications, is doomed to be unstable: any changes in this structure at higher levels of taxonomic hierarchy inevitably yield respective changes in the system of classifying designators. This threatens to give rise to a cumbersome system of "digital synonymy," the manipulation with which would require the creation of a constantly updated huge distributed database, probably with the elements of blockchain technology (see [Tapscott and Tapscott 2016] on it) to trace and correlate all modifications of numerical N-designators. The recent "cladistic revolution" in systematics shows that such changes can be both rather unpredictable (in a large time-scale) and quite radical, and should be followed by just as radical restructuring of the whole nominative numericlature (see Section 7.2).

At the same time, the above-mentioned examples using numericlature allow rather positive assessment of its potential as an auxiliary nomenclature system complementing the traditional one. So, the development of digital toolkits to resolve the significant task of making these systems compatible seems to be very promising [Ytow et al. 2001; Kennedy et al. 2005; Hussey et al. 2008; Pyle and Michel 2008; Patterson et al. 2010; Franz et al. 2016]. So the further development of numericlature will certainly require specific codification [Minelli 2017, 2019].

7.2 PHYLOGENETIC NOMENCLATURE

The first purposeful attempt to substantiate the new nomenclature systems metaphysically by direct reference to evolutionary theory was made by the adherents of classification Darwinism, who proposed legitimizing the trinomial and quadrinomial nomenclature instead of (or in addition to) the binomial one [Coues et al. 1886; Banks and Caudell 1912]. This trend was continued by developers of botanical biosystematics in the first half of the 20th century. One of its early leaders, the American botanist Wendell Holmes Camp (1904–1963), emphasized that in biosystematics "will arise something wholly different from Linnaean nomenclature […] The new system of nomenclature will have to be cut of a different cloth from the old and tailored so as to express, in descriptive terms, a vast array of most dynamic and involved genetic situations" [Camp 1951: 126]. Correspondingly, within this taxonomic theory, both the structure and terminology of intraspecific units appeared to be significantly complicated in order to reflect their place in ecological and evolutionary processes: their number reached several dozens by the mid 20th century [Du Rietz 1930; Camp and Gilly 1943; Sylvester-Bradley 1952; Valentine and Löve 1958]. However, no specific nomenclature system was formalized as a kind of Biosystematic Code, so it did not officially compete with traditional botanical nomenclature.

At the end of the 20th century, the active development of the cladistic version of phylogenetic systematics[3] gave rise to the *phylogenetic nomenclature* (phylonomenclature). Its main objective was to bring descriptive language of systematics in line with the content of phylogenetic theory with its central concept of a hierarchically arranged phylogenetic pattern, whose hierarchy was presumed rankless [de Queiroz and Gauthier 1990, 1992, 1994; de Queiroz 1992, 1997, 2005, 2007; Minelli 1995; Ereshefsky 1997, 2001a,b; Knapp et al. 2004; Lee and Skinner 2007; Cantino and Queiroz 2010, 2020]. The developers of this nomenclature system stated that it should be "a system of nomenclature [...] that is more concordant with evolutionary concepts of taxa [and] more closely conform to the manner in which they are conceptualized" [de Queiroz and Cantino 2001: 269]. One of the main ideologists of phylonomenclature is the American zoologist Kevin de Queiroz, so following tradition, this nomenclature system is suggested to be called *Queirauthian* [Dubois 2005].

The reform of nomenclature undertaken by phylocoders was initially stimulated by the conviction that: (a) cladistic taxonomic theory is dominating and will dominate in systematics; and (b) its specific conceptual basis, viz., the ontology of historical groups instead of the traditional ontology of classes, required specific principles of recognizing and naming taxa [de Queiroz and Gauthier 1990, 1992, 1994; Minelli 1995; Ereshefsky 1997, 2001b; Laurin 2007]. The key innovation of this nomenclature system became a proposal to recognize and designate only clades (strictly defined monophyletic groups) and exclude any taxa and their names that do not fit this criterion [de Queiroz and Gauthier 1990, 1992, 1994; de Queiroz 1997, 2007; de Queiroz and Cantino, 2001; Cantino and de Queiroz 2010, 2020]. The traditional and phylogenetic nomenclature systems were proposed to be distinguished terminologically: the first was designated as *taxonomy* (not in the sense of a theoretical section of systematics, see [Simpson 1961; Pavlinov 2018, 2021] and Section 1.1), and the second as *cladonomy* [Brummitt 1997; Dubois et al. 2021]. The latter was called *post-Linnaean* nomenclature [Ereshefsky 2001a,b], which is not quite correct (see beginning of Chapter 6). Taxon designators are proposed to be called *phylonyms* or *phyloreferences* [Cantino and de Queiroz 2010, 2020; Cellinese et al. 2021].

Phylonomenclature was supposed from the very beginning to become a unified nomenclature system for all subject areas of biological systematics. Its first version, after discussion at the meeting of its developers at Harvard University (1998), was published in 2000; it was followed by several revisions (2004, 2006, 2007, 2010),

---

[3] It is usually identified with the whole of phylogenetics, which is not correct [Pavlinov 2018, 2021].

which were officially published on the Internet. The International Society for Phylogenetic Nomenclature was established at the Second International Congress on Phylogenetic Nomenclature (Paris, 2004) to develop a draft of the respective rulebook. The *Committee on Phylogenetic Nomenclature* was established at the above-mentioned Society to regulate the development and application of phylogenetic nomenclature. The current **International Code of Phylogenetic Nomenclature** (*PhyloCode*), Version 6, was ratified by this Committee in 2019 [Cantino and de Queiroz 2020].

The Preamble of *PhyloCode* states that it applies to all clades of both extant and extinct organisms; it is used concurrently with rank-based Codes; it relies on the latter in recognizing the acceptability of preexisting names, but governs the application of those names independently from rank-based Codes; contrary to the provisions of *PhyloCode*, the names have no official standing under it. A separate clause indicates that this Code comes into force simultaneously with the publication of the specific nomenclator entitled *Phylonyms: A Companion to the PhyloCode* which also happened in 2020 [de Queiroz et al. 2020].

The following guiding principles are stated: (1) *reference*—names are intended to designate taxa, but not to indicate their characteristics; (2) *clarity*—names should unambiguously indicate taxa, which is achieved by explicit definitions of the latter; (3) *uniqueness*—each taxon should be designated by a single accepted name; (4) *stability*—the names of taxa should not change over time; (5) *phylogenetic context*—this Code is concerned with the naming of phylogenetically conceptualized taxa; (6) *taxonomic freedom*—this Code does not restrict freedom of opinions with regard to hypotheses about relationships; (7) *no "case law"*—nomenclature problems are resolved by this Code through direct application of its articles. As seen, *PhyloCode* proclaims adherence to the nominalist concept of nomenclature and limits its scope of application to a phylogenetically specified context. Some of the principles occurring in the introductory sections of the traditional Codes (Romanization, precedence, etc.), are rightly relegated to particular rules considered in the respective chapters. Since this Code claims to be of general biological application, it lacks the principle of independence of the subject-area Codes.

Taxon is defined as a group of organisms or a species to which a name is assigned; and clade is defined as a unique ancestor (organism, population, or species) and all its descendants. It is specified that only clades are named, but not all of them are to be named on a mandatory basis. Because of rank independence of phylonomenclature, the consideration of synonymy, homonymy, and preexistence of the clade names does not depend on their position in the hierarchy. With this, it is stipulated that, although the Code does not require the fixation of ranks, it does not prohibit them.

The clade names are regulated very strictly and rather sophisticatedly, so the chapter considering them is the longest. It describes specific requirements for the formation of phylogenetically correct names [Art. 9] and for their choice as established ones [Art. 10], and defines the specifiers and qualifying clauses used in the definitions of clades [Art. 11] (see also below). The name of a newly recognized clade is converted from a preexisting one or established as a new one. Article 9 contains the list of admissible phylogenetic definitions of clade; either of them or some combinations thereof must be indicated in the protologue accompanying the establishment of a new clade name. It is specially stipulated [Art. 10] that the choice of name for a newly recognized clade should minimize its discrepancy with existing nomenclature. Therefore, a new name should be introduced only if the respective clade differs significantly in its circumscription from previously identified groups.

It is inadmissible to form clade names on the basis of preexisting species names (similar to the suppression of tautonymy in the genus-species binomen). In order to distinguish between clades with different phylogenetic statuses, various prefixes (*Pan-*, *Apo-*) are introduced into their names. The homonymy of names established under different subject-area Codes (botanical, zoological, etc.) is eliminated by adding group-specific prefixes to coinciding preexisting names: *Phyto-* for higher plants, *Phyco-* for algae (not including cyanophytes), *Myco-* for fungi, *Zoo-* for animals, *Protisto-* for non-photosynthetic protists, *Bacterio-* for prokaryotes.

An important Chapter IV considers conditions for the *N*-definitions of clades by means of *specifiers* and *qualifying clauses*. The specifiers refer to ancestral species, specimens (mainly the types indicated with the establishment of preexisting species name), and apomorphies (indicated by the phylogenetic definition of clade); it is noteworthy that the clades themselves are not considered as the specifiers. The specifiers may be either *internal* (explicitly included in the clade) or *external* (explicitly excluded from the clade). Qualifying clauses serve to consider preexisting names with respect to their possible phylogenetic meaning (the interpretation of the name of the mammal group *Pinnipedia* is analyzed as an example). The *reference phylogeny* provides a substantive context for the application of clade names by means of their phylogenetic definition. The lengthy explanations and examples in this chapter show how problematic the resolution of certain nomenclatural tasks becomes if substantive judgments about the phylogenetic status of groups are involved.

Two categories of authorship are introduced, *nominal* (authorship of the name as such) and *definitional* (authorship of the definition of a clade bearing this name) [Art. 19]. They coincide in the case of a newly recognized clade with a new name. In the case of a converted name, the authorship of a preexisting name is nominal, and of an accepted name is definitional. In the case of a replacement clade name, the authorship of both replacement and replacing names are nominal.

The traditional names of species-group taxa are admissible to use if they meet the requirements of the corresponding subject-area Codes, but they are not regulated as such by *PhyloCode* [Art. 21]. By this, the principle of binomiality is partly retained, albeit not in a fundamental way [Cantino et al. 1999].

Criticism of phylonomenclature is quite extensive (e.g., [Brummitt 1997; Benton 2000; Nixon and Carpenter 2000; Stuessy 2000; Keller et al. 2003; Monsch 2006; Dyke and Sigwart 2007; Rieppel 2006, 2008; Dubois 2007; Kraus 2008]). From a traditional point of view, its most serious drawback is its strong conceptualization, according to which the freedom of taxonomic decisions proclaimed in *PhyloCode* is allowed only insofar as it does not contradict the main idea of phylonomenclature

[Rasnitsyn 1996, 2002; Rieppel 2006; Pavlinov 2014, 2015a].[4] Added to this might be logical inconsistency of at least some of its provisions; thus, a suggestion to exclude species category from its scope [Cellinese et al. 2021] seems to make clade definition with the reference to its ancestral species logically confused. Besides, as far as the judgment of such a species is always conjectural, its inclusion in the clade N-definition makes the latter non-operational [Pavlinov 2015a]. Thus, "rather than solving those problems with which the traditional Codes have struggled for more than a century, the PhyloCode simply perpetuates them, because it is subsidiary to and fundamentally dependent upon the traditional Codes" [Brower 2020: 625].

The claim of phylonomenclature to take a dominant position, on the one hand, and the rather strong position of traditional nomenclature, on the other hand, lead to simultaneous functioning of two essentially different nomenclature systems in contemporary systematics. This circumstance may entail another serious "schism" in taxonomic nomenclature, causing its fundamental instability. Therefore, calls for a search for opportunities to combine the most significant provisions of these nomenclature systems are quite understandable [Barkley et al. 2004; Kuntner and Agnarsson 2006; Naomi 2014]. The so-called *Kinman System* of cladisto-eclectic classification of organisms [Kinman 1994] seems to be one of the possible practical options.

---

[4] It is, however, stated that PhyloCode does not restrict the freedom of taxonomic decisions concerning content of phylogenetically defined taxa, as it is actually applied to the phylonyms and not to phylogenetic hypotheses [Bryant and Cantino 2002].

Instead of Conclusion
A General Outlook

What has been will be again, what has been done will be done again.
                                                                Ecclesiastes

A theoretical-historical analysis of taxonomic nomenclature based on its consideration as a specific linguistic system being shaped by a complex interaction of non-coinciding factors of various levels of generality appeared to be quite productive in revealing, in the first approximation, some general mechanisms of its developing, structuring, and functioning.

This analysis indicated that the nomenclature concepts and systems were developed following changes in the theoretical consideration of "taxonomic reality" as the subject area of biological systematics that was developed at a different stage of its conceptual history. It was these changes that made the descriptive language of systematics adequate for changing ideas about the structure of this "reality." This is because the latter's complex structure presumes the possibility of its consideration from various aspects and elaboration of more or less specific means for describing them. Due to this, the descriptive language of systematics split into "dialects," each with its particular interpretations of general principles of nomenclature.

Such inevitable global dynamic means the transitory and therefore "local" character of any particular nomenclature concepts and systems, which was emphasized by A. de Candolle one and a half centuries ago [de Candolle 1868]. This entails a kind of "nomenclature pluralism" as an immanent property of the professional language of systematics. Therefore, the main goals declared by the "founding fathers" of contemporary nomenclature in the 19th century were not achieved: the current professional language of biological systematics is neither universal nor stable; moreover, the goals seem to be unachievable.[1]

And yet, there are also quite clear elements of the preservation of structural stability in this background dynamic. They are easy to reveal by superimposing the conceptual history of taxonomic nomenclature on its logical structure (or *vice versa*). It turns out that the nomenclature systems proposed at different times, in one way or

---

[1] However, biology is not unique in this regard; nomenclature in chemistry faces the same problem [Leigh et al. 1998; Elk 2004].

another, reproduced several general patterns, specifying them in accordance with certain particular taxonomic theories [Pavlinov 2015a, 2019].

In fact, the most recent suggestion of replacing traditional nomenclature with a phylogenetic one repeats the most serious conceptual shift in the development of nomenclature from an essentialist to a nominalist one in the second half of the 18th century. Certain rank-dependent rules of contemporary traditional Codes (such as binomiality) remained as an evident "relic" of the old essentialist nomenclature preserved under a new nominalist "envelope." On the other hand, the rankless nomenclature system of cladistics reproduces the structure of the genus–species scheme of scholastics. Proposals to recognize and designate only "definite" genera by *Linnaean Canons* of the mid 18th century or "natural" groups by *Paris Laws* of the mid 19th century or clades by *PhyloCode* of the beginning of the 21th century are essentially the same with regard to their basic ontological consideration: in fact, they all refer to certain groups acknowledged to be "real" according to the respective background metaphysics. Finally, the most recent development of the rational-logical nomenclature that ascribes taxonyms with a basically classificatory function reproduces the same ideas of the rationalists of the 17th century.

All this gives the impression that the "space of logical possibilities" for the basic concepts shaping the regulative taxonomic nomenclature, as it was developed during the 17th–19th centuries, is close to exhaustion, and its historical fate is reminiscent, to an extent, of "walking in a circle" [Moore 2005; Pavlinov 2015a]. This reflects the rather conservative nature of the professional language of systematics, which, in turn, is a consequence of the "healthy conservatism" of systematics in general. Such conservatism seems to be caused largely by the peculiarities of the structure of "taxonomic reality," to which only a quite specifically structured descriptive language seems to be adequate. An important part of this language is the, hardly accidentally originated, traditional nomenclature, in which excessively abrupt transformations are probably contraindicated.

In this regard, a possible forecast of the nearest future of taxonomic nomenclature seems to be rather simple [Pavlinov 2015a]. Obviously, it will continue to develop following possible changes in the cognitive situation of systematics. With this, it will reproduce in this development, in one way or another, the basic structure of its professional language, which was laid at the beginning of its formation. Based on certain general considerations, it can be assumed that the historical dynamics of this language, like any sufficiently developed linguistic system, will combine the same two opposite trends, integration and diversification, just as in the 19th–20th centuries.

This means that theorists will continue to fight for improvement of the regulative nomenclature by further developing its basic principles and enriching its thesaurus with new concepts and notions. The detailing of the existing "precedent" nomenclature regulators will be supplemented by their more comprehensive interdisciplinary (logical, linguistic, juridical, etc.) analyses. The latter will hopefully provide a clearer understanding of the functioning, structuring, and development of taxonomy nomenclature and the basic causes responsible for them. Various theory-dependent nomenclature systems will continue to develop, each with its own conceptual basis corresponding to particular visions of the nature of "taxonomic reality" and the ways of its descriptions. All this will contribute to the maturation of onymology (taxonymy), which now finds itself in its infancy. On the other hand, practitioners will continue to call for simplification of the rules for naming organisms and for greater adaptation of these rules to the everyday needs of practical and applied systematics.

Among these needs, the most recent challenges are shaped by the expanding practice of: (a) describing new species based on the analysis of molecular genetic data; and (b) the active development of digital technologies, including databases with taxonomic designators and other relevant information. Active participants in this practice are "techies," who usually have no basic taxonomic knowledge, but are forced to employ its elements when describing "taxonomic reality" as they understand it "technically." They tend to consider the traditional language of taxonomic descriptions untenable and suggest their own designative means in parallel to the traditional ones or just to replace them. The crucial problem here is that these new suggestions coming from essentially different sources are not seriously considered with respect to their compatibility both between themselves and with the traditional ones. So, all this taken together threatens to spawn a new wave of chaos in nominative taxonomic nomenclature [Minelli 2013, 2017, 2019].

One can see in these challenges and responses to them another manifestation of the "repetition of the past": the current situation reproduces in outlines those, in which taxonomic nomenclature found itself twice, first, at the beginning of the 18th and second, at the beginning of the 19th centuries, and which demanded its stronger codification. Thus, the above-mentioned new challenges, like previous ones, will make nomenclaturists searching for new means to make taxonomic nomenclature more adequate to the most recent ideas about both the structure of "taxonomic reality" and the methods for studying and describing it.

References

Abebe, E., T. Mekete, and W. Decraemer. 2014. *E*-typing for nematodes: An assessment of type specimen use by nematode taxonomists with a summary of types deposited in the Smithsonian Nematode Collection. *Nematology* 16:879–88. DOI:10.1163/15685411-00002826.

Adams, M. J., E. J. Lefkowitz, A. M. Q. King, et al. 2017. 50 years of the International Committee on Taxonomy of Viruses: Progress and prospects. *Archives of Virology* 162:1441–6. DOI:10.1007/s00705-016-3215-y.

Adanson, M. 1757. *Histoire naturelle du Sénégal. Coquillages* […]. Paris: Bauche. www.biodiversitylibrary.org/bibliography/47118.

Adanson, M. 1763. *Familles des plantes*, Pt. 1. Paris: Vincent. ISBN:9780913196250 (1964 reprint).

Aescht, E. 2018. Reflecting on a theory of biological nomenclature (onymology). *Alytes* 36:212–37. DOI:10.1099/ijsem.0.000778.

Agassiz, A. 1871. Systematic zoology and nomenclature. *The American Naturalist* 5:353–6. https://archive.org/details/jstor-2447062/mode/2up.

Agrawal, A. 2002. Indigenous knowledge and the politics of classification. *International Social Science Journal* 54:287–97. DOI:10.1111/1468-2451.00382.

Aguiar, M. J. J., J. C. Santos, and M. V. Urso-Guimarães. 2017. On the use of photography in science and taxonomy: How images can provide a basis for their own authentication. *Bionomina* 12:44–7. DOI:10.11646/bionomina.12.1.4.

Ahrens, D., S. T. Ahyong, A. Ballerio, et al. 2021. Is it time to describe new species without diagnoses?—A comment on Sharkey et al. (2021). *Zootaxa*. https://zenodo.org/record/4899151#.YMCe_iQYAz0.

Aime, M. C., A. N. Miller, T. Aoki, et al. 2021. How to publish a new fungal species, or name, version 3.0. *IMA Fungus* 12:1–15. DOI:10.1186/s43008-021-00063-1.

Alekseev, E. B., I. A. Gubanov, and V. N. Tikhomirov. 1989. [*Botanical nomenclature*.] Moscow: Moscow University Publishing (in Russian).

Alinei, M. 2021. *Names of animals, animals as names: Synthesis of a research.* www.academia.edu/11751209/Names_of_animals_animals_as_names_synthesis_of_a_research.

Allard-Kropp, M. 2020. *Languages and worldview*. St. Louis (MO): University of Missouri. https://irl.umsl.edu/cgi/viewcontent.cgi?article=1016&context=oer.

Allen, J. A. 1905. A new code of nomenclature. *Science, New Series* 21:428–33. DOI:10.1126/science.21.533.428-a.

Allen, J. A. 1906. The "elimination" and "first species" methods of fixing the types of genera. *Science, New Series* 24:773–9. DOI:10.1126/science.24.624.773-b.

Allen, J. A. 1907. The first species rule for determining types of genera: How it works in ornithology. *Science, New Series* 25:546–54. DOI:10.1126/science.25.640.546.

Alonso-Zarazaga, M. A. 2005. Nomenclature of higher taxa: A new approach. *The Bulletin of Zoological Nomenclature* 62:189–99. www.biodiversitylibrary.org/item/107011#page/195/mode/1up.

Amadon, D. 1966. Another suggestion for stabilizing nomenclature. *Systematic Biology* 15:54–8. DOI:10.2307/sysbio/15.1.54.

Amorim, D. S., C. M. D. Santos, F.-T. Krell, et al. 2016. Timeless standards for species delimitation. *Zootaxa* 4137:121–8. DOI:10.11646/zootaxa.4137.1.9.

Amyot C.-G.-B. 1848. *Entomologie française. Rhynchotes. Méthode mononymique.* Paris: J.-B. Baillière. www.biodiversitylibrary.org/bibliography/34231.

Andino, R., and E. Domingo. 2015. Viral quasispecies. *Virology* 479–480:46–51. DOI:10.1016/j.virol.2015.03.022

Anonymous. 1847. The rules of American pomology. *The Horticulturist and Journal of Rural Art and Rural Taste* 2:273–5. www.biodiversitylibrary.org/item/30187#page/7/mode/1up.

Anonymous. 1848. The rules of American pomology—As adopted by the Horticultural Societies [...]. *The Horticulturist and Journal of Rural Art and Rural Taste* 3:480–1. www.biodiversitylibrary.org/item/30512#page/7/mode/1up.

Anonymous. 1904. Code of botanical nomenclature. *Bulletin of the Torrey Botanical Club* 31:249–90. www.biodiversitylibrary.org/page/13175642#page/299/mode/1up.

Anonymous. 1907. American code of botanical nomenclature. *Bulletin of the Torrey Botanical Club* 34:167–78. www.biodiversitylibrary.org/item/45479#page/208/mode/1up.

Anonymous. 1909. Propositions relating to the amendment and completion of the international rules of botanical nomenclature [...]. *Bulletin of the Torrey Botanical Club* 36:55–63. www.biodiversitylibrary.org/item/46477#page/70/mode/1up.

Anonymous. 1928. Report. Rules of entomological nomenclature [...]. *Proceedings of the Entomological Society of London* 3:1R–13R. DOI:10.1111/j.1365-3032.1928.tb00473.x.

Anonymous. 1934. *International rules of botanical nomenclature adopted by the Fifth International Botanical Congress, Cambridge, 1930.* London: Taylor and Francis. www.iapt-taxon.org/historic/Congress/IBC_1930/prepub.pdf.

Anonymous. 1961. *International Code of Zoological Nomenclature*, 1st ed. London: International Trust for Zoological Nomenclature. www.biodiversitylibrary.org/item/107561#page/5/mode/1up.

Anonymous. 1964. *International Code of Zoological Nomenclature adopted by the XV International Congress of Zoology.* London: International Trust for Zoological Nomenclature. www.biodiversitylibrary.org/item/107127#page/5/mode/1up.

Anonymous. 1982. The rules of nomenclature of viruses. *Intervirology* 17:23–5. DOI:10.1159/000149283.

Anonymous. 1985. *International Code of Zoological Nomenclature adopted by the XV International Congress of Zoology.* London: International Trust for Zoological Nomenclature. ISBN:9780853010067.

Anonymous. 1999. *International Code of Zoological Nomenclature*, 4th ed. London: International Trust for Zoological Nomenclature. ISBN:9780853010067.

Anonymous. 2001. International Committee on Bionomenclature. *The Bulletin of Zoological Nomenclature* 58:6–7. www.biodiversitylibrary.org/item/105441#page/8/mode/1up.

Anonymous. 2005. Turkey renames "divisive" animals. *BBC News* 8 March. http://news.bbc.co.uk/2/hi/europe/4328285.stm.

Anonymous. 2011. *The draft BioCode.* www.bionomenclature.net/biocode2011.html.

Anonymous. 2012. *International Code of Nomenclature for algae, fungi, and plants (Melbourne Code)* [...]. Königstein: Koeltz Scientific Books. www.iapt-taxon.org/nomen/main.php.

Anonymous. 2013. *The International Code of Virus Classification and Nomenclature.* https://talk.ictvonline.org/information/w/ictv-information/383/ictv-code.

Arber, A. 1938. *Herbals: Their origin and evolution. A chapter in the history of botany, 1470–1670.* London: Cambridge University Press. ISBN:9781108016711.

Aref'ev, V. A., and L. A. Lisovenko. 1995. [*English–Russian defining dictionary of genetic terms.*] Moscow: VNIRO. ISBN:5853821326 (in Russian).

References

Ascherson, P., and A. Engler. 1895. Versammlung deutscher Naturforscher und Aerzte in Wien. *Osterreichische Botanische Zeitschrift* 45:27–35. www.biodiversitylibrary.org/item/91441#page/33/mode/1up.

Astrin, J. J., X. Zhou, and B. Misof. 2013. The importance of biobanking in molecular taxonomy, with proposed definitions for vouchers in a molecular context. *Zookeys* 365:67–70. DOI:10.3897/zookeys.365.5875.

Atran, S. 1990. *The cognitive foundations of natural history: Towards an anthropology of science.* New York: Cambridge University Press. ISBN:9780521438711.

Atran, S. 1998. Folk biology and the anthropology of science: Cognitive universals and cultural particulars. *Behavioral and Brain Sciences* 21:547–609. DOI:10.1017/S0140525X98001277.

Atran, S. 1999. Itzaj Maya folkbiological taxonomy. In *Cognitive universals and cultural particulars*, eds. D. L. Medin, and S. Atran, 119–205. Folkbiology. Cambridge (MA): MIT Press. www.researchgate.net/publication/290812064_Itzaj_Maya_folkbiological_taxonomy_Cognitive_universals_and_cultural_particulars.

Atran, S. 2002. Modular and cultural factors in biological understanding: an experimental approach to the cognitive basis of science. In *The cognitive basis of science*, eds. P. Carruthers, S. P. Stich, and M. Siegal, 41–72. Cambridge (UK): Cambridge University Press. ISBN:9780511613517.

Atran, S., and D. Medin. 2008. *The native mind and the cultural construction of nature.* Cambridge (MA): MIT Press. ISBN:9780262514088.

Aubert, D. 2016. Doit-on parler de "nomenclature binomiale" ou bien de "nomenclature binominale"? *La Banque des Mots* 91:7–14. https://hal.archives-ouvertes.fr/hal-01343212/document.

Azarkin, N. M. 2003. [*General history of jurisprudence.*] Moscow: Juridical Literature Publishing. ISBN:5726009932 (in Russian).

Bachmann, H. 1896. Karl Nikolaus Lang Dr. Phil. et Med. 1670–1741. *Der Geschichtsfreund* 51:167–280. ISSN:1421-2919.

Balch, F. N. 1909. A lawyer on the nomenclature question. *Science, New Series* 29:998–1000. www.jstor.org/stable/1634623?seq=1#metadata_info_tab_contents.

Banks, N., and A. N. Caudell. 1912. *The Entomological Code. A code of nomenclature for use in entomology.* Washington (DC): Judd & Detweiler. www.biodiversitylibrary.org/item/16541#page/1/mode/1up.

Barkley, T., P. DePriest, V. Funk, et al. 2004. Linnaean nomenclature in the 21st century: A report from a workshop on integrating traditional nomenclature and phylogenetic classification. *Taxon* 53:153–8. DOI:10.2307/4135501.

Barnhart, J. H. 1895. Family nomenclature. *Bulletin of the Torrey Botanical Club* 22:1–24. www.biodiversitylibrary.org/item/8000#page/14/mode/1up.

Barrett, J. A., and T. LaCroix. 2020. Epistemology and the structure of language. *Erkenntnis*. https://doi.org/10.1007/s10670-020-00225-4.

Barron, E. S., C. Sthultz, D. Hurley, et al. 2015. Names matter: Interdisciplinary research on taxonomy and nomenclature for ecosystem management. *Progress in Physical Geography* 39:640–60. https://journals.sagepub.com/toc/ppga/39/5.

Barskov, I. S., B. T. Yanin, and T. V. Keznetsova. 2004. [*Paleontological descriptions and nomenclature.*] Moscow: Moscow University Publishing. ISBN:5211049217 (in Russian).

Bartlett, H. 1940. The concept of the genus. 1. History of the generic concept in botany. *Bulletin of the Torrey Botanical Club* 67:319–62. DOI:10.2307/2481068.

Bather, F. A. 1897. A postscript on the terminology of types. *Science, New Series* 5:843–4. DOI:10.1126/science.5.126.843-a.

Bauhin, C. 1596. *ΦΥΤΟΠΙΝΑΞ, seu Enumeratio Plantarum ab Herbarijs nostro seculo descriptarum, cum earum differentijs* [...]. Basileae: Sebastianum. www.biodiversity library.org/item/30648#page/1/mode/1up.

Bauhin, C. 1623. *ΠΙΝΑΞ theatri botanici* [...] *Index in Theophrasti, Dioscoridis, Plinii et Botanicorum* [...] *Methodicè secundum earum & genera & species proponens* [...]. Basileae: Sumptibus & typis Ludovici Regii. ISBN:9785881618476 (2012 reprint).

Beaudreau, A. H., P. S. Levin, and K. C. Norman. 2011. Using folk taxonomies to understand stakeholder perceptions for species conservation. *Conservation Letters* 4:451–63. DOI:10.1111/j.1755-263X.2011.00199.x.

Bengtson, S. 1985. Taxonomy of disarticulated fossils. *Journal of Paleontology* 59:1350–8. www.jstor.org/stable/1304949.

Bengtson, S. 1988. Open nomenclature. *Palaeontology* 31:223–7. www.palass.org/sites/default/files/media/publications/palaeontology/volume_31/vol31_part1_pp223-227.pdf.

Bentham, G. 1858. Memorandum on the principles of generic nomenclature in botany, as referred to in the preceding paper. *Journal of the Proceedings of the Linnean Society, Botany* 2:30–3. DOI:10.1111/j.1095-8339.1857.tb02465a.x.

Bentham, G. 1879. On some points in botanical nomenclature. *Journal of Botany, British and Foreign* 17:45–8. www.biodiversitylibrary.org/item/35880#page/55/mode/1up.

Bentham, G. 1880. Notes on Euphorbiaceae. *Journal of the Linnean Society of London, Botany* 17:185–267. www.biodiversitylibrary.org/item/8370#page/193/mode/1up.

Benton, M. J. 2000. Stems, nodes, crown clades. and rank-free lists: Is Linnaeus dead? *Biological Review* 5:633–45. DOI:10.1111/j.1469-185x.2000.tb00055.x.

Berg, L. S. 1932. [*Freshwater fish of the USSR and neighboring countries*], Part 1, 3rd ed. Leningrad: VIORCh Publishing (in Russian).

Bergeret, J. P. 1783–1785. *Phytonomatotechnie Universelle, c'est-à-dire, l'art de donner aux plantes des noms tirés de leurs caractères* [...], Vols. 1–3. Paris: Didot et Poisson. https://gallica.bnf.fr/ark:/12148/bpt6k10575844.image.

Berlin, B. 1973. Folk systematics in relation to biological classification and nomenclature. *Annual Review of Ecology and Systematics* 4:259–71. DOI:10.1146/annurev.es.04.110173.001355.

Berlin, B. 1992. *Ethnobiological classification: Principles of categorization of plants and animals in traditional societies*. Princeton (NJ): Princeton University Press. ISBN: 9780691631004.

Berlin, B. 2007. 'Just another fish story?' Size-symbolic properties of fish names. In *Animal names*, eds. A. Minelli, G. Ortalli, and G. Sanga, 9–20. Venice: Istituto Veneto di Scienze, Lettere ed Arti. ISBN:8888143386.

Berlin, B., D. E. Breedlove, and P. H. Raven. 1973. General principles of classification and nomenclature in folk biology. *American Anthropologist* 75:214–42. DOI:10.1525/aa.1973.75.1.02a00140.

Bertrand, Y., and M. Härlin. 2008. Phylogenetic hypotheses, taxonomic sameness and the reference of taxon names. *Zoologica Scripta* 37:337–47. DOI:10.1111/j.1463-6409.2007.00323.x.

Bessey, C. E. 1892. Rochester Rules. *The American Naturalist* 26:860–1. www.biodiversitylibrary.org/item/129412#page/334/mode/1up.

Bessey, C. E. 1895. A protest against "Rochester Rules". *The American Naturalist* 29:666–8. www.biodiversitylibrary.org/item/129692#page/55/mode/1up.

Béthoux, O. 2007. Cladotypic taxonomy revisited. *Arthropod Systematics & Phylogeny* 65:127–33. www.researchgate.net/publication/228677180_Cladotypic_taxonomy_revisited.

References

Béthoux, O. 2010. Optimality of phylogenetic nomenclatural procedures. *Organisms Diversity & Evolution* 10:173–91. DOI.org/10.1007/s13127-010-0005-3.

Bílý, S., M. G. Volkovitsh, and T. C. Macrae. 2018. Case 3769—Proposed use of the plenary power to declare the pamphlet "Procrustomachia" as an unavailable work. *The Bulletin of Zoological Nomenclature* 75:220–3. www.biotaxa.org/bzn/article/view/44159.

Blackwelder, R. E. 1967. *Taxonomy. A text and reference book.* New York: John Wiley. ISBN:9780471078005.

Blackwood. W., T. Cadell, and W. Davies. 1808. [Review of] "The principles of botany and of vegetable physiology [...]". *The Edinburgh Review, or Critical Journal* 11:75–88. ISSN:1751-8482.

Blanchard, R. 1889. De la nomenclature des êtres organisés. In *Compte-Rendu des Séances du Congrès International de Zoologie*, ed. R. Blanchard, 333–424. Paris: Société Zoologique de France. https://babel.hathitrust.org/cgi/pt?id=hvd.32044106262256&view=1up&seq=359.

Blanchard, R. 1892. Deuxième rapport sur la nomenclature des êtres organisés [...]. In *Congrès International de Zoologie* [...], Part 1, 303–314. https://gallica.bnf.fr/ark:/12148/bpt6k1417923p.

Blanchard, R. 1893. Deuxième rapport sur la nomenclature des êtres organisés [...]. *Mémoires de la Société Zoologique de France* 6:126–201. www.biodiversitylibrary.org/item/38553#page/150/mode/1up.

Blumenbach, J. F. 1782. *Handbuch der Naturgeschichte.* Göttingen: Johann Christian Dieterich. ISBN:9780274455188 (2018 reprint).

Bobrov, E. G. 1970. [*Carl Linnaeus. 1707–1778.*] Leningrad: Nauka (in Russian).

Bock, H. 1546. *Neu Kreuterbuch von Underscheidt, Würckung und Namen der Kreuter, so in teutschen Landen wachsen* [...]. Strasbourg: Wendel Rihel. www.biodiversitylibrary.org/item/33579#page/2/mode/1up.

Bodson, L. 2005. Naming the exotic animals in ancient Greek and Latin. In *Animal names*, eds. A. Minelli, G. Ortalli, and G. Sanga, 441–80. Venice: Istituto Venetio di Scienze, Lettere ed Arti. ISBN:8888143386.

Boerman, A. J. 1953. Carolus Linnaeus. A psychological study. *Taxon* 2:145–56. DOI:10.2307/1216487.

Bolton, C. J. 1996. Proper names, taxonomic names and necessity. *The Philosophical Quarterly* 46:145–57. DOI:10.2307/2956383.

Bonneuil, C. 2002. The manufacture of species: Kew gardens, the empire, and the standardisation of taxonomic practices in late 19th century botany. In *Instruments, travel and science* [...], eds. M.-N. Bourguet, C. Licoppe, and O. Sibum, 189–215. London: Routledge. ISBN:9780415272957.

Borrell, B. 2007. The big name hunters. *Nature* 446:253–5. DOI:10.1038/446253a.

Bos, L. 2000. Structure and typography of virus names. *Archives of Virology* 145:429–32. DOI:10.1007/s007050050035.

Bos, L. 2002. International naming of viruses: A digest of recent developments. *Archives of Virology* 147:1471–7. DOI:10.1007/s007050050035.

Bowker, G. C. 1999. The game of the name: Nomenclatural instability in the history of botanical informatics. In *Proceedings of the 1998 Conference on the History and Heritage of Science Information Systems*, eds. M. E. Bowden, T. B. Hahn, and R. V. Williams, 74–83. Medford: Information Today. ISBN:1573870803.

Bradley, J. C. 1958. Draft of the English text of the Règles. *Bulletin of Zoological Nomenclature* 14:7–370. www.biodiversitylibrary.org/item/44295#page/11/mode/1up.

Brickell, C. D., B. R. Baum, W. L. A. Hetterscheid, et al. (eds.). 2004a. *International Code of Nomenclature for Cultivated Plants.* Utrecht: Bohn, Scheltema & Holkema. www.ishs.org/scripta-horticulturae/international-code-nomenclature-cultivated-plants.

Brickell, C. D., B. R. Baum, W. L. A. Hetterscheid, et al. 2004b. International code of nomenclature for cultivated plants: Glossary. *Acta Horticulturae* 647:85–123. DOI:10.17660/ActaHortic.2004.647.13.

Brickell, C. D., C. Alexander, J. C. David et al. (eds.). 2009. International code of nomenclature for cultivated plants [...], 8th ed. *Scripta Horticulturae* 10:1–184. www.ishs.org/news/icncp-international-code-nomenclature-cultivated-plants-9th-edition.

Brickell, C. D., C. Alexander, J. C. David et al. (eds.). 2016. *International Code of Nomenclature for Cultivated Plants [...], 9th ed.* Katwijk aan Zee (Netherlands): ISHS. ISBN:9789462611160.

Briquet, J. 1897. Règles de nomenclature pour les botanistes attachés au Jardin et au Musée royaux de Botanique de Berlin [...]. *Bulletin de l'Herbier Boissier* 5:768–79. www.biodiversitylibrary.org/item/105007#page/812/mode/1up.

Briquet, J. 1905. *Texte synoptique des documents destinés à servir de base aux débats du Congrès International de Nomenclature Botanique de Vienne 1905.* Berlin: R. Friedlinder. www.amazon.ca/-/fr/John-Briquet/dp/B00HQJHW9A.

Briquet, J. 1906. *Règles internationales de la nomenclature botanique adoptées par le Congrès International de Botanique de Vienne 1905.* Jena: Gustav Fischer. ISBN:9780265300763 (2019 reprint).

Briquet, J. 1912. *Règles internationales de la nomenclature botanique adoptées par le Congrès International de Botanique de Vienne 1905 [...] de Bruxelles 1910. Deuxième édition [...].* Jena: Gustav Fischer. www.biodiversitylibrary.org/item/78379#page/5/mode/1up.

Britten, J. 1888. Recent tendencies in American botanical nomenclature. *The Journal of Botany, British and Foreign* 26:257–62. www.biodiversitylibrary.org/item/109870#page/266/mode/1up.

Britten, J. 1895. American nomenclature. *The Journal of Botany, British and Foreign* 33:19–23, 149–52. www.biodiversitylibrary.org/item/33747#page/159/mode/1up.

Britten, J. 1898. The fifty years' limit in nomenclature. *The Journal of Botany, British and Foreign* 36:90–4. www.biodiversitylibrary.org/item/36199#page/104/mode/1up.

Britton, N. L. 1893. Proceedings of the Botanical Club, A.A.A.S., Madison meeting, August 18–22, 1893. *Bulletin of the Torrey Botanical Club* 20:360–5. www.biodiversitylibrary.org/item/7998#page/462/mode/1up.

Britton, N. L., F. V. Coville, J. M. Coulter, et al. 1892. [Rochester rules.] *Bulletin of the Torrey Botanical Club* 19:290–2. www.biodiversitylibrary.org/item/7997#page/367/mode/1up.

Brower, A. V. Z. 2010. Alleviating the taxonomic impediment of DNA barcoding and setting a bad precedent [...]. *Systematics and Biodiversity* 8:485–91. DOI:10.1080/14772000.2010.534512.

Brower, A. V. Z. 2020. Dead on arrival: A postmortem assessment of "phylogenetic nomenclature", 20+ years on. *Cladistics* 36:627–37. DOI:10.1111/cla.12432.

Brown, C. H. 1984. *Language and living things: Uniformities in folk classification and naming.* New Brunswick: Rutgers University Press. ISBN:9780813510088.

Brown, C. H. 1986. The growth of ethnobiological nomenclature. *Current Anthropology* 27:1–19. DOI:10.1086/203375.

Brown, D. E. 1991. *Human universals.* New York: McGraw-Hill. ISBN:9780070082090.

Bruford, E. A., B. Braschi, P. Denny, et al. 2020. Guidelines for human gene nomenclature. *Nature Genetics* 52:754–8. DOI:10.1038/s41588-020-0669-3.

Brummitt, R. K. 1997. Taxonomy versus cladonomy: A fundamental controversy in biological systematics. *Taxon* 46:723–34. DOI:10.2307/1224478.

References

Brummitt, R. K. and C. E. Powell. 1992. *Authors of plant names*. Kew: Royal Botanic Gardens. ISBN:0947643443.

Brunfels, O. 1530. *Herbarum vivae eicones ad naturae imitationem summa cum diligentia et artificio effigiatse [...]*. T. I. Argentorati: Ioannem Scottii. ISBN:9789333327398 (2015 reprint).

Bryant, H. N., and P. D. Cantino. 2002. A review of criticisms of phylogenetic nomenclature: Is taxonomic freedom the fundamental issue? *Biological Review* 77:39–55. DOI:10.1017/S1464793101005802.

Brzozowski, J. A. 2020. Biological taxon names are descriptive names. *History and Philosophy of the Life Sciences* 42:29. DOI:10.1007/s40656-020-00322-1.

Buchanan, R. E. 1959. The international code of nomenclature of the bacteria and viruses. *Systematic Zoology* 8:27–39. DOI:10.2307/sysbio/8.1.27.

Buchanan, R. E., R. St. John-Brooks, and R. S. Breed. 1948. International bacteriological code of nomenclature. *Journal of Bacteriology* 55:287–306. DOI:10.1128/jb.55.3.287-306.1948.

Bull, M. J., J. R. Marchesi, P. Vandamme, et al. 2012. Minimum taxonomic criteria for bacterial genome sequence depositions and announcements. *Journal of Microbiological Methods* 89:18–21. DOI:10.1016/j.mimet.2012.02.008.

Burling, L. D. 1912. The nomenclature of types. *Journal of the Washington Academy of Sciences* 2:519–20. www.biodiversitylibrary.org/item/18375#page/5/mode/1up.

Cain, A. J. 1959. The post-Linnaean development of taxonomy. *Proceedings of the Linnean Society of London* 170:234–44. DOI:10.1111/j.1095-8312.1959.tb00857.x.

Cain, A. J. 1994. Rank and sequence in Caspar Bauhin's Pinax. *Botanical Journal of the Linnean Society* 114:311–56. DOI:10.1111/j.1095-8339.1994.tb01839.x.

Camp, W. H. 1951. Biosystematy. *Brittonia* 7:113–27. DOI:10.2307/2804701.

Camp, W. H., and C. L. Gilly. 1943. The structure and origin of species, with a discussion of intraspecific variability and related nomenclatural problems. *Brittonia* 4:323–85. DOI:10.2307/2804896.

Cantino, P. D., and K. de Queiroz. 2010. *International Code of Phylogenetic Nomenclature (PhyloCode), Version 4c*. www.ohio.edu/phylocode/PhyloCode4c.pdf.

Cantino, P. D., and K. de Queiroz. 2020. *International Code of Phylogenetic Nomenclature (PhyloCode), Version 6*. Boca Raton (FL): CRC Press. ISBN:9781138332829. http://phylonames.org/code/.

Cantino, P. D., H. N. Bryant, K. de Queiroz, et al. 1999. Species names in phylogenetic nomenclature. *Systematic Biology* 48:790–807. DOI:10.1080/106351599260012.

Caprini, R. 2007. Meaning, semantics, taboo, onomasiology and etymology. In *Animal names*, eds. A. Minelli, G. Ortalli, and G. Sanga, 235–44. Venice: Istituto Veneto di Scienze, Lettere ed Arti. ISBN:8888143386.

Carnap, R. 1969. *The logical structure of the world: And pseudoproblems in philosophy*. Chicago (IL): Open Court. ISBN:9780812695236.

Carus, J. V., L. Döderlein, and K. Möbius. 1894. Berathung des zweiten Entwurfes von Regeln für die zoologische Nomenclatur im Auftrage der deutschen zoologischen Gesellschaft. *Verhandlungen der deutschen zoologischen Gesellschaft* 3:84–8. www.biodiversitylibrary.org/item/182004#page/88/mode/1up.

Cellinese, N., D. A. Baum, and B. D. Mishler. 2012. Species and phylogenetic nomenclature. *Systematic Biology* 61:885–91. DOI:10.1093/sysbio/sys035.

Cellinese, N., S. Conix, and H. Lapp. 2021. Phyloreferences: Tree-native, reproducible, and machine-interpretable taxon concepts. *EcoEvoRxiv Preprints*. https://ecoevorxiv.org/57yjs/.

Cesalpino, A. 1583. *De plantis libri XVI Andreae Cesalpini Aretini [...]* Florentiae: Georgium Marescottum. ISBN:9789333481038 (2015 reprint).

Chaikovsky, Yu. V. 2007. [The natural system and the taxonomic names.] In *The Linnaenan miscellanea*, ed. I. Ya. Pavlinov, 381–436. Moscow: Moscow University Publishing. ISSN:0134-8647 (in Russian).

Chakrabarty, P. 2010. Genetypes: A concept to help integrate molecular phylogenetics and taxonomy. *Zootaxa* 2632:67–8. DOI:10.11646/zootaxa.2632.1.4.

Chamberlain, J. R. 1992. Biolinguistic systematics and marking. In *Third International Symposium on Language and Linguistics*, 1279–93. Bangkok: Chulalongkorn University. http://sealang.net/sala/archives/pdf8/chamberlain1992biolinguistic.pdf.

Chamberlin, W. J. 1952. *Entomological nomenclature and literature*, 3rd ed. Dubuque (IA): W. C. Brown. ASIN:B0006ATD70.

Chaper, M. 1881. *De la Nomenclature des êtres organisés*. Paris: Société Zoologique de France. ISBN:9781169582378 (2010 reprint).

Chaper, M. 1889. Rapport fait au nom de la Commission de nomenclature de la Société Zoologique de France. *Compte-Rendu des Séances du Congrès International de Zoologie*, ed. R. Blanchard, 437–66. Paris: Société Zoologique de France. https://babel.hathitrust.org/cgi/pt?id=hvd.32044106262256&view=1up&seq=463.

Chebanov, S. V., and G. Ya. Martynenko. 1998. [*Semiotics of descriptive texts (typological aspect)*.] Saint Petersburg: Saint Petersburg University Publishing (in Russian).

Cheeseman, T. F. 1908. Notes on botanical nomenclature, with remarks on the rules adopted by the International Botanical Congress of Vienna. *Transactions and Proceedings of the New Zealand Institute* 40:447–65. ISSN:1176-6158.

Choi, J. H., H. J. Lee, and A. Shipunov. 2015. All that is gold does not glitter? Age, taxonomy, and ancient plant DNA quality. *PeerJ* 3:e1087. DOI:10.7717/peerj.1087.

Chomsky, N. 1987. *Language and problems of knowledge*. Cambridge (MA): The MIT Press. ISBN:9780262530705.

Chuang-tzu. 1964. *Basic writings*. Translated by Burton Watson. Washington (DC): Columbia University Press. ISBN:9780231105958.

Cianferoni, F., and L. Bartolozzi. 2016. Warning: Potential problems for taxonomy on the horizon? *Zootaxa* 4139:128–30. DOI:10.11646/zootaxa.4139.1.8.

Ciferri, R., and R. Tomaselli. 1955. The symbiotic fungi of lichens and their nomenclature. *Taxon* 4:190–2. http://www.jstor.org/stable/1216802.

Clements, F. E. 1902. Greek and Latin in biological nomenclature. *University Studies (University Nebraska)* 3:1–85. ISBN:9780548619018 (2007 reprint).

Clerck, C. 1757. *Svenska spindlar uti sina hufvudslågter indelte samt under några [...] / Aranei Svecici, descriptionibus et figuris æneis illustrate [...]*. Stockholmiae: Laurentius Salvius. www.biodiversitylibrary.org/item/209583#page/7/mode/1up.

Clusius, C. 1601. *Caroli Clusi atrebatis Rariorum Plantarum Historia [...]*. Antverpiae: Ioannem Moretum. www.botanicus.org/title/b12075048.

Cogniaux, A. 1911. Un complément aux règles de nomenclature botanique—Nomenclature horticole. *Bulletin de la Société royale de botanique de Belgique* 47:363–424. ISSN: 0037-9557.

Collar, N. J. 2000. Collecting and conservation: Cause and effect. *Bird Conservation International* 10:1–15. DOI:10.1017/S0959270900000010.

Cook, L. G., R. D. Edwards, M. D. Crisp, et al. 2010. Need morphology always be required for new species descriptions? *Invertebrate Systematics* 24:322–6. https://core.ac.uk/download/pdf/335002345.pdf.

Cook, O. F. 1898. Stability in generic nomenclature. *Science, New Series* 8:186–90. DOI:10.1126/science.8.189.186.

Cook, O. F. 1900. The method of types in botanical nomenclature. *Science, New Series* 12:475–81. DOI:10.1126/science.12.300.475.

References

Cook, O. F. 1901. Priority of place and the method of types. *Science, New Series* 13:712–13. DOI:10.1126/science.13.331.712.

Cook, O. F. 1902a. Types and synonyms. *Science, New Series* 15:646–56. DOI:10.1126/science.15.382.646.

Cook, O. F. 1902b. Types versus residues. *Science, New Series* 16:311–12. DOI:10.1126/science.16.399.311.

Cope, E. D. 1878. The report of the committee of the American Association of 1876 on biological nomenclature. *The American Naturalist* 12:517–25. www.biodiversitylibrary.org/item/128840#page/527/mode/1up.

Coquillett, D. W. 1907. The first reviser and elimination. *Science, New Series* 25:625–6. DOI:10.1126/science.25.642.625-c.

Cordus, V. 1561. *Simesusij annotationes in Pedacij Dioscoridis Anazarbei de Medica materia libros V* [...]. Argentorati: Iosias Rihelius. https://archive.org/details/mobot31753000817848?ref=ol&view=theater.

Corliss, J. O. 1962. Taxonomic-nomenclatural practices in protozoology and the new International Code of Zoological Nomenclature. *The Journal of Protozoology* 9:307–24. DOI:10.1111/j.1550-7408.1962.tb02626.x.

Corliss, J. O. 1995. The ambiregnal protists and the codes of nomenclature: A brief review of the problem and of proposed solutions. *The Bulletin of Zoological Nomenclature* 52:11–17. www.biodiversitylibrary.org/item/44798#page/29/mode/1up.

Cornelius, P. F. S. 1987. Use versus priority in zoological nomenclature: A solution for an old problem. *The Bulletin of Zoological Nomenclature* 44:79–85. ISSN:0007-5167.

Costa, C. M., and R. P. Roberts 2014. Techniques for improving the quality and quantity of DNA extracted from herbarium specimens. *Phytoneuron* 48:1–8. http://www.phytoneuron.net/2014Phytoneuron/48PhytoN-DNAextraction.pdf.

Cotterill, F. P. D., P. J. Taylor, S. Gippoliti, et al. 2014. Why one century of phenetics is enough: Response to "Are there really twice as many bovid species as we thought?" *Systematic Biology* 63:819–32. DOI:10.1093/sysbio/syu003.

Coues, E., and J. A. Allen. 1897. The Merton rules. Science, New Series 6:9–19. DOI:10.1126/science.6.131.9.

Coues, E., J. A. Allen, R. Ridgway, et al. 1886. *The code of nomenclature and check-list of North American birds* [...]. New York: American Ornithologists' Union. 392 p. www.biodiversitylibrary.org/item/16484#page/7/mode/1up.

Cowie, A. P. (ed.). 1998. *Phraseology: Theory, analysis, and applications*. Oxford: Oxford University Press. ISBN:9780198299646.

Croft, W., and A. Cruse. 2004. *Cognitive linguistics*. Cambridge (UK): Cambridge University Press. ISBN:9780511803864.

Cronquist, A. 1991. Do we know what we are doing? In *Improving the stability of names: needs and options*, ed. D. L. Hawksworth, 301–11. Königstein: Koeltz Scientific Books. ISBN:3874293289.

Crotch, M. A. 1870. On the generic nomenclature of Lepidoptera. *Cistula Entomologica* 2:59–70.

Cruz, H., and J. Smedt. 2007. The role of intuitive ontologies in scientific understanding: The case of human evolution. *Biology and Philosophy* 22:351–68.

Culberson, W. L. 1961. Proposed changes in the international code governing the nomenclature of lichens. *Taxon* 10:161–5. http://www.jstor.org/stable/1216004.

Cuvier, G. 1817. *Le règne animal distribué d'après son organisation* [...], Vols. 1–4. Paris: Chez Déterville. ISBN:9781139567107 (2012 reprint).

Dadi, T. H., B. Y. Renard, L. H. Wieler, et al. 2017. SLIMM: Species level identification of microorganisms from metagenomes. *PeerJ* 5:e3138. https://peerj.com/articles/3138/.

Dall, W. H. 1877. *Nomenclature in zoology and botany* [...]. Salem (MA): Salem Press. ISBN:9781125466865.

Dana, J. D. 1846. *Report of scientific nomenclature* [...]. New Haven (CT): B. L. Hamlen. https://bit.ly/3yGjezC.

Daston, L. 2004. Type specimens and scientific memory. *Critical Inquiry* 31:153–82. DOI:10.1086/427306.

David, P., G. Vogel, and A. Dubois. 2011. On the need to follow rigorously the rules of the code for the subsequent designation of a nucleospecies (type species) for a nominal genus which lacked one [...]. *Zootaxa* 2992:1–51. DOI:10.11646/zootaxa.2992.1.1.

Davis, K., and A. Borisenko. 2017. *Introduction to access and benefit-sharing and the Nagoya Protocol: What DNA barcoding researchers need to know*. Sofia: Pensoft. ISBN:9789546429056.

Davis, L. 2015. These scientific names were chosen purely to insult certain people. *Gizmodo*. https://gizmodo.com/these-scientific-names-were-chosen-purely-to-insult-cer-1691360201.

Dayrat, B. 2003. *Les botanistes et la flore de France: Trois siècles de découvertes*. Paris: Muséum National d'Histoire Naturelle. ISBN:2856535488.

Dayrat, B. 2010. Celebrating 250 dynamic years of nomenclatural debates. In *Systema naturae 250. The Linnaean ark*, ed. A. Polaszek, 186–239. Boca Raton (FL): CRC Press. ISBN:9781420095012.

Dayrat, B., C. Schander, and K. Angielczyk. 2004. Suggestions for a new species nomenclature. *Taxon* 53:485–91. DOI:10.2307/4135627.

Dayrat, B., P. D. Cantino, J. A. Clarke, et al. 2008. Species names in the *PhyloCode*: The approach adopted by the International Society for Phylogenetic Nomenclature. *Systematic Biology* 57:507–14. DOI:10.1080/10635150802172176.

de Candolle, A.-P. 1813. *Théorie élémentaire de la botanique, oe exposition des principes de la classification naturelle* [...]. Paris: Deterville. www.biodiversitylibrary.org/item/88297#page/7/mode/1up.

de Candolle, A.-P. 1819. *Théorie élémentaire de la botanique, oe exposition des principes de la classification naturelle* [...], 2nd ed. Paris: Deterville. ISBN:9781142242220 (2010 reprint).

de Candolle, A. 1867. *Lois de la nomenclature botanique adoptées par le Congrès international de botanique tenu à Paris en août 1867* [...]. Paris: J. B. Baillière. https://gallica.bnf.fr/ark:/12148/bpt6k981450.image.

de Candolle, A. 1868. *Laws of botanical nomenclature adopted by the International Botanical Congress held at Paris in August, 1867* [...]. London: L. Reeve. www.biodiversitylibrary.org/item/117841#page/5/mode/1up.

de Candolle, A. 1869. Réponse à diverses questions et critiques faites sur le recueil des lois de nomenclature botanique, tel que le Congrès international de 1867 l'a publié. *Bulletin de Société Botanique de France* 16:64–81. DOI:10.1080/00378941.1869.10825237.

de Candolle, A. 1880. *La phytographie, l'art de décrire les végétaux considérés sous différents points de vue*. Paris: G. Masson. www.biodiversitylibrary.org/item/69577#page/7/mode/1up.

de Candolle, A. 1883. *Nouvelles remarques sur la nomenclature botanique. Supplément au commentaire du même auteur qui accompagnait le texte des lois*. Geneva: H. Georg. www.biodiversitylibrary.org/item/221146#page/7/mode/1up.

de Candolle, A., and C. A. Cogniaux. 1876. Quelques points de nomenclature botanique. *Bulletin de la Société Royale de Botanique de Belgique* 15:477–85. www.jstor.org/stable/pdf/20790642.pdf.

References

de Queiroz, K. 1992. Phylogenetic definitions and taxonomic philosophy. *Biology and Philosophy* 4l:295–313. DOI:10.1007/BF00129972/.
de Queiroz, K. 1997. The Linnaean hierarchy and the evolutionization of taxonomy, with emphasis on the problem of nomenclature. *Aliso* 15:125–44. ISSN:0065-6275.
de Queiroz, K. 2005. Linnaean, rank-based, and phylogenetic nomenclature: Restoring primacy to the link between names and taxa. *Symbolae Botanicae Upsalienses* 33:127–40. https://repository.si.edu/bitstream/handle/10088/4506/VZ_2005deQueirozSymBotUps.pdf.
de Queiroz, K. 2007. Toward an integrated system of clade names. *Systematic Biology* 56:956–74. DOI:10.1080/10635150701656378.
de Queiroz, K. 2012. Biological nomenclature from Linnaeus to the *PhyloCode*. *Bibliotheca Herpetologica* 9:135–45. https://repository.si.edu/bitstream/handle/10088/17640/vz_2012deQueirozBibHerp.pdf?sequence=1&isAllowed=y.
de Queiroz, K., and J. Gauthier. 1990. Phylogeny as a central principle in taxonomy: Phylogenetic definitions of taxon names. *Systematic Zoology* 39:307–22. DOI:10.2307/2992353.
de Queiroz, K., and J. Gauthier. 1992. Phylogenetic taxonomy. *Annual Review of Ecology and Systematics* 23:449–80. DOI:10.1146/annurev.es.23.110192.002313.
de Queiroz, K., and J. Gauthier. 1994. Toward a phylogenetic system of biological nomenclature. *Trends in Ecology and Evolution* 9:27–31. DOI:10.1016/0169-5347(94)90231-3.
de Queiroz, K., and P. D. Cantino. 2001. Phylogenetic nomenclature and the *PhyloCode*. *The Bulletin of Zoological Nomenclature* 58:254–71. DOI:10.1201/9780429446320.
de Queiroz, K., P. D. Cantino, and J. A. Gauthier (eds.). 2020. *Phylonyms: A companion to the PhyloCode*. Boca Raton (FL): CRC Press. ISBN:9781138332935.
De Smet, W. 1973. *Initiation à la Nomenclature Biologique Nouvelle* (N.B.N.). Kalmthout: Association pour l'Introduction de la Nomenclature Biologique Nouvelle. https://uia.org/s/or/en/1100020068.
De Smet, W. 1991a. La sistemo N.B.N. (Nova Biologia Nomenklaturo). Kalmthout: Asocio por la Enkonduko de Nova Biologia Nomenklaturo. ISBN:8085853523 (2001 reprint).
De Smet, W. 1991b. Meeting user needs by an alternative nomenclature. In *Improvement of the stability of names: Needs and options*, ed. D. L. Hawksworth, 179–81. Königstein: Koeltz Scientific Books. ISBN:9783874293280.
De Vos, P., and H. G. Truper. 2000. Judicial Commission of the International Committee on Systematic Bacteriology [...]. *International Journal of Systematic and Evolutionary Microbiology* 50:2239–44. DOI:10.1099/00207713-50-6-2239.
DeCandolle, A. P., and K. Sprengel. 1821. *Elements of the philosophy of plants, containing the principles of scientific botany* [...]. London: T. Cadell. ISBN:9781108037464 (2011 reprint).
Declaration 45. Addition of recommendations to Article 73 and of the term "specimen, preserved" to the glossary. *International Commission on Zoological Nomenclature*. www.iczn.org/the-code/declaration-45-addition-of-recommendations/#:~:text=Establishing%20new%20species-group%20taxa,or%20when%20specimens%20must%20be.
Demoulin, V., D. L. Hawksworth, R. P. Korf, et al. 1981. A solution of the starting point problem in the nomenclature of fungi. *Taxon* 30:52–63. DOI:10.2307/1219390.
Devitt, M., and K. Sterelny. 1999. *Language and reality. An introduction to the philosophy of language*, 2nd ed. Cambridge (MA): MIT Press. ISBN:9780262540995.
Dioscorides. 2000. *De materia medica. Being an herbal with many other medicinal materials written in Greek in the first century of the common era* [...]. Johannesburg: IBI DIS Press. ISBN:0620234350.

Dodoens, R. 1553. *Trium priorum de stirpium historia commentariorum imagines ad vivum expressae* [...]. Antverpiae: Joanni Leoi. www.biodiversitylibrary.org/item/30650#page/1/mode/1up.

Dodoens, R. 1557. *Histoire des plantes: En laquelle est contenue la description entière des herbes herbes, c'est-à-dire leurs espèces* [...]. Antverpiae: Joanni Leoi. https://gallica.bnf.fr/ark:/12148/bpt6k534046/f4.item.

Dompere, K. K. 2009. *Fuzzy rationality. A critique and methodological unity of classical, bounded and other rationalities.* Berlin: Springer-Verlag. ISBN:9783540880837.

Donegan, T. M. 2008. New species and subspecies descriptions do not and should not always require a dead type specimen. *Zootaxa* 1761:37–48. DOI:10.11646/zootaxa.1761.1.4.

Douvillé, H. 1882a. Règles à suivre pour établir la nomenclature des espèces. In *Congrès Géologique International, compte rendu de la 2-me session, Bologne, 1881*, 592–5. Bologne: Fava et Garaniani. https://archive.org/details/compterendudela00conggoog/page/n649/mode/2up.

Douvillé, H. 1882b. Rapport fait à la chargée d'étudier la question des Règles a suivre pour établir la nomenclature des espèces In *Congrès Géologique International, compte rendu de la 2-me session, Bologne, 1881*, 596–608. Bologne: Fava et Garaniani. https://archive.org/details/compterendudela00conggoog/page/n653/mode/2up.

du Plessis, N. M. 2017. A rule for naming objects. *Biological Theory* 12:39–49. DOI:10.1007/s13752-016-0256-0

Du Rietz, G. E. 1930. The fundamental units of biological taxonomy. *Svensk Botanisk Tidskrift* 24:333–428. https://ru.scribd.com/doc/248585975/Du-Rietz-1930-the-Fundamental-Units-of-Biological-Taxonomy.

Dubois, A. 2000. Synonymies and related lists in zoology: General proposals, with examples in herpetology. *Dumerilia* 4:33–98. ISSN:1256-7779.

Dubois, A. 2005. Proposed rules for the incorporation of nomina of higher-ranked zoological taxa in the International Code of Zoological Nomenclature. 1. Some general questions, concepts and terms of biological nomenclature. *Zoosystema* 27:365–426. https://sciencepress.mnhn.fr/sites/default/files/articles/pdf/z2005n2a8.pdf.

Dubois, A. 2006a. New proposals for naming lower-ranked taxa within the frame of the International Code of Zoological Nomenclature. *Comptes Rendus Biologies* 329:823–40. DOI:10.1016/j.crvi.2006.07.003.

Dubois, A. 2006b. Proposed rules for the incorporation of nomina of higher-ranked zoological taxa in the International Code of Zoological Nomenclature. 2. The proposed rules and their rationale. *Zoosystema* 28:165–258. www.researchgate.net/publication/265399833_Proposed_Rules_for_the_incorporation_of_nomina_of_higher-ranked_zoological_taxa_in_the_International_Code_of_Zoological_Nomenclature_2_The_proposed_Rules_and_their_rationale.

Dubois, A. 2007. Phylogeny, taxonomy and nomenclature: The problem of taxonomic categories and of nomenclatural ranks. *Zootaxa* 1519:27–68. DOI:10.11646/zootaxa.1519.1.3.

Dubois, A. 2008a. Zoological nomenclature: Some urgent needs and problems. In *Future Trends of Taxonomy*, 15–18. Carvoeiro (Portugal): EDIT.

Dubois, A. 2008b. A partial but radical solution to the problem of nomenclatural taxonomic inflation and synonymy load. *Biological Journal of the Linnean Society* 93:857–63.

Dubois, A. 2008c. Phylogenetic hypotheses, taxa and nomina in zoology. *Zootaxa* 1950:51–86. DOI:10.11646/zootaxa.1950.1.7.

Dubois, A. 2010a. Bionomina, a forum for the discussion of nomenclatural and terminological issues in biology. *Bionomina* 1:1–10. DOI:10.11646/bionomina.1.1.1.

Dubois, A. 2010b. Retroactive changes should be introduced in the Code only with great care: Problems related to the spellings of nomina. *Zootaxa* 2426:1–42. DOI:10.11646/zootaxa.2426.1.1.

Dubois, A. 2010c. Zoological nomenclature in the century of extinctions: Priority *vs.* "usage". *Organisms Diversity & Evolution* 10:259–74. DOI:10.1007/s13127-010-0021-3.

Dubois, A. 2011a. The International Code of Zoological Nomenclature must be drastically improved before it is too late. *Bionomina* 2:1–104. www.zin.ru/animalia/coleoptera/pdf/dubois_2011_bionomina_iczn.pdf.

Dubois, A. 2011b. The rich but confusing terminology of biological nomenclature: A first step towards a comprehensive glossary. *Bionomina* 3:1–23. DOI:10.11646/bionomina.3.1.6.

Dubois, A. 2011c. Describing a new species. *Taprobanica: The Journal of Asian Biodiversity* 2:6–24. DOI:10.4038/tapro.v2i1.2703.

Dubois, A. 2012. The distinction between introduction of a new nomen and subsequent use of a previously introduced nomen in zoological nomenclature. *Bionomina* 5:57–80. www.biotaxa.org/Bionomina/article/view/171.

Dubois, A. 2013. Zygoidy, a new nomenclatural concept. *Bionomina* 6:1–26. DOI:10.11646/bionomina.6.1.1.

Dubois, A. 2015. The Duplostensional Nomenclatural System for higher zoological nomenclature. *Dumerilia* 5:1–108. www.researchgate.net/publication/287800820_The_Duplostensional_Nomenclatural_System_for_higher_zoological_nomenclature.

Dubois, A. 2017a. Diagnoses in zoological taxonomy and nomenclature. *Bionomina* 12:63–85. DOI:10.11646/bionomina.12.1.8.

Dubois, A. 2017b. A few problems in the generic nomenclature of insects and amphibians, with recommendations for the publication of new generic nomina in zootaxonomy [...]. *Zootaxa* 4237:1–16. DOI: 10.11646/zootaxa.4237.1.1.

Dubois, A. 2017c. The need for reference specimens in zoological taxonomy and nomenclature. *Bionomina* 12:4–38. DOI:10.11646/bionomina.12.1.2.

Dubois, A. 2020a. Allocation of nomina to taxa in zoological nomenclature. *Bionomina* 18. www.mapress.com/j/bn/article/view/bionomina.18.1.1. DOI:10.11646/bionomina.18.1.1.

Dubois, A. 2020b. Nomenclatural consequences of the *Oculudentavis khaungraae* case, with comments on the practice of "retraction" of scientific publications. *Zoosystema* 42:475–82. DOI:10.5252/zoosystema2020v42a23.

Dubois, A., and A. Nemésio. 2007. Does nomenclatural availability of nomina of new species or subspecies require the deposition of vouchers in collections? *Zootaxa* 1409:1–22. DOI:10.11646/zootaxa.1409.1.1.

Dubois, A., A. M. Bauer, L. M. P. Ceríaco, et al. 2019. The *Linz ZooCode* project: A set of new proposals regarding the terminology, the principles and rules of zoological nomenclature [...]. *Bionomina* 17:1–111. www.mapress.com/j/bn.

Dubois, A., A. Nemésio, and R. Bour. 2014. Primary, secondary and tertiary syntypes and virtual lectotype designation in zoological nomenclature, with comments on the recent designation of a lectotype for *Elephas maximus* Linnaeus, 1758. *Bionomina* 7:45–64. http://citeseerx.ist.psu.edu/viewdoc/download?doi=10.1.1.643.6938&rep=rep1&type=pdf.

Dubois, A., A. Ohler, and R. A. Pyron. 2021. New concepts and methods for phylogenetic taxonomy and nomenclature in zoology, exemplified by a new ranked cladonomy of recent amphibians (Lissamphibia). *Megataxa* 5:1–738. DOI:10.11646/MEGATAXA.5.1.1.

Dubois, A., E. Aescht, and E. C. Dickinson. 2016. Burning questions and problems of zoological nomenclature. The Linz International Workshop of Zoological Nomenclature (9–10 July 2014). *Dumerilia* 6:24–34. DOI:10.11646/bionomina.17.1.1.

Dubois, A., P.-A. Crochet, E. C. Dickinson, et al. 2013. Nomenclatural and taxonomic problems related to the electronic publication of new nomina and nomenclatural acts in zoology, with brief comments on optical discs and on the situation in botany. *Zootaxa* 3735:1–94. DOI:10.11646/zootaxa.3735.1.1.

Dunning, J. W. 1872. On the relation between generic and specific names. *The Entomologist's Monthly Magazine* 8:290–294. www.biodiversitylibrary.org/item/102841#page/624/mode/1up.

Dupérré, N. 2020. Old and new challenges in taxonomy: What are taxonomists up against? *Megataxa* 1:59–62. DOI: 10.11646/megataxa.1.1.12.

Dupuis, C. 1974. Pierre André Latreille (1762–1833). The foremost entomologist of his time. *Annual Review of Entomology* 19:1–14. DOI:10.1146/annurev.en.19.010174.000245.

Durand, E. J. 1909. A discussion of some of the principles governing the interpretation of pre-Persoonian names [...]. *Science, New Series* 29:670–6. DOI:10.1126/science.29.747.670.

Dyke, G. J., and J. D. Sigwart. 2007. A search for a 'smoking gun'. No need for an alternative to the Linnaean system of classification. In *Animal names*, eds. A. Minelli, G. Ortalli, and G. Sanga, 49–65. Venice: Istituto Veneto di Scienze, Lettere ed Arti. ISBN:8888143386.

Earle, F. S. 1904. The necessity for reform in the nomenclature of the fungi. *Science, New Series* 19:508–10. DOI:10.1126/science.19.482.508.

Ebach, M. C., J. Morrone, L. R. Parenti, et al. 2008. International Code of Area Nomenclature. *Journal of Biogeography* 35:1153–7. DOI:10.1111/j.1365-2699.2008.01920.x.

Editorial. 2012. Amendment of Articles 8, 9, 10, 21 and 78 of the International Code of Zoological Nomenclature to expand and refine methods of publication. Zootaxa 3450:1–7. www.mapress.com/zootaxa/2012/f/zt03450p007.pdf.

Editorial. 2014. Zoological nomenclature and electronic publication—A reply to Dubois et al. (2013). *Zootaxa* 3779:3–5. DOI:10.11646/zootaxa.3779.1.2.

Edwards, W. H. 1873. Some remarks on entomological nomenclature. *The Canadian Entomologist* 5:21–36. www.biodiversitylibrary.org/item/22227#page/31/mode/1up.

Ehrlich, P. R. 1961. Systematics in 1970: Some unpopular predictions. *Systematic Zoology* 10:157–8. DOI:10.2307/2411612.

Elk, S. B. 2004. *A new unifying biparametric nomenclature that spans all of chemistry*. Amsterdam: Elsevier. ISBN:0444516859.

Ellen, R. F. 1993. *The cultural relations of classification. An analysis of Nuaulu animal categories from Central Seram.* Cambridge (UK): Cambridge University Press. ISBN:9780521431149.

Ellen, R. F. 2008. *The categorical impulse: Essays on the anthropology of classifying behavior.* Oxford (UK): Berghahn Books. ISBN:9781845451554.

Ellerman, J. R., and T. C. S. Morrison-Scot. 1951. *Checklist of Palaearctic and Indian mammals: 1758 to 1946.* London: Trustees of the British Museum (Natural History). ISBN:9781174907746 (2011 reprint).

Ellis, R. 2008. Rethinking the value of biological specimens: Laboratories, museums and the Barcoding of Life Initiative. *Museum and Society* 6:172–91. www.lancaster.ac.uk/fass/projects/taxonomy/docs/ellis.pdf.

Endersby, J. 2008. *Imperial nature, Joseph Hooker and the practice of Victorian science.* Chicago (IL): University of Chicago Press. ISBN:9780226207926.

Engler, A., I. Urban, A. Garcke, et al. 1897. Nomenclaturregeln für die Beamten des Königlichen Botanischen Gartens und Museums zu Berlin. *Notizblatt des Königlichen botanischen Gartens und Museums zu Berlin* 1:245–50. www.biodiversitylibrary.org/item/91425#page/291/mode/1up.

References

Engler, A., I. Urban, K. Schumann, et al. 1902. Zusätze zu den Berliner Nomenclatur-Regeln. *Botanische Jahrbücher für Systematik, Pflanzengeschichte und Pflanzengeographie* 31:24–5. www.biodiversitylibrary.org/item/693#page/1/mode/1up.

Ereshefsky, M. 1994. Some problems with the Linnaean hierarchy. *Philosophy of Science* 61:186–205. www.jstor.org/stable/188208.

Ereshefsky, M. 1997. The evolution of the Linnaean hierarchy. *Biology and Philosophy* 12:493–519. DOI:1 0.1023/A:1006556627052.

Ereshefsky, M. 2001a. *The poverty of the Linneaean hierarchy: A philosophical study of biological taxonomy.* New York: Cambridge University Press. ISBN:9780521038836.

Ereshefsky, M. 2001b. Names, numbers and indentations: A guide to post-Linnaean taxonomy. *Studies in History and Philosophy of Science* 32:361–83. DOI:10.1016/S1369-8486(01)00004-8.

Ereshefsky, M. 2002. Linnaean ranks: Vestiges of a bygone era. *Philosophy of Science* 9:305–15. DOI:10.1086/341854.

Ereshefsky, M. 2007a. Foundational issues concerning taxa and taxon names. *Systematic Biology* 56:295–301. DOI:10.1080/10635150701317401.

Ereshefsky, M. 2007b. The evolution of the Linnaean hierarchy. *Philosophy of Science* 12:493–519. DOI:10.1023/A:1006556627052.

Ereshefsky, M. 2017. *Species.* In *The Stanford encyclopedia of philosophy*, ed. E. N. Zalta. https://plato.stanford.edu/archives/fall2017/entries/species/.

Erikssont, M., L. Jeppsson, C. F. Bergman, et al. 2000. Paranomenclature and the rules of zoological nomenclature, with examples from fossil polychaete jaws (scolecodonts). *Micropaleolntology* 46:186–8. www.jstor.org/stable/1486156.

Erxleben, J. C. P. 1777. *Systema regni animalis per classes* [...]. *Classis I. Mammalia.* Lipsiae: Weygandianis. www.biodiversitylibrary.org/item/53898#page/7/mode/1up.

Evans, K. 2020. *Change species names to honor indigenous peoples, not colonizers.* www.scientificamerican.com/article/change-species-names-to-honor-indigenous-peoples-not-colonizers-researchers-say/.

Evans, K. M., and D. G. Mann. 2009. A proposed protocol for nomenclaturally effective DNA barcoding of microalgae. *Phycologia* 48:70–4. DOI:10.2216/08-70.1.

Evenhuis, N. L. 2008a. A compendium of zoological type nomenclature: A reference source. *Bishop Museum Technical Report* 41:1–23. www.semanticscholar.org/paper/A-Compendium-of-Zoological-Type-Nomenclature%3A-a-Evenhuis/98db1b6d6a7cbe662959959a177bcc708b108b38.

Evenhuis, N. L. 2008b. The "Mihi itch": A brief history. *Zootaxa* 1890:59–68. www.mapress.com/zootaxa/2008/f/zt01890p068.pdf,

Evenhuis, N. L., and T. Pape. 2010. Alfonso L. Herrera (1868–1942) and his little-known new system of naming animals and plants, with special reference to Diptera genus-group names. *Historical Dipterology*:17–26. www.researchgate.net/publication/283348235_Alfonso_L_Herrera_1868-1942_and_his_little-known_new_system_of_naming_animals_and_plants_with_special_reference_to_Diptera_genus-group_names.

Fabricius, I. C. 1778. *Philosophia entomologica sistens scientiae fundamenta* [...]. Hamburgi et Kilonii: Carol. Ernest. Bonnii. www.biodiversitylibrary.org/item/41699#page/7/mode/1up.

Fabricius, I. C. 1798. *Supplementum entomologiae systematicae.* Hafniae: Proft & Storck. www.biodiversitylibrary.org/item/132638#page/5/mode/1up.

Fabricius, I. C. 1801. *Systema eleutheratorum secundum ordines, genera, species* [...], Vol. 1. Kiliae: Bibliopol. Acad. Novi. https://books.google.ru/books/about/Systema_Eleutheratorum.html?id=-z6AAQAACAAJ&redir_esc=y.

Farber, R. L. 1976. The type-concept in zoology in the first half of the nineteenth century. *The Journal of the History of Biology* 9:93–119. DOI:10.1007/BF00129174.

Federhen, S. 2014. Type material in the NCBI Taxonomy Database. *Nucleic Acids Research* D1:D1086–D1098.

Felt, E. P. 1934. Classifying symbols for insects. *Journal of the New York Entomological Society* 42:373–92. www.biodiversitylibrary.org/item/205825#page/405/mode/1up.

Felt, E. P., T. D. A. Cockerell, and E. L. Troxell. 1930. Scientific names. *Science, New Series* 71:215–18. DOI:10.1126/science.71.1834.215-a.

Fernald, M. L. 1901a. Some recent publications and the nomenclatorial principles they represent. *Botanical Gazette* 31:183–97. www.jstor.org/stable/2465281.

Fernald, M. L. 1901b. The instability of the Rochester nomenclature. *Botanical Gazette* 32:359–67. www.jstor.org/stable/2465236.

Ficetola, G. F., C. Miaud, F. Pompanon, et al. 2008. Species detection using environmental DNA from water samples. *Biology Letters* 4:423–5. DOI:10.1098/rsbl.2008.0118.

Fischer, H. 1966. Conrad Gessner (1516–1565) as bibliographer and encyclopedist. *The Library, Fifth Series* 21:269–81. DOI:10.1093/library/s5-XXI.4.269.

Fischer, M. 1892. Rules of nomenclature adopted by the International Zoological Congress held in Paris, France, 1889. *The American Naturalist* 26:383–8. www.biodiversitylibrary.org/item/127888#page/424/mode/1up.

Fisher, W. K. 1905. A new Code of Nomenclature. *The Condor* 7:28–30. DOI:10.2307/1361355.

Flann, C., J. McNeill, F. R. Barri, et al. 2015. Report on botanical nomenclature—Vienna 2005 […]. *PhytoKeys* 45:1–341. DOI:10.3897/phytokeys.45.9138.

Flesness, N. R. 2003. International Species Information System (ISIS): Over 25 years of compiling global animal data to facilitate collection and population management. *International Zoo Yearbook* 38:53–61. DOI:10.1111/j.1748-1090.2003.tb02064.x.

Flores, R., J. W. Randles, M. Bar-Joseph, et al. 1998. A proposed scheme for viroid classification and nomenclature. *Archives of Virology* 143:623–9. DOI:10.1007/s007050050318.

Foucault, M. 1970. *The order of things. An archaeology of the human sciences.* New York: Pantheon Books. ISBN:9780679753353.

Francki, K. I. B., C. M. Fauquet, D. L. Knudson, et al. 1991. Classification and nomenclature of viruses—Fifth report of the international committee on taxonomy of viruses. *Archives of Virology*, Suppl. 2. ISBN:9783709191637.

Fransson, M. 2008. Report on the Life Science Identifier (LSID). *BBMRI WP5 team meeting, deCODE, Reykjavik, Iceland.* http://old.bbmri-eric.eu/documents/10181/68479/WP5_2008_09_Reykjavik_LSID_MF.pdf/1f2481f6-b1f4-4c8b-9c69-4b33fa156a44?version=1.0.

Franz, N. M., C. Zhang, and J. Lee. 2017. A logic approach to modeling nomenclatural change. *Cladistics* 34:336–57. DOI:10.1111/cla.12201.

Franz, N. M., M. Chen, P. Kianmajd, et al. 2016. Names are not good enough: Reasoning over taxonomic change in the *Andropogon* complex. *Semantic Web* 7:645–67. DOI:10.3233/SW-160220.

Friedmann, E. I., and L. J. Borowitzka. 1982. The symposium on taxonomic concepts in blue-green algae: Towards a compromise with the Bacteriological Code? *Taxon* 31:673–83. DOI:10.2307/1219683.

Fries, E. M. 1821–1829 (1832). *Systema micologicum: Sistens fungorum ordines, genera et species* […], Vols. 1–3. Lundae: Berlingiana; Gryphiswaldae: Mauritii. www.biodiversitylibrary.org/bibliography/5378.

Friis, I. 2019. G. C. Oeder's conflict with Linnaeus and the implementation of taxonomic and nomenclatural ideas in the monumental Flora Danica project (1761–1883). *Gardens' Bulletin Singapore* 71 (Suppl. 2):53–85. DOI:10.26492/gbs71(suppl.2).2019-07.

References

Frizzell, D. L. 1933. Terminology of types. *American Midland Naturalist* 14:637–68. DOI:10.2307/2420124.

Frolov, A. O., and A. O. Kostygov. 2013. [Protozoa, protists and protoctists in the eukaryotic system.] In *Contemporary problems of biological systematics*, eds. A. F. Alimov, and S. D. Stepanyanz, 250–71. Saint Petersburg: Zoological Institute. ISBN:9785873175895 (in Russian).

Gadow, H. 1893. *Vögel, Vol. II. Systematische Teil*. Amsterdam: T. J. Van Holkema. https://archive.org/details/vgel02gado.

Garrity, G. M., and C. Lyons. 2003. Future-proofing biological nomenclature. *Journal of Integrative Biology* 7:31–3. DOI:10.1089/153623103322006562.

Gaydenko, P. P. 2003. [*Scientic rationality and philosophical mind.*] Moscow: Progress-Traditsia. ISBN:5898261427 (in Russian).

Gentry, A., J. Clutton-Brock, and C. P. Groves. 2004. The naming of wild animal species and their domestic derivatives. *Journal of Archaeological Science* 31:645–51. DOI:10.1016/j.jas.2003.10.006.

Gesner, C. 1560. *Nomenclator aquatilium animantium. Icones animalium aquatilium in mari* […]. Zurich: Christoph. Froschoverum. ISBN:9785519152907 (2015 reprint).

Ghiselin, M. T. 1995. Ostensive definitions of the names of species and clades. *Biology and Philosophy* 10:219–22. DOI:10.1007/BF00852246.

Gibbs, A. J. 2000. Virus nomenclature descending into chaos. *Archives of Virology* 145:1505–7. DOI:10.1007/s007050070108.

Gibbs, A. J. 2003. Virus nomenclature, where next? *Archives of Virology* 148:1645–53. DOI:10.1007/s00705-003-0150-5.

Gibbs, A. J., B. D. Harrison, D. H. Watson, et al. 1966. What's in a virus name? *Nature* 209:450–4. DOI:10.1038/209450a0.

Gill, T. 1896. Some questions of nomenclature. *Science, New Series* 4:581–601. DOI:10.1126/science.4.95.581.

Gillman, L. N., and S. D. Wright. 2020. Restoring indigenous names in taxonomy. *Communications Biology* 3:609. DOI:10.1038/s42003-020-01344-y.

Gmelig-Nijboer, C.A. 1977. *Conrad Gessner's "Historia animalium": An inventory of Renaissance zoology*. Meppel (Netherlands): Krips Repro. ASIN:B0024TMTF4.

Gmelin, J. F. 1792. *The animal kingdom, or zoological system of the celebrated Sir Charles Linnaeus. Class I. Mammalia* […]. Edinburgh: A. Strahan, T. Cadell, & W. Creech. www.biodiversitylibrary.org/item/119041#page/9/mode/1up.

Gontier, N. 2009. The origin of the social approach in language and cognitive research exemplified by studies into the origin of language. In *Language and social cognition*, ed. H. Pishwa, 25–46. Berlin: De Gruyter Mouton. https://doi.org/10.1515/9783110216080.1.25.

Gould, M. D. 1843. Notice of some works, recently published, on the nomenclature of zoology. *The American Journal of Science and Arts* 45:1–12. www.biodiversitylibrary.org/item/52474#page/19/mode/1up.

Gradstein, S. R., M. Sauer, W. Braun, et al. 2001. TaxLink, a program for computer-assisted documentation of different circumscriptions of biological taxa. *Taxon* 50:1075–84. DOI:10.2307/1224722.

Gray, A. 1864. Nomenclature. *The American Journal of Science and Arts* 37:278–81. www.biodiversitylibrary.org/item/113611#page/286/mode/1up.

Gray, A. 1879. *Structural botany, or organography on the basis of morphology* […], 6th ed. New York: Blakeman, Taylor. www.biodiversitylibrary.org/item/62493#page/7/mode/1up.

Gray, A. 1883. Some points in botanical nomenclature: A review of "Nouvelle remarques sur la nomenclature botanique, par M. Alph. de Candolle", Geneva, 1888. *The American Journal of Science, 3rd Series* 26:417–437. www.biodiversitylibrary.org/item/120103#page/426/mode/1up.

Gray, S. F. 1821. *Natural arrangement of British plants, according to their relations to each other* [...], Vol. 1. London: Baldwin, Cradock & Joy. www.biodiversitylibrary.org/item/196374#page/7/mode/1up.

Graybeal, A. 1995. Naming species. *Systematic Biology* 44:237–50. DOI:10.1093/sysbio/44.2.237.

Green, M. L. 1927. History of plant nomenclature. *Bulletin of Miscellaneous Information (Royal Gardens, Kew)* 10:403–15. DOI:10.2307/2399589.

Greene, E. L. 1891a. Dr. Kuntze and his reviewers. *Pittonia* 2:263–81. www.biodiversitylibrary.org/item/52475#page/269/mode/1up.

Greene, E. L. 1891b. The Berlin protest. *Pittonia* 2:283–7. www.biodiversitylibrary.org/item/52475#page/290/mode/1up.

Greene, E. L. 1896. Some fundamentals of nomenclature. *Science, New Series* 3:13–16. DOI:10.1126/science.3.53.13.

Greene, E. L. 1909. *Landmarks of botanical history. A study of certain epochs in the development of the science of botany*, Part 1: Prior to 1562. Washington (DC): Smithsonian Institution. ISBN:9780804710756.

Greuter, W. 2004. Recent developments in international biological nomenclature. *Turkish Journal of Botany* 28:17–26. https://journals.tubitak.gov.tr/botany/issues/bot-04-28-1-2/bot-28-1-2-3-0211-13.pdf.

Greuter, W., D. L. Hawksworth, J. McNeill, et al. 1996. Draft *BioCode*: The prospective international rules for the scientific names of organisms. *Taxon* 45:349–72. DOI:10.2307/1224691.

Greuter, W., D. L. Hawksworth, J. McNeill, et al. 1998. Draft *BioCode* (1997): The prospective international rules for the scientific names of organisms. *Taxon* 47:127–50. DOI:10.2307/1224030.

Greuter, W., G. Garrity, D. L. Hawksworth, et al. 2011. Draft *BioCode*: Principles and rules regulating the naming of organisms. *Taxon* 60:201–12. DOI:10.1002/tax.601019.

Griffiths, A. J. 1981. A numericlature of the yeasts. *Antonie Van Leeuwenhoek* 47:547–63. DOI:10.1007/BF00443241.

Griffiths, G. C. D. 1973. Some fundamental problems in biological classification. *Systematic Biology* 22:338–43. DOI:10.2307/2412942.

Griffiths, G. C. D. 1976. The future of Linnean nomenclature. *Systematic Zoology* 25:168–73. DOI:10.2307/2412743.

Guérin-Méneville, F. E. 1843. Rapport d'une commission nommée par l'Association britannique pour l'avancement de la science [...]. *Revue Zoologique (1843)*:202–10. www.biodiversitylibrary.org/item/19446#page/210/mode/1up.

Gumperz, J. J., and S. C. Levinson. 1996. *Rethinking linguistic relativity*. Cambridge (UK): Cambridge University Press. ISBN:9780521448901.

Haber, M. H. 2012. How to misidentify a type specimen. *Biology and Philosophy* 27:767–84. DOI:10.1007/s10539-012-9336-0.

Hahn, G. 1981. Comment on the proposed amendments to the International Code of Zoological Nomenclature regarding ichnotaxa. *The Bulletin of Zoological Nomenclature* 38:93–4. https://archive.org/details/biostor-76394/mode/2up.

Hallier, H. 1900. Über Kautschuklianen und andere Apocyneen, nebst Bemerkungen über Hevea und einem Versuch zur Lösung der Nomenklaturfrage. *Jahrbücher der Hamburgischen wissenschaftlichen Anstalten* 17:19–216. ISBN:9781012203610 (2019 reprint).

Härlin, M. 2005. Definitions and phylogenetic nomenclature. *Proceedings of the California Academy of Sciences* 56, Suppl. I:216–24. www.researchgate.net/publication/263352844_Definitions_and_Phylogenetic_Nomenclature_Definitions_and_Phylogenetic_Nomenclature.

References

Härlin, M., and P. Sundberg. 1998. Taxonomy and philosophy of names. *Biology and Philosophy* 13:233–44. DOI:10.1023/A:1006583910214.
Harrison, B. 1973. *Form and content.* Hoboken: Blackwell. ISBN:9780631150305.
Harrison, P. 2006. The Bible and the emergence of modern science. *Science & Christian Belief* 18:115–32. DOI:10.1017/CHO9781139048781.029.
Harrison, P. 2009. Linnaeus as a second Adam? Taxonomy and the religious vocation. *Zygon* 44:879–93. DOI:10.1111/j.1467-9744.2009.01039.x.
Hart, C. R., and J. H. Long. 2011. Animal metaphors and metaphorizing animals: An integrated literary, cognitive, and evolutionary analysis of making and partaking of stories. *Evolution: Education and Outreach* 4:52–63. DOI:10.1007/s12052-010-0301-6.
Harting, P. 1871. Skizze eines rationellen Systems der zoologisches Nomenclature. *Archive für Naturgeschichte* 1:25–41. www.biodiversitylibrary.org/item/30468#page/37/mode/1up.
Häuser, C. L., A. Steiner, J. Holstein, et al. (eds.). 2005. *Digital imaging of biological type specimens. A manual of best practice* [...]. Stuttgart: Staatliches Museum für Naturkunde. ISBN:300017240-8.
Hawksworth, D. L. 1974. *Mycologist's handbook: An introduction to the principles of taxonomy and nomenclature in the fungi and lichens.* Kew: Commonwealth Mycological Institute. ISBN:9780851983066.
Hawksworth, D. L. (ed.). 1991. *Improvement the stability of names: Needs and options.* Königstein: Koeltz Scientific Books. ISBN:9783874293280.
Hawksworth, D. L. 2001. The naming of fungi. In: *Systematics and Evolution. The Mycota,* Vol. 7B, eds. D. J. McLaughlin, E. G. McLaughlin, and P. A. Lemke, 171–92. Berlin: Springer. DOI:10.1007/978-3-662-10189-6_6.
Hawksworth, D. 2002. The names behind the names. *Field Mycology* 3:15–19. DOI:10.1016/s1468-1641(10)60123-5.
Hawksworth, D. L. 2010. *Terms used in bionomenclature. The naming of organisms (and plant communities), including terms used in botanical* [...] *zoological nomenclature.* Copenhagen: Global Biodiversity Information Facility. www.gbif.org/ru/document/80577/terms-used-in-bionomenclature-the-naming-of-organisms-and-plant-communities.
Hawksworth, D. 2011. A new dawn for the naming of fungi: impacts of decisions made in Melbourne in July 2011 on the future publication and regulation of fungal names. *IMA Fungus* 2:155–62. DOI:10.5598/imafungus.2011.02.02.06.
Hawksworth, D. L., D. S. Hibbett, P. M. Kirk, et al. 2016. Proposals to permit DNA sequence data to serve as types of names of fungi. *Taxon* 65:899–900. DOI:10.12705/654.31.
Hawksworth, D. L., J. McNeill, P. H. A. Sneath, et al. 1994. Towards a harmonized bionomenclature for life on Earth [...]. *The Bulletin of Zoological Nomenclature* 51:188–216. www.biodiversitylibrary.org/item/44552#page/214/mode/1up.
Hawksworth, D. L., P. W. Crous, S. A. Redhead, et al. 2011. The Amsterdam Declaration on Fungal Nomenclature. *IMA Fungus* 2:105–12. DOI:10.5598/imafungus.2011.02.01.14. Epub 2011 Jun 7.
Hawksworth, D. L., T. W. May, and S. A. Redhead. 2017. Fungal nomenclature evolving: Changes adopted by the 19th International Botanical Congress in Shenzhen 2017, and procedures for the Fungal Nomenclature Session at the 11th International Mycological Congress in Puerto Rico 2018. *IMA Fungus* 8:211–18. DOI:10.5598/imafungus.2017.08.02.01.
Hedwig, J. 1801. *Species muscorum frondosorum: Descriptae et tabulis aeneis lxxvii coloratis illustratae.* Lipsiae: J. A. Barthi. https://archive.org/details/mobot31753002081708.
Heise, H., and M. P. Starr. 1968. Nomenifers: Are they christened or classified? *Systematic Zoology* 17:458–67. DOI:10.2307/2412043.

Heller, J. L. 1964. The early history of binomial nomenclature. *Huntia* 1:33–70. www.huntbotanical.org/admin/uploads/06hibd-huntia-1-pp33-70.pdf.
Hemming, F. 1957. *Copenhagen decisions on zoological nomenclature* […]. London: International Trust for Zoological Nomenclature. www.biodiversitylibrary.org/item/105616#page/7/mode/1up.
Hemming, F. 1958. Official text of the "Règles Internationales de la Nomenclature Zoologique" (International Code of Zoological Nomenclature) […]. *The Bulletin of Zoological Nomenclature* 14:i–xxviii. www.biodiversitylibrary.org/item/44295#page/27/mode/1up.
Hennebert, G. L., and W. Gams. 2003. Possibilities to amend or delete Article 59 of the International Code of Botanical Nomenclature to achieve a unified nomenclature and classification of the fungi. *Mycotaxon*. www.mycotaxon.com/resources/HennebertGams2003.pdf.
Hennig, W. 1969. *Die Stammesgeschichte der Insekten*. Frankfurt: Verlag Waldemar Kramer. ISBN:9780471278481.
Heppel, D. 1981. The evolution of the Code of Zoological Nomenclature. In *History in the service of systematics* […], eds. A. Wheeler, and J. H. Price, 135–41. London: Society for the Bibliography of Natural History. ISBN:9780901843050.
Herrando-Pérez, S., B. W. Brook, and C. J. A. Bradshaw. 2014. Ecology needs a convention of nomenclature. *BioScience* 64:311–21. DOI:10.1093/biosci/biu013.
Herrera, A. L. 1899. About a reform in nomenclature. *Science, New Series* 10:120–1. DOI:10.1126/science.10.239.120.
Herschel, J. F. 1830. *A preliminary discourse on the study of natural philosophy*. London: Longman. ISBN:9780511692727 (2009 reprint).
Hetterscheid, W. L. A., and W. A. Brandenburg. 1995. The culton concept: Setting the stage for an unambiguous taxonomy of cultivated plants. *Acta Horticulturae* 413:29–34. DOI:10.17660/ActaHortic.1995.413.5.
Hetterscheid, W. L. A., R. G. van der Berg, and W. A. Brandenburg. 1996. An annotated history of the principles of cultivated plant classification. *Acta Botanica Neerlandica* 45:123–34. DOI:10.1111/j.1438-8677.1996.tb00504.x.
Heywood, V. H. 1991. Needs for stability of nomenclature in conservation. In *Improving the stability of names: Needs and options*, ed. D. L. Hawksworth, 53–8. Königstein: Koeltz Scientific Books. ISBN:3874293289.
Hiern, W. P. 1878. On a question of botanical nomenclature. *Journal of Botany, British and Foreign* 16:72–4. www.biodiversitylibrary.org/item/35887#page/84/mode/1up.
Hill, J. H., and B. Mannheim. 1992. Language and world view. *Annual Review of Anthropology* 21:381–406. DOI:10.1146/annurev.an.21.100192.002121.
Hitchcock, A. S. 1905. Nomenclatorial type specimens of plant species. *Science, New Series* 21:828–32. DOI:10.1126/science.21.543.828.
Hitchcock, A. S. 1921. The type concept in systematic botany. *American Journal of Botany* 8:251–5. www.biodiversitylibrary.org/item/181550#page/311/mode/1up.
Hitchcock, A. S. 1922. [Type-basis code.] *The Journal of Botany, British and Foreign* 60:316–18. www.biodiversitylibrary.org/item/33755#page/372/mode/1up.
Hoffmann, H., and A. Roggenkamp. 2003. Population genetics of the nomen species *Enterobacter cloacae*. *Applied and Environmental Microbiology* 69:5306–18. DOI:10.1128/aem.69.9.5306-5318.2003.
Holland, P. 2015. *Pliny's natural history in thirty-seven books*. Salt Lake City (UT): Andesite Press. ISBN:9781297491566.
Holman, E. W. 2007. How comparable are categories in different phyla? *Taxon* 56:179–84. DOI: 10.2307/25065749.

References

Hołyński, R. B. 2020. Strict nomenclatural rules or subjective "best taxonomic practices": Is the Code a confusing factor? *Procrustomachia* 5:61–6. ISSN:2543-7747.

Hong, S.-B., S.-W. Kwon, and W.-G. Kim. 2012. Introductions of the New Code of Fungal Nomenclature and recent trends in transition into one fungus/one name system. *The Korean Journal of Mycology* 40:73–7. DOI:10.4489/KJM.2012.40.2.73.

Hopkinson, J. 1907. Dates of publication of the separate parts of Gmelin's edition (13th) of the "Systema Naturae" of Linneus. *Proceedings of the Zoological Society of London* 69:1035–7. www.biodiversitylibrary.org/item/98530#page/715/mode/1up.

Horton, T., L. Marsh, B. J. Bett, et al. 2021. Recommendations for the standardisation of open taxonomic nomenclature for image-based identifications. *Frontiers in Marine Science* 8:62. DOI:10.3389/fmars.2021.620702.

Hughes, N. F. 1989. *Fossils as information*. Cambridge (UK): Cambridge University Press. ISBN:0521366569.

Hull, D. L. 1966. Phylogenetic numericlature. *Systematic Zoology* 15:14–17. DOI:10.2307/sysbio/15.1.14.

Hull, D. L. 1968. The syntax of numericlature. *Systematic Zoology* 17:472–4. DOI:10.1093/sysbio/17.4.472.

Hull, D. L. 1983. Exemplars and scientific change. *Proceedings of the Biennial Meeting of the Philosophy of Science Association (PSA 18)* 2:479–503. ISBN:9780917586194.

Hussey, C., Y. de Jong, and D. Remsen. 2008. Actual usage of biological nomenclature and its implications for data integrators; a national, regional and global perspective. *Zootaxa* 1950:5–8. DOI:10.11646/zootaxa.1950.1.3.

International Commission. 2008. Proposed amendment of Articles 8, 9, 10, 21 and 78 of the International Code of Zoological Nomenclature to expand and refine methods of publication. *Zoological Journal of the Linnean Society* 154:848–55. DOI:10.1111/j.1096-3642.2008.00518.x.

International Commission. 2014. Zoological nomenclature and electronic publication—A reply to Dubois et al. (2013). *Zootaxa* 3779:003–005. DOI:10.11646/zootaxa.3779.1.2.

Issak, M. 2020. Curiosities of biological nomenclature. www.curioustaxonomy.net/puns/puns.html.

Jackson, B. D. 1887. A new "Index of plant-names." *Journal of Botany, British and Foreign* 25:66–71, 150–1. www.biodiversitylibrary.org/item/109218#page/71/mode/1up.

Jackson, J. A., L. Laikre, C. S. Baker, et al. 2012. Guidelines for collecting and maintaining archives for genetic monitoring. *Conservation Genetics Resources* 4:527–36. DOI:10.1007/s12686-011-9545-x.

Jahn, T. L. 1961. Man versus machine: A future problem in protozoan taxonomy. *Systematic Zoology* 10:179–92. DOI:10.2307/2411616.

Janick, J. 2003. Herbals: The connection between horticulture and medicine. *HortTechnology* 13:229–38. DOI:10.21273/HORTTECH.13.2.0229.

Jardine, W. 1858. *Memoirs of Hugh Edwin Strickland*. London: John van Roost. ISBN:9781108037693 (2011 reprint).

Jardine, W. 1866. Report of a Committee "appointed to report on the changes which they may consider desirable to make desirable, if any, in the Rules of Zoological Nomenclature […]". In *Report of the thirty-fifth meeting of the British Association for the Advancement of Science (1865)*, 25–42. London: John Murray. www.biodiversitylibrary.org/item/93098#page/96/mode/1up.

Jarvis, C. E. 1992. The Linnaean Plant Name Typification Project. *Botanical Journal of the Linnean Society* 109:503–13. DOI:10.1111/j.1095-8339.1992.tb01447.x.

Jeffrey, C. 1992. *Biological nomenclature*, 3rd ed. Cambridge (UK): Cambridge University Press. ISBN:9780521427753.

Jeffrey, C. 2003. Theoretical and practical problems in the classification and nomenclature of cultivated plants, with examples from Cucurbitaceae and Compositae. In *Rudolf Mansfeld and plant genetic resources*, ed. H. Knüpffer, and J. Ochsmann, 51–9. Bonn: ZADI. www.genres.de/fileadmin/SITE_MASTER/content/Schriftenreihe/Band22_ Gesamt.pdf.

Jordan, D. S. 1900. The first species named as the type of the genus. *Science, New Series* 12:785–7. DOI:10.1126/science.12.308.785.

Jordan, D. S. 1905. The method of elimination in fixing generic types in zoological nomenclature. *Science, New Series* 22:598–601. DOI:10.1126/science.22.567.598.

Jordan, D. S. 1907. The "first species" and the "first reviser". *Science, New Series* 25:467–9. DOI:10.1126/science.25.638.467-b.

Jordan, D. S. 1911. The use of numerals for specific names in systematic zoology. *Science, New Series* 33:370–3. DOI:10.1126/science.22.567.598.

Jørgensen, P. M. 1991. Difficulties in lichen nomenclature. *Mycotaxon* 40:497–501. ISSN:0093-4666.

Jørgensen, P. M. 1997. Lichen phototypes, nature's unmanageable misprints? *Taxon* 46:721–2. DOI:10.2307/1224477.

Jörger, K. M., and M. Schrödl. 2013. How to describe a cryptic species? Practical challenges of molecular taxonomy. *Frontiers in Zoology* 10:59. DOI:10.1186/1742-9994-10-59.

Jungius, J. 1662. *Doxoscopiae physicae minores, sive isagoge physica doxoscopica* [...]. Hamburgi: Johannis Naumanni. ASIN:B07R1YHBX3 (2019 reprint).

Jungius, J. 1747. *Opuscula botanico-physica* [...]. Coburgi: Georgii Ottonis. ISBN: 9781173363604 (2011 reprint).

Jussieu, A. L. 1773. Examen de la famille des renoncules. *Histoire de l'Académie Royale des Siences. Année 1773*: 214–40. https://gallica.bnf.fr/ark:/12148/bpt6k3572b/ f362.item.

Jussieu, A. L. 1789. *Genera plantarum secundum ordines naturales disposita* [...] Parisiis: Herissant et Theophilum Barrois. www.biodiversitylibrary.org/item/7125#page/ 1/mode/1up.

Kaiser, H., B. I. Crother, C. M. R. Kelly, et al. 2013. Best practices: In the 21st century, taxonomic decisions in herpetology are acceptable only when supported by a body of evidence and published via peer review. *Herpetological Review* 44:8–23. http:// hdl.handle.net/2436/621767.

Keller, R. A., R. N. Boyd, and Q. D. Wheeler. 2003. The illogical basis of phylogenetic nomenclature. *Botanical Review* 69:93–110. DOI:10.1663/0006-8101(2003)069[0093:TIBOP N]2.0.CO;2.

Kennedy, J. B., R. Kukla, and T. Paterson. 2005. Scientific names are ambiguous as identifiers for biological taxa: Their context and definition are required for accurate data integration. In *Data integration in the life sciences*, eds. B. Ludäscher, and L. Raschid, 80–95. Berlin: Springer-Verlag. DOI:10.1007/11530084_8.

Ketchum, M. S., P. W. Wojtkiewicz, A. B. Watts, et al. 2012. Novel North American hominins, next generation sequencing of three whole genomes and associated studies. https:// heavy.com/wp-content/uploads/2013/10/novel-north-american-hominins-final-pdf-download.pdf.

Kiesenwetter, E. 1858. Gesetze der entomologischen Nomenclatur. *Berliner Entomologische Zeitschrift* 2:xi–xvi. DOI:10.1002/mmnd.18580020316.

Kim, J. Y., K. S, Kim, M. G. Lockley, et al. 2008. Hominid ichnotaxonomy: An exploration of a neglected discipline. *Ichnos* 15:126–39. DOI:10.1080/10420940802467868.

King, A. M. Q., M. J. Adams, E. B. Carstens, et al. (eds.). 2011. *Virus taxonomy. Classification and nomenclature of viruses* [...]. London: Academic Press. ISBN:9780123846846.

References

Kinman, K. E. 1994. *The Kinman system: Toward a stable cladisto-eclectic classification of organisms, living and extinct, 48 phyla, 269 classes, 1,719 orders.* Hays (KA): K. E. Kinman. ASIN:B0000EHVS4.

Kirby, W. 1802. *Monographia apum Angliae* […], Vol. 2. Ipswich: J. Raw. www.biodiversity library.org/item/41176#page/7/mode/1up.

Kirby, W. F. 1892. On "type-specimens" and "type-figures" in entomology. *Science, New Series* 20:244–5. DOI:10.1126/science.ns-20.508.244-a.

Klein, J. Th. 1753. *Tentamen methodi ostracologicae sive dispositio naturalis Cochlidum et Concharum* […]. Lugdini Batavorum: G. J. Wishoff. www.biodiversitylibrary.org/item/203339#page/9/mode/1up.

Kluge, N. Yu. 1996. Myths in insect systematics and principles of zoological nomenclature. *Entomologicheskoye Obozrenie* 75:939–44. ISSN:0367-1445.

Kluge, N. Yu. 1999a. [A system of alternative nomenclatures of supraspecific taxa.] *Entomologicheskoye Obozrenie* 78:224–43. ISSN:0367-1445 (in Russian).

Kluge, N. Yu. 1999b. [Linnean and post-Linnean systematics of supraspecific taxa and new principles of nomenclature.] *Russian Ornithological Journal, Express Issue*:3–21. ISSN:0869-4362 (in Russian).

Kluge, N. Yu. 2020. [*Systematics of insects and principles of cladoendesis*], Vol. 1. Moscow: KMK Science Press (in Russian). ISBN:9785907213715.

Knapp, M., and M. Hofreiter. 2010. Next generation sequencing of ancient DNA: Requirements, strategies and perspectives. *Genes* (Basel) 1:227–43. DOI:10.3390/genes1020227.

Knapp, S., G. Lamas, E. N. Lughadha, et al. 2004. Stability or stasis in the names of organisms: The evolving codes of nomenclature. *Philosophical Transactions: Biological Sciences* 359:611–22. DOI:10.1098/rstb.2003.1445.

Koerner, L. 1996. Carl Linnaeus in his time and place. In *Cultures of natural history*, eds. N. Jardine, J. A. Secord, and E. C. Spary, 145–63. Cambridge (UK): Cambridge University Press. ISBN:9780521558945.

Kolosova, V. B. 2009. [*Lexicon and symbolism of Slavic folk botany. Ethnolinguistic aspect*]. Moscow: Indrik. ISBN:9785916740264 (in Russian).

Komárek J., and S. Golubić. 2004. *Guide to the nomenclature and formal taxonomic treatment of oxyphototroph prokaryotes (Cyanoprokaryotes)*. http://www.cyanodb.cz/files/CyanoGuide.pdf.

Korf, R. P. 1983. Sanctioned epithets, sanctioned names, and cardinal principles in "Pers." and "Fr." citations. *Mycotaxon* 16:341–52. ISSN:0093-4666.

Kosko, B. 1993. *Fuzzy thinking: The new science of fuzzy logic*. New York: Hyperion. ISBN:9780786880218.

Kraus, O. 2008. The Linnean foundations of zoological and botanical nomenclature. *Zootaxa* 1950:9–20. DOI:10.11646/zootaxa.

Kravetz, A. S. 2001. [Rigid designator.] *Vestnik Voronezh State University, Ser. 1* 2:94–127 (in Russian). ISSN:1995-5502.

Krell, F.-T. 2016. Preserve specimens for reproducibility. *Nature* 538:168. DOI:10.1038/539168b.

Krell, F.-T. 2020. Comment (Case 3769)—The journal "Procrustomachia" is available for nomenclatural purposes and should not be suppressed. *The Bulletin of Zoological Nomenclature* 77:89–91. DOI:10.21805/bzn.v77.a029.

Krell, F.-T., and S. A. Marshall. 2017. New species described from photographs: Yes? no? sometimes? A fierce debate and a new declaration of the ICZN. *Insect Systematics and Diversity* 1:3–19. DOI:10.1093/isd/ixx004.

Kripke, S. A. 1972. *Naming and necessity*. Cambridge (MA): Harvard University Press. ISBN:9780631128014.

Kronestedt, T. 2010. Carl Clerck and what became of his spiders and their names. In *European arachnology 2008*, eds. W. Nentwig, M. Entling, and C. Kropf, 105–117. Bern: Natural History Museum. ISSN:1660-9972.

Kubanin, A. A. 2001. [Analysis of the basic principles of the nomenclature of higher taxa by example of Briozoa.] *Paleontological Journal* 35:157–65. ISSN:0031-031X (in Russian).

Kubryakova, E. S. 2004. [*Language and knowledge. The role of language in the knowing the world.*] Moscow: Institute of Language RAS. ISBN:5944571748 (in Russian).

Kuntner, M., and I. Agnarsson. 2006. Are the Linnean and phylogenetic nomenclatural systems combinable? Recommendations for biological nomenclature. *Systematic Biology* 55:774–84. DOI:10.1080/10635150600981596.

Kuntze, O. 1891. *Revisio generum plantarum vascularium omnium* [...], Part 1. Leipzig: A. Felix.

Kuntze, O. 1893. *Revisio generum plantarum secundum leges nomenclature internationales* [...], Part 3. Leipzig: A. Felix. www.biodiversitylibrary.org/item/7553#page/2/mode/1up.

Kuntze, O. 1900. The advantages of 1737 as a starting-point of botanical nomenclature. *The Journal of Botany, British and Foreign* 38:7–10. www.biodiversitylibrary.org/item/108946#page/14/mode/1up.

Kupriyanov, A. V. 2005. [*Prehistory of biological systematics.*] Saint Petersburg: Saint Petersburg European University. ISBN:5943800433 (in Russian).

Lamarck, J.-B. 1778. *Flore française, ou description succinte de toutes les plantes* [...], Vols. I–III. Paris: l'Imprimerie Royale. www.biodiversitylibrary.org/item/197779#page/11/mode/1up.

Lamarck, J.-B. 1798. *Encyclopédie méthodique botanique*, Vol. 4, Part 2, 498–9. Paris: H. Agasse. https://babel.hathitrust.org/cgi/pt?id=hvd.32044102813847&view=1up&seq=512.

Lamarck, J.-B. 1809. *Philosophie zoologique, ou exposition* [...], Vol. 1, 2. Paris: Dentu & Auteur. ISBN:9780226468105 (2011 reprint).

Lamarck, J.-B. 1815. *Histoire naturelle des animaux sans vertèbres*, Vol. 1. Paris: Verdière. ISBN:9781139567411 (2013 reprint).

Lampman, A. M. 2010. How folk classification interacts with ethnoecological knowledge: A case study from Chiapas, Mexico. *Journal of Ecological Anthropology* 14:39–51. DOI:10.5038/2162-4593.14.1.3.

Lane, R. P., and J. Marshall. 1981. Geographic variation, races and subspecies. In *The evolving biosphere*, ed. P. Forey, 9–19. London: British Museum (Natural History).

Lang, C. N. 1722. *Methodus nova et facilix testacea marina* [...]. Lucernae: Wyssing. ISBN:9780656448159.

Lanham, U. 1965. Uninominal nomenclature. *Systematic Zoology* 14:144. DOI:10.2307/2411739.

Lapage, S. P., P. H. A. Sneath, E. F. Lessel, et al. (eds.). 1992. *International Code of Nomenclature of Bacteria* [...]. Washington (DC): ASM Press. ISBN:155581039X.

LaPorte, J. 2003. Does a type specimen necessarily or contingently belong to its species? *Biology and Philosophy* 18:583–8. DOI:10.1023/A:1025559319279.

LaPorte, J. 2018. Rigid designators. In *The Stanford encyclopedia of philosophy*, ed. E. N. Zalta. https://plato.stanford.edu/archives/spr2018/entries/rigid-designators/.

Larson, J. L. 1967. Linnaeus and the natural method. *Isis* 58:304–20. DOI:10.1086/350265.

Larson, J. L. 1971. *Reason and experience: The representation of natural order in the work of Carl von Linné*. Berkeley (CA): University California Press. ISBN:9780520018341.

Latham, J. 1790. *Index ornithologicus, sive, Systema ornithologiae* [...], Vol. 1. Londini: Sumptibus authoris. www.biodiversitylibrary.org/item/226470#page/5/mode/1up.

References

Latreille, P. A. 1801. *Histoire naturelle, générale et particulière des crustacés et des insectes*, Vol. 1. Paris: F. Dufart. www.biodiversitylibrary.org/item/80055#page/5/mode/1up.

Latreille, P. A. 1806. *Genera crustaceorum et insectorum secundum ordinem naturalem* [...], Vol. 1. Parisiis & Argentorati: Amand Koenig. www.biodiversitylibrary.org/item/132620#page/7/mode/1up.

Laurin, M. 2007. The advantages of phylogenetic nomenclature over Linnean nomenclature. In *Animal names*, eds. A. Minelli, G. Ortalli, and G. Sanga, 67–98. Venice: Istituto Veneto di Scienze, Lettere ed Arti. ISBN:8888143386.

Lee, M. S. Y., and A. Skinner. 2007. Stability, ranks, and the *PhyloCode*. *Acta Palaeontologica Polonica* 52:643–50. http://citeseerx.ist.psu.edu/viewdoc/download?doi=10.1.1.720.399&rep=rep1&type=pdf.

Legré, M. L. 1897. La botanique en provence au XVIE siècle: Mathias De Lobel et Pierre Pena. *Bulletin de la Société Botanique de France* 44, suppl. 1: xi–xlvii. DOI:10.1080/00378941.1897.10839641.

Leigh, G. J., H. A. Favre, and W. V. Metanomski. 1998. *Principles of chemical nomenclature. A guide to IUPAC recommendations*. London: Blackwell Science. ISBN:0865426856.

Lendemer, J. C. 2011. Changes to the International Code for Botanical Nomenclature passed in Melbourne: A lichenological explainer. *Opuscula Philolichenum* 10:6–13. http://sweetgum.nybg.org/images3/415/759/OP10_p2.pdf.

Leske, N. G. 1788. *Anfangsgründe der Naturgeschichte des Thierreichs*. Vienna: Christian Friedrich Wappler. https://reader.digitale-sammlungen.de/de/fs1/object/display/bsb10308030_00005.html.

Levine, A. 2001. Individualism, type specimens, and the scrutability of species membership. *Biology and Philosophy* 16:325–38. DOI:10.1023/A:1010674915907.

Lévi-Strauss, C. 1966. *The savage mind*. London: Weidenfeld & Nicolson. ISBN:9780226474847.

Lewis, W. A. 1871. On the application of the maxim "communis error facit jus" to scientific nomenclature. *The Entomologist's Monthly Magazine* 8:1–5. www.biodiversitylibrary.org/item/102841#page/335/mode/1up.

Lewis, W. A. 1872. *A discussion of the law of priority in entomological nomenclature* [...]. London: Williams & Norgate. ISBN:9781230008592 (2013 reprint).

Lewis, W. A. 1875. On entomological nomenclature and the rule of priority. *Transactions of the Entomological Society of London (1875)*, Appendix: i–xlii. www.biodiversitylibrary.org/item/50986#page/367/mode/1up.

Lindbeck, A. 1975. The changing role of the national state. *Kyklos* 28:23–46. DOI:10.1111/j.1467-6435.1975.tb01932.x.

Lindley, J. 1830. *An outline of the first principles of botany*. London: Longman. ISBN:9781165261482 (2010 reprint).

Lindley, J. 1832. *An introduction to botany*, Vol. 1. London: Longman. www.biodiversitylibrary.org/item/61746#page/7/mode/1up.

Link, H. F. 1798. *Philosophiae botanicae novae, seu institutionum phytographicorum prodromus*. Gottingae: Christ. Dietrich. https://gallica.bnf.fr/ark:/12148/bpt6k98423h?rk=21459;2.

Linnaeus, C. 1735. *Systema naturae, sive, regna tria naturae* [...]. Lugdini Batavorum: Theodorum Haak. ASIN:B07QXCN4B1 (2016 reprint).

Linnaeus, C. 1736. *Fundamenta botanica: quae majorum operum prodromi instar theoriam scientiae botanices* [...]. Amstelodami: Salomonem Schouten. ISBN:9789060460641 (2010 reprint).

Linnaeus, C. 1737a. *Critica botanica in qua nomina plantarum generica, specifica, & variantia* [...]. Lugduni Batavorum: Conradum Wishoff. ISBN:9781165930630 (2010 reprint).

Linnaeus, C. 1737b. *Genera plantarum eorumque characteres naturales secundum numerum, figuram, situm* [...]. Lugduni Batavorum: C. Wishoff; G. J. Wishoff. ISBN:9781294082880 (2013 reprint).

Linnaeus, C. 1751. *Philosophia botanica in qua explicantur fundamenta botanica cum definitionibus partium* [...]. Holmiae: Godofr. Kiesewetter. ISBN:9781104629731 (2009 reprint).

Linnaeus, C. 1753. *Species plantarum exhibentes plantas rite cognitas ad genera relatas* [...], Vols. I, II. Holmiae: Laurentii Salvii. ISBN:9785519064231 (2014 reprint).

Linnaeus, C. 1758–1759. *Systema naturae per regna tria naturae* [...] *Editio decima reformata*, Vols. I, II. Holmiae: Laurentii Salvii. www.biodiversitylibrary.org/item/10277#page/3/mode/1up.

Linnaeus, C. 1766–1767. *Systema naturae per regna tria naturae* [...] *Editio duodecima reformata*, Vols. I, II. Holmiae: Laurentii Salvii. www.biodiversitylibrary.org/item/137337#page/5/mode/1up.

Linnaeus, C. 1788. *Amoenitas academiae, seu, dissertationes variae physicae, medicae botanicae* [...], Vol. Quintum, 2nd ed. Arlangae: Iacobi Palm. www.biodiversitylibrary.org/item/15497#page/1/mode/1up.

Linnaeus, C. 1789. *Entomologia, fauna suecicae descriptionibus aucta* [...]. Vol. 1. Lugduni: Piestre et Delamolliere. www.biodiversitylibrary.org/item/47834#page/11/mode/1up.

Linnaeus, C. 1824. *Species plantarum exhibentes plantas rite cognitas ad genera relatas* [...], 4th ed., Vol. VI, Part I. Berolini: G. C. Nauk. www.biodiversitylibrary.org/item/14571#page/1/mode/1up.

Linnaeus, C. 1938. *The "Critica Botanica" of Linnaeus. Translated by the late Sir Arthur Hort* [...]. London: Ray Society. ASIN:B001172O78.

Linnaeus, C. 2003. *Linnaeus' Philosophia botanica. English edition, translated by Stephen Freer*. Oxford: Oxford University Press. ISBN:9781104629731.

Linné, C. 1775. *The elements of botany: Containing the history of the science: With accurate definitions of all the terms of art* [...]. London: T. Cadel & M. Hingestone. ISBN:9781332974092 (2018 reprint).

Linsley, E. G., and R. L. Usinger. 1959. Linnaeus and the development of the International Code of Zoological Nomenclature. *Systematic Zoology* 8:39–47. DOI:10.2307/sysbio/8.1.39.

List. 2020. *List of long species names*. https://en.wikipedia.org/wiki/List_of_long_species_names.

Little, F. J. 1964. The need for a uniform system of biological numericlature. *Systematic Zoology* 13:191–4. DOI:10.2307/sysbio/13.1-4.191.

Löbl, I. 2017. Assessing biodiversity: A pain in the neck. *Bionomina* 12:39–43. DOI:10.11646/bionomina.12.1.3.

Loftin, R. W. 1992. Scientific collecting. *Environmental Ethics* 14:253–64. DOI:10.5840/enviroethics199214320.

Long, K. J. 1996. Botany in medieval Latin. In *Medieval Latin: An introduction and bibliographical guide*, eds. F. A. C. Mantello, and A. G. Rieg, 401–6. Washington (DC): The Catholic University of America Press. ISBN:9780813208428.

Losev, A. F. 1990. [*The philosophy of name.*] Moscow: Moscow University Publishing (in Russian).

Ludwig, D. 2017. Indigenous and scientific kinds. *The British Journal for the Philosophy of Science* 68:187–212. DOI:10.1093/bjps/axv031.

Lwoff, A. 1964. The new provisional committee on nomenclature of viruses. *International Journal of Systematic and Evolutionary Microbiology* 14:53–6. DOI:10.1099/0096266X-14-1-53.

Lyubarsky, G. Yu. 2018. [*Origins of hierarchy: The history of taxonomic rank.*] Moscow: KMK Sci. Press. ISBN 9785950082962 (in Russian).

Maat, J. 2004. *Philosophical languages in the seventeenth century: Dalgarno, Wilkins, Leibniz.* Dordrecht: Kluwer. ISBN:9781402017582.

Magnol, P. 1689. *Prodromus historiae generalis plantarum in quo familiae plantarum [...].* Monspelij: Gabrielis & Honorati Pech. https://gallica.bnf.fr/ark:/12148/bpt6k6244284d.

Maiden, J. H. 1906. International rules for botanical nomenclature, adopted by the International Botanical Congress, Vienna, 1905. *The Journal of Botany, British and Foreign* 44:74–94. ISBN:9781230146560 (2013 reprint)

Maiden, J. H. 1907. International rules for botanical nomenclature (adopted by the International Botanical Congress, Vienna, 1905). *Journal and Proceedings Royal Society of New South Wales for 1906* 40:74–94. www.biodiversitylibrary.org/page/41577712#page/106/mode/1up.

Malécot, V. 2008. Les règles de nomenclature. In *Biosystema 25. Linnaeus, systématique et biodiversité*, eds. M. Veuille, J.-M. Drouin, P. Deleporte, et al., 41–76. Paris: Société Française de Systématique. https://hal-agrocampus-ouest.archives-ouvertes.fr/hal-00729760.

Mandrioli, M. 2008. Insect collections and DNA analyses: How to manage collections? *Museum Management and Curatorship* 23:193–9. DOI:10.1080/09647770802012375.

Markova, E. M. 2008. [Proto-Slavic names of trees as a reflection of a fragment of the linguistic world picture of the Slavs.] *Acta Linguistica* 2:37–45. ISSN:1313-2296 (in Russian).

Marsh, O. C. 1898. The value of type specimens and importance of their preservation. *Geological Magazine* 5:548–52. www.cambridge.org/core/journals/geological-magazine/article/abs/vthe-value-of-typespecimens-and-importance-of-their-preservation1/462B6CBB172B74F50FC5474B7904A88F.

Marshall, S. A., and N. L. Evenhuis. 2015. New species without dead bodies: A case for photo-based descriptions, illustrated by a striking new species of *Marleyimyia* Hesse (Diptera, Bombyliidae) from South Africa. *Zookeys* 525:117–27. DOI:10.3897/zookeys.525.6143.

Martin, G. 2006. The impact of frozen tissue and molecular collections on natural history museum collections. *NatSCA News* 10:31–47. www.natsca.org/sites/default/files/publications/NatSCA%20News%20Issue%2010-12.pdf.

Matthews, R. E. F. 1983. The history of viral taxonomy. In *Critical appraisal of viral taxonomy*, ed. R. E. F. Matthews, 219–45. Boca Raton (FL): CRC Press. ISBN:9781315892122.

Matthews, S. C. 1973. Notes on open nomenclature and on synonymy lists. *Palaeontology* 16:713–19. www.palass.org/sites/default/files/media/publications/palaeontology/volume_16/vol16_part4_pp713-719.pdf.

Maxwell, S. J. 2019. A description of a new endemic carnivorous Marsupialia in Myrtoideae Forests of Australia: A taxonomic misadventure with phototypes. *Research in Zoology* 9:12–15. DOI:10.5923/j.zoology.20190901.03.

May, T. W., S. A. Redhead, L. Lombard, et al. 2018. XI International Mycological Congress: Report of Congress action on nomenclature proposals relating to fungi. *IMA Fungus* 9:xxii–vii. DOI:10.1007/BF03449448.

Mayo, M. A., and M. C. Horzinek. 1998. A revised version of the International Code of Virus Classification and Nomenclature. *Archive of Virology* 143:1645–54. DOI:10.1007/s007050050406.

Mayr, E. 1942. *Systematics and the origin of species, from the viewpoint of zoologist.* New York: Columbia University Press. ISBN:9780674862500.

Mayr, E. 1963. *Animal species and evolution.* Cambridge (MA): Harvard University Press. ISBN:9780674865327.

Mayr, E. 1969. *Principles of systematic zoology.* New York: McGraw Hill. ISBN:9780070411432.

Mayr, E. 1988. *Toward a new philosophy of biology*. New York: Cambridge University Press. ISBN:9780674896666.
McNeill, J. 2004. Nomenclature of cultivated plants: A historical botanical standpoint. *Acta Horticulturae* 634:29–36. wwwlib.teiep.gr/images/stories/acta/Acta%20634/634_2.pdf.
McNeill, J. 2008. The taxonomy of cultivated plants. *Acta Horticulturae* 799:21–8. DOI:10.17660/ActaHortic.2008.799.1.
McNeill, J., and N. J. Turland. 2011. Synopsis of proposals on botanical nomenclature Melbourne 2011: A review of the proposals [...] to the XVIII International Botanical Congress. *Taxon* 60:243–86. DOI:10.1002/tax.601033.
McOuat, G. R. 1996. Species, rules and meaning: The politics of language and the ends of definitions in 19th century natural history. *Studies in the History and Philosophy of Science* Pt. A, 21:413–519. DOI:10.1016/0039-3681(95)00060-7.
Measey, J. 2013. Taxonomic publishing, vandalism and best practice: *African Journal of Herpetology* makes changes that will safeguard authors. *African Herp News* 60:2–4. ISSN:1017-6187.
Meierotto, S., M. J. Sharkey, D. H. Janzen, et al. 2019. A revolutionary protocol to describe understudied hyperdiverse taxa and overcome the taxonomic impediment. *Deutsche Entomologische Zeitschrift* 66:119–45. DOI:10.3897/dez.66.34683.
Meiklejohn, K. A., N. Damaso, and J. M. Robertson. 2019. Assessment of BOLD and GenBank—Their accuracy and reliability for the identification of biological materials. *PLoS ONE* 14:e0217084. DOI:10.1371/journal.pone.0217084.
Melville, R. V. 1981. International Code of Zoological Nomenclature. Deferment of proposal to introduce provisions to regulate paranomenclature. *The Bulletin of Zoological Nomenclature* 38:166–7. www.biodiversitylibrary.org/item/44480#page/200/mode/1up.
Melville, R. V. 1995. *Toward stability in the names of animals. A history of the Commission on Zoological Nomenclature, 1895–1995*. London: Internat. Trust for Zoological Nomenclature. ISBN:9780853010050.
Melville, R. V., and J. D. D. Smith. 1987. Official lists and indexes of names and works in zoology. London: International Trust for Zoological Nomenclature. www.biodiversitylibrary.org/item/20019#page/5/mode/1up.
Mequignon, A. 1932. Latreille et le génotype. In *Société Entomologique de France. Le livre du Centenaire*, 149–56. Paris: Firmin & Didot. ASIN:B0000DTMHQ.
Merriam, C. H. 1895. Unity of nomenclature in zoölogy and botany. *Science, New Series* 1:161–2. DOI:10.1126/science.1.6.161.
Merriam, C. H. 1897. Type specimens in natural history. *Science, New Series* 5:731–2. DOI:10.1126/science.5.123.731-a.
Meyen, S. V., and Yu. A. Shreyder. 1976. [Methodological issues of the of theory of classification.] *Voprosy Philosophii* 12:67–79. ISSN:0042-8744 (in Russian).
Michener, C. D. 1963. Some future developments in taxonomy. *Systematic Zoology* 12:151–72. DOI:10.2307/2411757/.
Michener, C. D. 1964. The possible use or uninominal nomenclature to increase the stability of names in biology. *Systematic Zoology* 13:182–90. DOI:10.2307/sysbio/13.1-4.182.
Mielke, A., and K. Zuberbühler. 2013. A method for automated individual, species and call type recognition in free-ranging animals. *Animal Behaviour* 86:475–82. DOI:10.1016/j.anbehav.2013.04.017.
Miguel, V. 2020. The promise of next-generation taxonomy. *Megataxa* 1:35–8. DOI:10.11646/megataxa.1.1.6.
Mikhailov, K. E. 1991. Classification of fossil egg shells of amniotic vertebrates. *Acta Palaeontologica Polonica* 36:193–238. www.researchgate.net/publication/285159952_Classification_of_fossil_eggshells_of_amniotic_vertebrates.

References

Mikhailov, K. E., E. S. Bray, and K. F. Hirsch. 1996. Parataxonomy of fossil egg remains (Veterbrata): Principles and applications. *Journal of Vertebrate Paleontology* 16:763–9. DOI:10.1080/02724634.1996.10011364.

Miller, W. 1899. A practical reform in the nomenclature of cultivated plants. *Botanical Gazette* 28:264–8. www.jstor.org/stable/2465408.

Minelli, A. 1995. The changing paradigms of biological systematics: New challenges to the principles and practice of biological nomenclature. *The Bulletin of Zoological Nomenclature* 52:303–9. www.biodiversitylibrary.org/item/44798#page/333/mode/1up.

Minelli, A. 2008. Zoological *vs.* botanical nomenclature: A forgotten "BioCode" experiment from the times of the Strickland Code. *Zootaxa* 1950:21–38. DOI:10.11646/ZOOTAXA.1950.1.5.

Minelli, A. 2013. Zoological nomenclature in the digital era. *Frontiers in Zoology* 10:4. http://www.frontiersinzoology.com/content/10/1/4.

Minelli, A. 2017. Grey nomenclature needs rules. *Ecologica Montenegrina* 7:654–66. DOI:10.37828/em.2016.7.31.

Minelli, A. 2019. The galaxy of the non-Linnaean nomenclature. *History and Philosophy of Life Sciences* 41:31. DOI:10.1007/s40656-019-0271-0.

Minelli, A., G. Ortalli, and G. Sanga (eds.). 2005. *Animal names*. Venice: Istituto Veneto di Scienze, Lettere ed Arti. ISBN:8888143386.

Mirkin, B. M. 1985. [*Theoretical foundations of contemporary phytocenology.*] Moscow: Nauka (in Russian).

Monsch, K. A. 2006. The *PhyloCode*, or alternative nomenclature: Why it is not beneficial to palaeontology, either. *Acta Palaeontologica Polonica* 51:521–4. DOI:10.1080/02724634.2018.1528450.

Montfort, D. 1810. Conchyliologie systématique et classification méthodique des Coquilles, Vol. 2. Paris: Shoell. www.biodiversitylibrary.org/page/11065017#page/9/mode/1up.

Moore, G. 1998. A comparison of traditional and phylogenetic nomenclature. *Taxon* 47:561–79. DOI:10.2307/1223578.

Moore, G. 2001. A review of the nomenclatural difficulties associated with misplaced rank-denoting terms. *Taxon* 50:495–505. DOI:10.2307/1223897.

Moore, G. 2003. Should taxon names be explicitly defined? *The Botanical Review* 69:2–21. DOI:10.1663/0006-8101(2003)069[0002:STNBED]2.0.CO;2.

Moore, G. 2005. A review of past and current debates in nomenclature: 250 years of progress or going around in a circle? *Acta Universitatis Upsaliensis* 33:109–17. ISSN:0562-2719.

Moore, M., M. E. Jameson, and A. Paucar-Cabrera. 2014. Taxonomic vandalism is an emerging problem for biodiversity science: A case study in the Rutelini (Coleoptera: Scarabaeidae: Rutelinae). *Entomological Society of America Annual Meeting.* https://esa.confex.com/esa/2014/webprogram/Paper87442.html.

Moore, R. T. 1974. Proposal for the recognition of super ranks. *Taxon* 23:650–2. DOI:10.2307/1218807.

Morard, R., G. Escarguel, A. K. M. Weiner, et al. 2016. Nomenclature for the nameless: A proposal for an integrative molecular taxonomy of cryptic diversity exemplified by planktonic Foraminifera. *Systematic Biology* 65:925–40. DOI:10.1093/sysbio/syw031.

Morison, R. 1672. *Plantarum umbelliferarum distributio nova* […]. Oxonii: Theatro Sheldoniano. https://bibdigital.rjb.csic.es/records/item/12179-redirection.

Müller, F. 1884. Einige Bemerkungen zu den Regeln der Pflanzen-Benennungen. *Botanisches Centralblatt* 18:118–22. https://archive.org/details/botanischeszentr0518bota/page/118/mode/2up.

Müller-Wille, S. 2007. Collection and collation: Theory and practice of Linnaean botany. *Studies in the History and Philosophy of Biology and Biomedical Sciences* 38:541–62. DOI:10.1016/j.shpsc.2007.06.010.

Mulligan, C. J. 2005. Isolation and analysis of DNA from archaeological, clinical, and natural history specimens. *Methods in Enzymology* 395:87–103. DOI:10.1016/S0076-6879(05)95007-6.

Mulsant, E. 1859–1860. *Opuscules Entomologiques*. Paris: Magnin et Blanchard. www.biodiversitylibrary.org/item/113994#page/7/mode/1up.

Murphy, E. A., C. M. Fauquet, D. H. L. Bishop, et al. (eds.). 1995. *Virus taxonomy. Classification and nomenclature of viruses. Sixth report of the International Committee on Taxonomy of Viruses*. Vienna: Springer-Verlag. ISBN:9783211825945.

Murray, J. A. 1782. *Vindiciae nominum trivialium stirpibus a Linneo equ impertitorum*. Gottingae: Jo. Christ. Dieterich. https://gdz.sub.uni-goettingen.de/id/PPN654995648.

Murray, J. A. 1784. *Caroli Linnaei equities systema vegitibilium secundum classes ordines genera species* […]. Gottingae: Jo. Christ. Dieterich. www.biodiversitylibrary.org/item/10309#page/1/mode/1up.

Nachman, M. W. 2013. Genomics and museum specimens. *Molecular Ecology* 22:5966–8. DOI:10.1111/mec.12563.

Nalimov, V. V. 1979. [*A probabilistic model of language. On relation between natural and artificial languages.*] Moscow: Nauka (in Russian).

Naomi, S.-I. 2014. Proposal of an integrated framework of biological taxonomy: A phylogenetic taxonomy, with the method of using names with standard endings in clade nomenclature. *Bionomina* 7:1–44. DOI:10.11646/bionomina.7.1.1.

Needham, J. G. 1911. The law that inheres in nomenclature. *Science, New Series* 33: 813–16. DOI:10.1126/science.33.856.813.

Needham, J. S. 1930. Scientific names. *Science*, 71:26–8. DOI:10.1126/science.71.1828.26.

Nemésio, A. 2009. Nomenclatural availability of nomina of new species should always require the deposition of preserved specimens in collections: A rebuttal to Donegan (2008). *Zootaxa* 2045:1–16. DOI:10.11646/zootaxa.2045.1.1.

Nicolson, D. H. 1977. Typification of names *vs*. typification of taxa; proposal on Article 48 and reconsideration of *Mitracarpus hirtus vs. M. villosus* (Rubiaceae). *Taxon* 26:569–74. DOI:10.2307/1219653.

Nicolson, D. H. 1991. A history of botanical nomenclature. *Annals of the Missouri Botanical Garden* 78:33–56. DOI:10.2307/2399589.

Nikishina, I. Yu. 2002. [The notion of "concept" in cognitive linguistics.] In [*Language, consciousness, communication*] eds. V. V. Krasnykh, and A. I. Isotov, 5–7. Moscow: MAKS. ISBN:5317005507 (in Russian).

Nixon, K. C., and J. M. Carpenter. 2000. On the other "Phylogenetic systematics". *Cladistics* 16:298–318. DOI:10.1006/clad.2000.0137.

Nonveiller, G. 1963. *Amphimallon solstitialis matutinalis* ssp. nov. (Scarabaeidae, Coleoptera). *Acta Entomologica Musei Nationalis Pragae* 35:171–6. www.aemnp.eu/data/article-846/827-35_0_171.pdf.

Norton, D. A., J. M. Lord, D. R. Given, et al. 1994. Over-collecting: An overlooked factor in the decline of plant taxa. *Taxon* 43:181–5. DOI:10.2307/1222876.

Oberholser, H. C. 1920. The nomenclature of families and subfamilies in zoology. *Science, New Series* 52:142–7. DOI:10.1126/science.52.1337.142.

Ochsmann, J. 2003. Some notes on problems of taxonomy and nomenclature of cultivated plants. In *Rudolf Mansfeld and plant genetic resources*, eds. H. Knüpffer, and J. Ochsmann, 42–50. Bonn: ZADI. www.genres.de/fileadmin/SITE_MASTER/content/Schriftenreihe/Band22_Gesamt.pdf.

References

Odening, K. 1979. Zur Taxonomie und Benennung der Haustier. *Zoologische Garten* N.F. 49:89–103. ISSN:0044-5169.

Ogilby, W. 1838. Further observations on "Rules for Nomenclature". *The Magazine of Natural History N.S.* 2:275–84. www.biodiversitylibrary.org/item/19511#page/293/mode/1up.

Ogilvie, B. W. 2003. The many books of nature: Renaissance naturalists and information overload. *Journal of the History of Ideas* 64:29–40. DOI:10.1353/jhi.2003.0015.

Ogilvie, B. W. 2006. *The science of describing. Natural history in Renaissance Europe.* Chicago (IL): University Chicago Press. ISBN:9780226620886.

Oldroyd, H. 1966. The future of taxonomic entomology. *Systematic Zoology* 15:253–60. DOI:10.2307/2411984.

Opinion 2027. 2003. Case 3010. Usage of 17 specific names based on wild species which are predated by or contemporary with those based on domestic animals […]. *The Bulletin of Zoological Nomenclature* 60:81–4. www.biodiversitylibrary.org/item/107012#page/97/mode/1up.

Oren, A. 2004. A proposal for further integration of the Cyanobacteria under the Bacteriological Code. *International Journal of Systematic and Evolutionary Microbiology* 54:1895–902. DOI:10.1099/ijs.0.03008-0.

Oren, A. 2005. Nomenclature of the cyanophyta/cyanobacteria/cyanoprokaryotes under the International Code of Nomenclature of Prokaryotes. *Algological Studies* 117:39–52. DOI:10.1127/1864-1318/2005/0117-0039.

Oren, A. 2008. Prokaryote nomenclature. *Encyclopedia of life sciences.* DOI:10.1002/9780470015902.a0021150.

Osbaldeston, T. A., and Wood R. P. 2000. Introduction. In *Dioscorides. De materia medica* […], xx–xxxviii. Johannesburg: IBI DIS Press. ISBN:9780620234351.

Oshanin V. F. 1911. [*Codes of international rules for systematic nomenclature.*] Saint Petersburg: Russian Entomological Society (in Russian).

Page, R. D. 2006. Taxonomic names, metadata, and the semantic web. *Biodiversity Informatics* 3:1–15. DOI:10.17161/BI.V3I0.25.

Page, R. D. 2016. DNA barcoding and taxonomy: Dark taxa and dark texts. *Philosophical Transactions of the Royal Society* B 371:20150334. DOI:10.1098/rstb.2015.0334.

Palii, V. M. 2013. The contribution of O. S. Vialov to the development of ichnological classification and nomenclature. *Stratigraphy and Geological Correlation* 21:249–51. DOI:10.1134/S0869593813030076.

Pallas, P. S. 1776. *Novae species quadrupedum e glirium ordine.* Erlangae: Wolfgangi Waltheri. https://gallica.bnf.fr/ark:/12148/bpt6k97040r.image.

Páll-Gergely, B., A. Hunyadi, and K. Auffenberg. 2020. Taxonomic vandalism in malacology: Comments on molluscan taxa recently described by N. N. Thach and colleagues (2014–2019). *Folia Malacologica* 28:35–76. DOI:10.12657/folmal.028.002.

Palmer, T. S. 1904. Index generum mammalium: A list of the genera and families of mammals. *North American Fauna* 23:5–984. www.biodiversitylibrary.org/item/88553#page/9/mode/1up.

Parés-Casanova, P. M. 2015. *A proposal on the scientific nomenclature of domestic animals.* Lleida (Spain): Universitat de Lleida. www.researchgate.net/publication/284182137_A_proposal_on_the_scientific_nomenclature_of_domestic_animals.

Parker, C. T., B. J. Tindall, and G. M. Garrity (eds.). 2019. International Code of Nomenclature of Prokaryotes […]. *International Journal of Systematic and Evolutionary Microbiology* 69:S1–S111. https://ccug.se/documents/taxonomy/prokaryotic_code/2019_Parker-Tindall-Garrity_Prokaryotic%20Code%20-%202008%20Revision.pdf.

Parkes, K. C. 1967. A qualified defense of traditional nomenclature. *Systematic Zoology* 16:268–73. DOI:10.2307/2412078.

Parkinson, P. G. 1975. The International Code of Botanical Nomenclature: An historical review and bibliography. *Tane* 21:153–74. http://www.thebookshelf.auckland.ac.nz/docs/Tane/Tane-21/23%20The%20International%20Code%20of%20Botanical.pdf.

Parkinson, P. G. 1984. The concept of nomenclatural illegitimacy, including 32 proposals to amend the Code. *Taxon* 33:469–92. DOI:10.1002/j.1996-8175.1984.tb03902.x.

Parkinson, P. G. 1990. *A Reformed Code of Botanical Nomenclature* [...]. Oberreifenberg: Koeltz Botanical Books. http://worldcat.org/entity/work/id/234734112.

Parra, L. A., J. C. Zamora, D. Hawksworth, et al. 2018. Proposals for consideration at IMC11 to modify provisions related solely to fungi in the International Code of Nomenclature for algae, fungi, and plants. *IMA Fungus* i–vii. DOI:10.1007/BF03449481.

Parte, A. C. 2014. LPSN—List of prokaryotic names with standing in nomenclature. *Nucleic Acids Research* 42:D613–D616. DOI:10.1093/nar/gkt1111.

Patterson, D. J., and J. Larsen. 1992. Perspective on protistan nomenclature. *The Journal of Protozoology* 39:125–31. DOI:10.1111/j.1550-7408.1992.tb01292.x.

Patterson, D. J., D. Remsen, W. A. Marino, et al. 2006. Taxonomic indexing—Extending the role of taxonomy. *Systematic Biology* 55:367–73. DOI:10.1080/10635150500541680.

Patterson, D. J., J. Cooper, P. M. Kirk, et al. 2010. Names are key to the big new biology. *Trends in Ecology and Evolution* 25:686–91. DOI:10.1016/j.tree.2010.09.004.

Pavlinov, I. Ya. 2011. [How it is possible to build a taxonomic theory.] *Zoologicheskie Issledovania* 10:45–100. ISSN:1025–532x (in Russian).

Pavlinov, I. Ya. 2013. [*Taxonomic nomenclature. 1. From Adam to Linnaeus.*] Moscow: KMK Sci. Press. ISBN:9785873178834 (in Russian).

Pavlinov, I. Ya. 2014. [*Taxonomic nomenclature. 2. From Linnaeus to first Codes.*] Moscow: KMK Sci. Press. ISBN:9785873179893 (in Russian).

Pavlinov, I. Ya. 2015a. [*Nomenclature in systematics. History, theory, practice.*] Moscow: KMK Sci. Press. ISBN:9785990715745 (in Russian).

Pavlinov, I. Ya. 2015b. [*Taxonomic nomenclature. 3. Contemporary Codes.*] Moscow: KMK Sci. Press. ISBN:9785990656437 (in Russian).

Pavlinov, I. Ya. 2016. [Biodiversity and biocollections—Problem of correspondence.] In *Aspects of biodiversity*, eds. I. Ya. Pavlinov, M. V. Kalyakin, and A. V. Sysoev, 733–86. Moscow: KMK Sci. Press. ISBN:9785990841666 (in Russian).

Pavlinov, I. Ya. 2018. [*Foundations of biological systematics: Theory and history.*] Moscow: KMK Sci. Press. ISBN:9785604074992 (in Russian).

Pavlinov, I. Ya. 2019. [*Nomenclature in systematics: Why do we call them that way?*] Moscow: KMK Sci. Press. ISBN:9785907213074 (in Russian).

Pavlinov, I. Ya. 2021. *Biological systematics: History and theory*. Boca Raton (FL): CRC Press. ISBN:9780367654450.

Pellegrin, P. 1987. Logical difference and biological difference: The unity of Aristotle's thought. In *Philosophical issues in Aristotle's biology*, eds. A. Gotthelf, and J. Lennox: 313–38. New York: Cambridge Univ. Press. ISBN:9780521310918/.

Pena, P., and M. Lobelius. 1570. *Stirpium adversaria nova* [...]. Londini: Thomae Purfoetii.

Penev, L., A. Paton, N. Nicolson, et al. 2016. A common registration-to-publication automated pipeline for nomenclatural acts for higher plants [...] fungi [...] animals [...]. *ZooKeys* 550:233–46. DOI:10.3897/zookeys.550.9551.

Pentinsaari, M., S. Ratnasingham, S. E. Miller, et al. 2020. BOLD and GenBank revisited—Do identification errors arise in the lab or in the sequence libraries? *PLoS ONE* 15:e0231814. DOI:10.1371/journal.pone.0231814.

Perminov, V. Ya. 2001. [*Philosophy and foundations of mathematics.*] Moscow: Progress-Traditsia. ISBN:5898260986 (in Russian).

Persoon, C. H. 1801. Synopsis methodica fungorum..., Pt. 1, 2. Gottingae: H. Dietrich.

Petersen, R. H. 1993. *A brief history of the type method.* http://www.biologie.uni-hamburg.de/b-online/library/tennessee/nom-hist.htm.

Petit-Thouars, L.-M. 1822. *Histoire particulière des plantes orchidées recuellies* […]. Paris: the author, Arthus Bertrand, Treuttel & Wurtz. www.biodiversitylibrary.org/item/9881#page/3/mode/1up.

Petri, M. J. 2001. Introduction. In *Carl von Linné. Nemesis divina*. Dordrecht: Springer Science + Business Media. ISBN:9789048156542.

Pine, R. H., and E. E. Gutiérrez. 2018. What is an "extant" type specimen? Problems arising from naming mammalian species—Group taxa without preserved types. *Mammal Review* 48:12–23. DOI:10.1111/mam.12108.

Pinevich, A. V. 2015. Proposal to consistently apply the International Code of Nomenclature of Prokaryotes (ICNP) to names of the oxygenic photosynthetic bacteria (cyanobacteria) […]. *International Journal of Systematic and Evolutionary Microbiology* 65:1070–4. DOI:10.1099/ijs.0.000034.

Pinker, S. 1993. *The language instinct: How the mind creates language.* New York: William Morrow. ISBN:9780061336461.

Pleijel, F., U. Jondelius, E. Norlinder, et al. 2008. Phylogenies without roots? A plea for the use of vouchers in molecular phylogenetic studies. *Molecular Phylogenetics and Evolution* 48:369–71. DOI:10.1016/j.ympev.2008.03.024.

Plumb, J. G., G. J. Kellogg, and I. Ya. G. P. Peffer. 1884. Report of the committee on nomenclature. *Transactions of the Wisconsin State Horticultural Society* 14:104–6. https://images.library.wisc.edu/WI/EFacs/USAIN/WSHS/WSHS1883/reference/wi.wshs1883.i0029.pdf.

Plumier, C. 1703. *Nova plantarum americanarum genera.* Parisiis: Johannem Boudot. www.biodiversitylibrary.org/item/120812#page/11/mode/1up.

Poche, F. 1911. Die Klassen und höheren Gruppen des Tierreichs. *Archiv für Naturgeschichte* 77, Supplement:63–136. ISSN:0365-6136.

Popova, Z. D., and I. A. Sternin. 2007. [*Cognitive linguistic.*] Moscow: AST "Vostok–Zapad". ISBN:9785170451036 (in Russian).

Prigogine, I., and G. Nicolis. 1985. Self-organisation in nonequilibrium systems: Towards a dynamics of complexity. In *Bifurcation analysis*, eds. M. Hazewinkel, R. Jurkovich, and J. H. P. Paelinck, 3–12. Dordrecht: Springer. DOI:10.1007/978-94-009-6239-2_1.

Puillandre, N., P. Bouchet, M.-C. Boisselier-Dubayle, et al. 2012. New taxonomy and old collections: Integrating DNA barcoding into the collection curation process. *Molecular Ecology Resources* 12:396–402. DOI:10.1111/j.1755-0998.2011.03105.x.

Pulteney, R. 1781. *A general view of the writings of Linnaeus.* London: T. Payne. www.biodiversitylibrary.org/item/175082#page/7/mode/1up.

Puton, A. 1880. Quelques mots sur la nomenclature entomologique. *Annales de la Société Entomologique de France* 10:33–40. www.biodiversitylibrary.org/item/34129#page/39/mode/1up.

Pyle, C. M. 2000. Conrad Gessner on the spelling of his name. *Archives of Natural History* 27:175–86. DOI:10.3366/anh.2000.27.2.175.

Pyle, R. L., and E. Michel. 2008. ZooBank: Developing a nomenclatural tool for unifying 250 years of biological information. *Zootaxa* 1950:39–50. DOI:10.11646/zootaxa.1950.1.6.

Quine, W. V. 1969. *Ontological relativity and other essays.* New York: Columbia University Press. ISBN:9780231083577.

R. L. 1905. Règles internationales de la nomenclature zoologique. *Nature* 71:534. DOI:10.1038/071534a0.

Raclavský, J. 2012. Semantic paradoxes and transparent intensional logic. In *The Logica yearbook 2011*, eds. M. Pelis, and V. Punchochar, 239–52. ISBN:9781848900714.

Ramsbottom, J. 1955. Linnaeus' nomenclature. *Proceedings of the Linnean Society of London* 165:164–6. DOI:10.1111/j.1095-8312.1955.tb00738.x.

Raposo, M. A., and G. M. Kirwan. 2017. What lies beneath the controversy as to the necessity of physical types for describing new species? *Bionomina* 12:52–6. DOI:10.11646/bionomina.12.1.6.

Rasnitsyn, A. P. 1982. Proposal to regulate the names of taxa above the family group. *The Bulletin of Zoological Nomenclature* 39:200–7. www.biodiversitylibrary.org/item/44481#page/234/mode/1up.

Rasnitsyn, A. P. 1986. [Parataxa and paranomenclature.] *Paleontological Journal* 3:11–21. ISSN:0031-031X (in Russian).

Rasnitsyn, A. P. 1992. [Principles of nomenclature and the nature of taxon.] *Zhurnal Obshchei Biologii* 55:307–13. ISSN:0044-4596 (in Russian).

Rasnitsyn, A. P. 1996. Conceptual issues in phylogeny, taxonomy, and nomenclature. *Contributions to Zoology* 66:3–41. DOI:10.1163/26660644-06601001.

Rasnitsyn, A. P. 2002. [Evolutionary process and taxonomy methodology.] *Trudy Russkogo Entomologicheskogo Obshchestva* 73:5–108. ISSN:1605-7678 (in Russian).

Raspail, X. 1899. A propos d'un projet de reforme à la nomenclature des êtres organisés et des corps inorganiques. *Memorias y Revista de la Sociedad Cientifica "Antonio Alzate"* 12:475–80. www.biodiversitylibrary.org/item/88926#page/517/mode/1up.

Ratnasingham, S., and P. D. N. Hebert. 2007. BARCODING, BOLD: The Barcode of Life Data System. *Molecular Ecology Notes* 7:355–64. DOI:10.1111/j.1471-8286.2007.01678.x.

Raven, C. E. 1986. *John Ray: Naturalist*, 2nd ed. Cambridge (UK): Cambridge University Press. ISBN:9780521310833.

Ray, J. 1682. *Methodus plantarum nova: brevitatis & perspicuitatis causa synoptice in tabulis exhibita* […]. Londini: Henrici Faithorne & Joannis Kersey. ISBN:9780903874465 (2014 reprint).

Ray, J. 1693. *Synopsis methodica animalium quadrupedum et serpentini* […]. Londini: S. Smith & B. Walford. ISBN:9785882974182 (2012 reprint).

Ray, J. 1696. *De variis plantarum methodus dissertation brevis* […]. Londini: S. Smith & B. Walford. http://digital.onb.ac.at/OnbViewer/viewer.faces?doc=ABO_%2BZ184733202.

Remsen, D. 2016. The use and limits of scientific names in biological informatics. *ZooKeys* 550:207–23. DOI:10.3897/zookeys.550.9546.

Remsen, J. V. 1997. Museum specimens: Science, conservation and morality. *Bird Conservation International* 7:363–6. DOI:1 0.1017/S0959270900001696.

Renner, S. S. 2016. A return to Linnaeus's focus on diagnosis, not description: The use of DNA characters in the formal naming of species. *Systematic Biology* 65:1085–95. DOI:10.1093/sysbio/syw032

Revel, N. 2007. The symbolism of animals' names: Analogy, metaphor, totemism. In *Animal names*, eds. A. Minelli, G. Ortalli, and G. Sanga, 339–44. Venice: Istituto Veneto di Scienze, Lettere ed Arti. ISBN:8888143386.

Reynier, A. 1893. Nouvelle proposition de réforme dans la nomenclature botanique. *Bulletin de la Société des Amis des Sciences Naturelles Rouen*, 3rd series, 239–41. https://sasnmr.fr/bulletins/bulletins-sasnmr-de-1865-a-1899/158-bulletin-sasnmr-1892.

Reynolds, D. R., and J. W. Taylor. 1991. DNA specimens and the "International Code of Botanical Nomenclature". *Taxon* 40:311–15. DOI:10.2307/1222985.

Rhodin, A. G. J., H. Kaiser, P. P. van Dijk, et al. 2015. Comment on Spracklandus Hoser, 2009 (Reptilia, Serpentes, Elapidae): Request for confirmation of the availability of the generic name […]. *The Bulletin of Zoological Nomenclature* 72:64–78. www.biotaxa.org/bzn/article/view/38247.

Rickett, H. W. 1953. Expediency *vs.* priority in nomenclature. *Taxon* 2:117–24. www.iapt-taxon.org/historic/Congress/IBC_1954/expediency.pdf.

References

Rickett, H. W. 1959. The status of botanical nomenclature. *Systematic Zoology* 8:22–7. DOI:10.2307/sysbio/8.1.22.

Ride, W. D. 1988. Towards a unified system of biological nomenclature. In *Prospects in systematics*, ed. D. L. Hawksworth, 332–53. Oxford: Clarendon Press.

Riedel, A., K. Sagata, S. Surbakti, et al. 2013a. One hundred and one new species of *Trigonopterus* weevils from New Guinea. *ZooKeys* 280:1–150. DOI:10.3897/zookeys.280.3906.

Riedel, A., K. Sagata, Y. K. Suhardjono, et al. 2013b. Integrative taxonomy on fast track—Towards more sustainability in biodiversity research. *Frontiers in Zoology* 10:15. DOI:10.1186/1742-9994-10-15.

Rieppel, O. 2006. The *PhyloCode*: A critical discussion of its theoretical foundation. *Cladistics* 22:186–97. DOI:10.1111/j.1096-0031.2006.00097.x.

Rieppel, O. 2008. Origins, taxa, names and meanings. *Cladistics* 24:598–610. DOI:10.1111/j.1096-0031.2007.00195.x.

Rindsberg, A. K. 2018. Ichnotaxonomy as a science. *Annales Societatis Geologorum Poloniae* 88:91–110. DOI:10.14241/asgp.2018.012.

Rivinus, A. 1690. *Introductio generalis in Rem Herbariam* [...]. Lipsiae: Johannis Heinichii. https://cdm21057.contentdm.oclc.org/digital/collection/coll13/id/144430.

Robinson, B. L. 1895. Recommendations regarding the nomenclature of systematic botany. *Botanical Gazette* 20:263. www.jstor.org/stable/2464772.

Robinson, B. L. 1897. The official nomenclature of the Royal Botanical Garden and Museum of Berlin. *Botanical Gazette* 24:107–10. www.jstor.org/stable/2464391.

Robinson, B. L. 1898. Some reasons why the Rochester nomenclature cannot be regarded as consistent or stable. *Botanical Gazette* 25:437–45. www.jstor.org/stable/2464530.

Rodendorf, B. B. 1977. [On the rationalization of the names of high-rank taxa in zoology.] *Paleontological Journal* 2:14–22. ISSN:0031-031X (in Russian).

Rookmaaker, L. C. 2011. The early endeavours by Hugh Edwin Strickland to establish a code for zoological nomenclature in 1842–1843. *The Bulletin of Zoological Nomenclature* 68:29–40. http://www.rhinoresourcecenter.com/pdf_files/135/1355800865.pdf.

Ross, R. 1966. The generic names published by N. M. von Wolf. *Acta Botanica Neerlandica* 15:147–61. DOI:10.1111/j.1438-8677.1966.tb00221.x.

Rowe, K. C., S. Singhal, M. D. Macmanes, et al. 2011. Museum genomics: Low cost and high-accuracy genetic data from historical specimens. *Molecular Ecology Resources* 11:1082–92. DOI:10.1111/j.1755-0998.2011.03052.x.

Rowley, D. L., J. A. Coddington, M. W. Gates, et al. 2007. Vouchering DNA-barcoded specimens: Test of a nondestructive extraction protocol for terrestrial arthropods. *Molecular Ecology Notes* 7:915–24. DOI:10.1111/j.1471-8286.2007.01905.x.

Ryberg, M., and R. H. Nilsson. 2018. New light on names and naming of dark taxa. *MycoKeys* 30:31–9. DOI:10.3897/mycokeys.30.24376.

Sachs, J. 1906. *History of botany, 1530–1860*. Oxford: Clarendon Press. ISBN:9781408603901 (2007 reprint).

Saint-Lager, J.-B. 1880. Réforme de la nomenclature botanique. *Annales de la Société Botanique de Lyon* 7:1–154. www.biodiversitylibrary.org/item/246333#page/25/mode/1up.

Saint-Lager, J.-B. 1881. Nouvelles remarques sur la nomenclature botanique. *Annales de la Société Botanique de Lyon* 8:149–203. www.biodiversitylibrary.org/item/246347#page/173/mode/1up.

Saint-Lager, J.-B. 1886. *Le procès de la nomenclature botanique et zoologique*. Paris: J. B. Baillière et fils. ISBN:9781120411433 (2009 reprint).

Sangster, G., and F. G. Rozendaal. 2004. Systematic notes on Asian birds. 41. Territorial songs and species-level taxonomy of nightjars of the *Caprimulgus macrurus* complex, with the description of a new species. *Zoologische Verhandelingen* 26:7–45. ISSN:0024-1652.

Santos, C., A. D. De Souza Amorim, B. Klassa, et al. 2016. On typeless species and the perils of fast taxonomy. *Systematic Entomology* 41:511–15. DOI: 10.1111/syen.12180.

Santos, L. M., and L. R. R. Faria. 2011. The taxonomy's new clothes: A little more about the DNA-based taxonomy. *Zootaxa* 3025:66–8. www.biotaxa.org/Zootaxa/article/view/zootaxa.3025.1.5.

Sarjeant, W. A. S., and W. J. Kennedy. 1973. Proposal of a code for the nomenclature of trace-fossils. *Canadian Journal of Earth Sciences* 10:460–475. DOI:10.1139/e73-046.

Särkinen, T., M. Staats, J. E. Richardson, et al. 2012. How to open the treasure chest? Optimising DNA extraction from herbarium specimens. *PLoS ONE* 7:e43808. DOI:10.1371/journal.pone.0043808.

Scharf, S. 2008. Multiple independent inventions of a non-functional technology. Combinatorial descriptive names in botany, 1640–1830. *Spontaneous Generations* 2:145–84. DOI:10.4245/sponge.v2i1.3552.

Schaum, H. 1862. On the restoration of obsolete names in entomology. *Transactions of the Entomological Society of London* 3rd series 1:323–7. DOI:10.1111/j.1365-2311.1862.tb01282.x.

Schenk, E. T., and J. H. McMasters. 1956. *Procedure in taxonomy*, 3rd ed. Stanford (CA): Stanford University Press. https://archive.org/details/procedureintaxon00sche/mode/2up.

Schindel, D. E., and S. E. Miller. 2010. Provisional nomenclature: The onramp to taxonomic names. In *Systema naturae 250—The Linnaean ark*, ed. A. Polaszek, 109–15. Boca Raton (FL): CRC Press. ISBN:9781420095012.

Schoch, C. L., B. Robbertse, V. Robert, et al. 2014. Finding needles in haystacks: Linking scientific names, reference specimens and molecular data for fungi. *Database* 2014:1–21. DOI:10.1093/database/bau061.

Schuchert, C. 1897. What is a type in natural history? *Science, New Series* 5:636–40. DOI:10.1126/science.5.121.636.

Schuchert, C., and S. S. Buckman. 1905. The nomenclature of types in natural history. *Science, New Series* 21:899–901. DOI:10.1126/science.21.545.899-b.

Schuh, R. T. 2003. The Linnaean system and its 250-year persistence. *The Botanical Review* 69:59–78. DOI:10.1663/0006-8101(2003)069[0059:TLSAIY]2.0.CO;2.

Sclater, P. L. 1878. *Rules for zoological nomenclature drawn up by the late H. E. Strickland* [...]. London: John Murray. https://archive.org/details/rulesforzoologi00sciegoog.

Sclater, P. L. 1896. Remarks on the divergencies between the "Rules for Naming Animals" of the German Zoological Society and the Stricklandian Code of Nomenclature. *The Auk* 13:306–19. DOI:10.2307/4068345.

Seberg, O. 1984. The hierarchy of types again. *Taxon* 33:496–7. DOI:10.1002/j.1996-8175.1984.tb03905.x.

Seilacher, A. 1964. Sedimentological classification and nomenclature of trace fossils. *Sedimentology* 3:253–6. DOI:10.1111/j.1365-3091.1964.tb00464.x.

Sereno, P. C. 2007. Logical basis for morphological characters in phylogenetics. *Cladistics* 23:565–87. DOI:10.1111/j.1096-0031.2007.00161.x.

Sharkey, M. J., D. H. Janzen, W. Hallwachs, et al. 2021. Minimalist revision and description of 403 new species in 11 subfamilies of Costa Rican braconid parasitoid wasps [...]. *ZooKeys* 1013:1–665. DOI:10.3897/zookeys.1013.55600.

Sharp, D. 1873. *The object and method of zoological nomenclature*. London: E. W. Janson, Williams & Norgate. https://archive.org/details/b22352259.

Shatalkin, A. I. 1999. [Semantic structure of taxonomic names.] *Zhurnal Obshchei Biologii* 60:150–63. ISSN:0044-4596 (in Russian).

Shatalkin, A. I., and T. V. Galinskaya. 2017. A commentary on the practice of using the so-called typeless species. *ZooKeys* 693:129–39. DOI:10.3897/zookeys.693.10945.

Shear, C. L. 1902. The starting point for generic nomenclature in botany. *Science, New Series* 16:1035–6. DOI:10.1126/science.16.417.1035.

Shear, C. L. 1910. Nomenclature at Brussels. *Science, New Series* 32:594–5. DOI:10.1126/science.32.826.594.

Shiffman, D. 2019. *Scientists should stop naming species after awful people.* https://blogs.scientificamerican.com/observations/scientists-should-stop-naming-species-after-awful-people/.

Shipunov, A. B. 1999. [*Basics of theory of systematics.*] Moscow: Open Lyceum VZMS (in Russian).

Shipunov, A. 2011. The problem of hemihomonyms and the on-line hemihomonyms database (HHDB). *Bionomina* 4:65–72. www.biotaxa.org/Bionomina/article/view/4.

Siddell, S. G., P. J. Walker, E. J. Lefkowitz, et al. 2020. Binomial nomenclature for virus species: A consultation. *Archives of Virology* 165:519–25. DOI:10.1007/s00705-019-04477-6.

Sigler, L., and D. L. Hawksworth. 1987. International Commission on the Taxonomy of Fungi (ICTF): Code of practice for systematic mycologists. *Mycopathologia* 4:83–6. DOI:10.1007/BF00436673.

Sigovini, M., E. Keppel, and D. Tagliapietra. 2016. Open nomenclature in the biodiversity era. *Methods in Ecology and Evolution* 7:1217–25. DOI:10.1111/2041-210X.12594.

Simpson, G. G. 1940. Types in modern taxonomy. *American Journal of Science* 238:413–31. DOI:10.2475/ajs.238.6.413.

Simpson, G. G. 1961. *Principles of animal taxonomy.* New York: Columbia University Press. ISBN:9780231024273.

Simpson, S. 1975. Classification of trace fossils. In *The study of trace fossils,* ed. R.W. Frey, 39–54. Berlin: Springer. DOI:10.1007/978-3-642-65923-2_3.

Singer, R. 1960. Persoon's Synopsis 1801 as starting point for all fungi? *Taxon* 9:35–7. DOI:10.2307/1217835.

Siu, R. G. H., and E. T. Reese. 1955. Proposal for a system of biological nomenclature, with special reference to microorganisms. *Farlowia: A Journal of Crypogamic Botany* 4:399–407. www.biodiversitylibrary.org/item/33577#page/425/mode/1up.

Skuncke, M.-C. 2008. Linnaeus: An 18th century background. In *The Linnaean legacy: Three centuries after his birth,* eds. M. J. Morris, and L. Berwick, 19–26. London: Wiley-Blackwell. ISSN:0950-1096.

Slaughter, M. 1982. *Universal languages and scientific taxonomy in the seventeenth century.* Cambridge (UK): Cambridge University Press. ISBN:9780521135443.

Sluys, R. 2021. Attaching names to biological species: The use and value of type specimens in systematic zoology and natural history collections. *Biological Theory* 16:49–61. DOI:10.1007/s13752-020-00366-3.

Smith, A. C. 1957. Fifty years of botanical nomenclature. *Brittonia* 9:2–8. DOI:10.2307/2804843.

Smith, E. F. 1895. Austro-German views on botanical nomenclature. *The American Naturalist* 29:585–6. www.biodiversitylibrary.org/item/129413#page/716/mode/1up.

Smith, H. M. 1962. The hierarchy of nomenclatural status of generic and specific names in zoological taxonomy. *Systematic Zoology* 11:139–42. DOI:10.2307/2411877.

Smith, H. M., and R. B. Smith. 1972. Chresonymy ex synonymy. *Systematic Zoology* 21:445. DOI:10.1093/sysbio/21.4.445.

Smith, J. A. 1789. *Plantarum icones hactenus inedita, plerumque ad plantas in Herbario Linneano* […]. Londoni: Benj. White et Filii. www.biodiversitylibrary.org/item/105103#page/2/mode/1up.

Smith, J. A. 1821. *A selection of the correspondence of Linnaeus and other naturalists from the original manuscripts*, Vol. 2. London: Longman, Hust, Rees & Brown. www.biodiversitylibrary.org/item/239177#page/7/mode/1up.

Smith, J. D. D. 2001. *Official lists and indexes of names and works in zoology. Supplement 1986–2000*. London: Internat. Trust for Zoological Nomenclature. www.biodiversitylibrary.org/item/107003#page/5/mode/1up.

Sneath, R. H. A., and R. R. Sokal. 1973. *Numerical taxonomy. The principles and methods of numerical classification*. San Francisco (CA): W. H. Freeman. ISBN:9780716706977.

Sokal, R. R., and R. H. A. Sneath. 1963. *Principles of numerical taxonomy*. San Francisco (CA): W. H. Freeman. ASIN:B0006AYNO8.

Somervuo, P., A. Harma, and S. Fagerlund. 2006. Parametric representations of bird sounds for automatic species recognition. *IEEE Transactions on Audio, Speech, and Language Processing* 14:2252–63. DOI: 10.1109/TASL.2006.872624.

Sosef, M. S. M., J. Degreef, H. Engledow, et al. 2020. *Botanical classification and nomenclature, an introduction*. Meise: Meise Botanic Garden. ISBN:9789492663207.

Speers, L. 2005. *E*-Types—A new resource for taxonomic research. In *Digital imaging of biological type specimens. A manual of best practice* […], eds. C. L. Häuser, A. Steiner, J. Holstein, et al., 19–24. Stuttgart: Staatliches Museum für Naturkunde. ISBN:3000172408.

Spencer, R. D., and R. G. Cross. 2007. The International Code of Botanical Nomenclature (ICBN), the International Code of Nomenclature for Cultivated Plants (ICNCP), and the cultigen. *Taxon* 56:938–40. DOI:10.2307/25065875.

Spencer, R., R. Cross, and P. Lumley. 2007. *Plant names. A guide to botanical nomenclature*, 3rd ed. Melbourne: Royal Botanic Gardens. ISBN:9780643094406.

Spooner, D. M., W. L. A. Hetterscheid, R. G. van den Berg, et al. 2003. Plant nomenclature and taxonomy: An horticultural and agronomic perspective. *Horticultural Reviews* 28:1–59. DOI:10.1002/9780470650851.ch1.

Sprague, T. A. 1921. Plant nomenclature: Some suggestions. *The Journal of Botany, British and Foreign* 59:153–60. www.biodiversitylibrary.org/item/34333#page/187/mode/1up.

Sprague, T. A. 1928. The Herbal of Otto Brunfels. *Journal of the Linnean Society of London, Botany* 48:79–124. DOI:10.1111/j.1095-8339.1928.tb02577.x.

Sprague, T.A., and E. Nelmes. 1931. The Herbal of Leonhart Fuchs. *Journal of the Linnean Society of London, Botany* 48:545–642. DOI:10.1111/j.1095-8339.1931.tb00596.x.

Sprengel, K. 1807–1808. *Historia rei herbariae*, Vols. 1–2. Amstelodami: Taberna librariae et artium. https://catalog.hathitrust.org/Record/011568938.

Stafleu, F. A. 1963. Adanson and the "Familles des plantes". In *Adanson: The bicentennial of Michel Adanson's "Familles des Plantes"*, part 1, ed. G. H. M. Lawrence, 123–263. Pittsburg (PA): Hunt Botanical Library. www.huntbotanical.org/admin/uploads/05-hibd-adanson-pt1-pp123-264.pdf.

Stafleu, F. A. 1971. *Linnaeus and Linnaeans*. Utrecht: A. Oosthoek. ISBN:9789060460641.

Stanier, R. Y., W. R. Sistrom, T. A. Hansen, et al. 1978. Proposal to place the nomenclature of the Cyanobacteria (blue-green algae) under the rules of the International Code of Nomenclature of Bacteria. *International Journal of Systematic Bacteriology* 28:335–6. DOI:10.1099/00207713-28-2-335.

Starobogatov, Ya. I. 1991. Problems in the nomenclature of higher taxonomic categories. *The Bulletin of Zoological Nomenclature* 48:6–18. www.biodiversitylibrary.org/item/44489#page/22/mode/1up.

Starr, M. P., and H. Heise. 1969. Regarding nomenclatural types (nomenifers): A proposal for amending principle 11 and rule 9 of the International Code of Nomenclature of Bacteria. *International Journal of Systematic Bacteriology* 19:173–81.

www.microbiologyresearch.org/docserver/fulltext/ijsem/19/2/ijs-19-2-173.pdf?expires=1616406824&id=id&accname=guest&checksum=1C238BE5BF32B7AC3E22BEF978A1EDD9.
Stearn, W. T. 1952. Proposed International Code of Nomenclature for Cultivated Plants. Historical introduction. *Journal of the Royal Horticultural Society of London* 77:157–73. ISSN:0035-8924.
Stearn, W. T. 1953. International Code of Nomenclature for Cultivated Plants. In *Report of the 13th International Horticultural Congress, 1952*, Vol. 1, ed. P. M. Synge, 42–68. London: Royal Horticultural Society. ASIN:B000Z6B2IS.
Stearn, W. T. 1955. Linnaeus's "Species Plantarum" and the language of botany. *Proceedings of the Linnean Society of London* 165:158–64. DOI:10.1111/j.1095-8312.1955.tb00737.x.
Stearn, W. T. 1959. The background of Linnaeus's contributions to the nomenclature and methods of systematic biology. *Systematic Zoology* 8:4–22. DOI:10.2307/sysbio/8.1.4.
Stearn, W. T. 1985. *Botanical Latin*, 3rd ed. London: David & Charles. https://archive.org/details/BOTANICALLATINWILLIAMSTEARN.
Stearn, W. T. 1986. Historical survey of the naming of cultivated plants. *Acta Horticulturae* 182:19–28. DOI:10.17660/ActaHortic.1986.182.1.
Stebbins, L. 1950. *Variation and evolution in plants*. New York: Columbia University Press. ISBN:9780231017336.
Stejneger, L. 1924. A chapter in the history of zoological nomenclature. *Smithsonian Miscellaneous Collections* 77:1–21. https://repository.si.edu/bitstream/handle/10088/23657/SMC_77_Stejneger_1924_1_1-21.pdf?sequence=1&isAllowed=y.
Stenzel, H. B. 1950. Proposed uniform endings for names of higher categories in zoological systematics. *Science* 112:94. DOI:10.1126/science.112.2899.94.
Sterns, E. E. 1888. The nomenclature question and how to settle it. *Bulletin of the Torrey Botanical Club* 15:230–5. www.biodiversitylibrary.org/item/7991#page/233/mode/1up.
Stevens, P. F. 1991. George Bentham and the Kew Rule. In *Improving the stability of names: needs and options*, ed. D. L. Hawksworth, 157–68. Königstein: Koeltz Scientific Books. ISBN:3874293289.
Stevens, P. F. 2002. Why do we name organisms? Some reminders from the past. *Taxon* 51:11–26. https://pdfs.semanticscholar.org/33c3/68749360e7e9659b11c0ae02a5465502c7c6.pdf.
Stiles, C. W. 1906. A plan to ensure the designation of generic types. An open letter to systematic zoologists. *Science, New Series* 23:700–1. DOI:10.1126/science.23.592.700-a.
Stiles, C. W. 1907. The "First species rule" *vs.* the "Law of priority" in determining types of genera. *Science, New Series* 25:145–7. DOI:10.1126/science.25.630.145-b.
Stiles, C. W., and A. Hassal. 1905. Determination of generic types, and a list of roundworm genera, with their original and type species. *Bulletin of Bureau of Animal Industry* 79:5–152. www.biodiversitylibrary.org/item/75518#page/3/mode/1up.
Stone, W. 1906. The relative merits of the "elimination" and "first species" method in fixing the types of genera, with special reference to ornithology. *Science, New Series* 24:560–5. DOI:10.1126/science.24.618.560-a.
Stone, W. 1907. The first species rule versus elimination. *Science, New Series* 25:147–51. DOI:10.1126/science.25.630.147.
Strand, M., and P. Sundberg. 2011. A DNA-based description of a new nemertean (phylum Nemertea) species. *Marine Biology Research* 7:63–70. DOI:10.1080/17451001003713563.
Strickland, H. E. 1835. On the arbitrary alteration of established terms in natural history. *The Magazine of Natural History* 8:36–40. www.biodiversitylibrary.org/item/19589#page/50/mode/1up.

Strickland, H. E. 1837. Rules for zoological nomenclature. *The Magazine of Natural History* N.S. 1:173–6. www.biodiversitylibrary.org/item/87352#page/189/mode/1up.

Strickland, H. E. 1838a. Remarks on Mr. Ogilby's "Further observations on rules for nomenclature". *The Magazine of Natural History N.S.* 2:326–331. www.biodiversitylibrary.org/item/19511#page/344/mode/1up.

Strickland, H. E. 1838b. Reply to Mr. Ogilby's "Observations on rules for nomenclature". *The Magazine of Natural History N.S.* 2:198–204. www.biodiversitylibrary.org/item/19511#page/216/mode/1up.

Strickland, H. E. 1845. Report on the recent progress and present state of ornithology. *Report of the British Association for the Advancement of Science (1844)* 170–221. London: John Murray. www.biodiversitylibrary.org/item/47344#page/224/mode/1up.

Strickland, H. E., J. S. Henslow, J. Phillips, et al. 1843a. Report of a committee appointed to "consider of the rules by which the nomenclature of zoology may be established on a uniform and permanent basis". In *Report of the twelfth meeting of the British Association for the Advancement of Science (1842)*, 105–21. London: John Murray. www.biodiversitylibrary.org/item/46628#page/145/mode/1up.

Strickland, H. E., J. S. Henslow, J. Phillips, et al. 1843b. Series of propositions for rendering the nomenclature of zoology uniform and permanent […]. *The Annals and Magazine of Natural History* 11:259–75. www.biodiversitylibrary.org/item/61799#page/273/mode/1up.

Stuessy, T. F. 2000. Taxon names are not defined. *Taxon* 49:231–3. DOI:10.2307/1223709.

Stuessy, T. F. 2008. *Plant taxonomy. The systematic evaluation of comparative data.* 2nd ed. New York: Columbia University Press. ISBN:9780231147125.

Sunagawa, S., D. Mende, G. Zeller, et al. 2013. Metagenomic species profiling using universal phylogenetic marker genes. *Nature Methods* 10:1196–9. DOI:10.1038/nmeth.2693.

Susov, I. P. 2006. [*Introduction to linguistics.*] Moscow: AST "Vostok–Zapad". ISBN: 5760901109 (in Russian).

Svenson, H. K. 1945. On the descriptive method of Linnaeus. *Rhodora* 47:273–302. www.jstor.org/stable/23304637.

Swainson, W. 1820–1821. *Zoological illustrations, or original figures and descriptions […].* London: Boldwin, Cradock & Joy. www.biodiversitylibrary.org/item/92614#page/7/mode/1up.

Swainson, W. 1836. *On the natural history and classification of birds*, Vol. 1. London: Longman. www.biodiversitylibrary.org/item/131541#page/17/mode/1up.

Swingle, W. T. 1913. Types of species in botanical taxonomy. *Science, New Series* 37:864–5. DOI:10.1126/science.37.962.864.

Sylvester-Bradley, P. C. 1952. *The classification and coordination of infraspecific categories.* London: Systematic Association.

Sytin, A. K. 1997. [Peter Simon Pallas as a botanist.] Moscow: KMK Science Press (in Russian). ISBN:9785873179626.

Talmy, L. 2000. *Toward a cognitive semantics,* Vol. 1. Cambridge (MA): MIT Press. ISBN:9780262700962.

Tapscott, D., and A. Tapscott. 2016. *Blockchain revolution: How the technology behind bitcoin is changing money, business, and the world.* London: Penguin Books. ISBN:9781101980149.

Taylor, J. W. 2011. One fungus = one name: DNA and fungal nomenclature twenty years after PCR. *IMA Fungus* 2:113–20. DOI:10.5598/imafungus.2011.02.02.01.

Taylor, P. M. 1990. *The folk biology of the Tobelo people. A study in folk classification.* Washington (DC): Smithsonian Inst. Press. ISBN:9780835743259.

Temminck, C. J. 1815. *Manuel d'ornithologie, ou tableau systèmatique des oiseaux qui se troevent en Europe.* Amsterdam: J. C. Sepp & fils. https://books.google.ru/books/about/Manuel_d_ornithologie_ou_tableau_systém.html?id=agoOAAAAQAAJ&redir_esc=y.

References

Theophrastus. 1916. *Enquiry into plants*, Vol. 1. Cambridge (MA): Harvard University Press. www.biodiversitylibrary.org/item/58434#page/9/mode/1up.

Thines, M., P. W. Crous, M. C. Aime, et al. 2018. Ten reasons why a sequence-based nomenclature is not useful for fungi anytime soon. *IMA Fungus* 9:177–83. DOI:10.5598/imafungus.2018.09.01.11.

Thines, M., T. Aoki, P. W. Crous, et al. 2020. Setting scientific names at all taxonomic ranks in italics facilitates their quick recognition in scientific papers. *IMA Fungus* 11:25. DOI:10.1186/s43008-020-00048-6.

Thomas, O. 1897. Types in natural history and nomenclature of rodents. *Science, New Series* 6:485–7. DOI:10.1126/science.6.143.485-a.

Timm, R. M., and R. R. Ramey. 2005. What constitutes a proper description? *Science* 309:2163–4. DOI:10.1126/science.309.5744.2163c.

Tindall, B. J. 2008. Are the concepts of legitimate and illegitimate names necessary under the current International Code of Nomenclature of Bacteria? [...] *International Journal of Systematic and Evolutionary Microbiology* 58:1979–86. DOI:10.1099/ijs.0.2008/006239-0.

Tindall, B. J., P. Kämpfer, J. P. Euzéby, et al. 2006. Valid publication of names of prokaryotes according to the rules of nomenclature: Past history and current practice. *International Journal of Systematic and Evolutionary Microbiology* 56:2715–20. DOI:10.1099/ijs.0.64780-0.

Tishechkin, D. Y. 2014. The use of bioacoustic characters for distinguishing between cryptic species in insects: Potentials, restrictions, and prospects. *Entmological Review* 94:289–309. DOI:10.1134/S0013873814030014.

Tobias, P. V. 1969. Bigeneric nomina: A proposal for modification of the rules of nomenclature. *American Journal of Physical Anthropology* 31:103–6. DOI:10.1002/ajpa.1330310115.

Tornier, G. 1898. Grundlagen einer wissenschaftlichen Tier- und Pflanzennomenclatur. *Zoologisches Anzeiger* 21:575–80. www.biodiversitylibrary.org/item/37581#page/593/mode/1up.

Tournefort, J. Pitton de. 1694. *Élémens de botanique, ou méthode pour connoître les plantes*, Vol. 1. Paris: De l'Imprimerie Royale. ISBN:9781272619824 (2012 reprint).

Tournefort, J. Pitton de. 1700. *Institutiones rei herbariae. Editio altera* [...], Vol. 1. Parisiis: Typ. Regia. www.biodiversitylibrary.org/item/14433#page/1/mode/1up.

Trehane, P. 2004. 50 years of the International Code of Nomenclature for Cultivated Plants: Future prospects for the Code. *Acta Horticulturae* 634:17–27. DOI:10.17660/ActaHortic.2004.634.1.

Troncoso-Palacios, J., M. Ruiz De Gamboa, R. Langstroth, et al. 2019. Without a body of evidence and peer review, taxonomic changes in Liolaemidae and Tropiduridae (Squamata) must be rejected. *Zookeys* 813:39–54. DOI:10.3897/zookeys.813.29164.

Tuominen, J., N. Laurenne, and E. Hyvönen. 2011. Biological names and taxonomies on the semantic web—Managing the change in scientific conception. *The Semantic Web: Research and Applications Lecture Notes in Computer Science* 6644: 255–269.

Turland, N. 2019. *The Code decoded. A user's guide to the International Code of Nomenclature for algae, fungi, and plants*, 2nd ed. Sofia: Pensoft Publishers. ISBN:9789546429636.

Tuxen, S. L. 1967. The entomologist J. C. Fabricius. *Annual Review of Entomology* 12:1–15. DOI:0.1146/annurev.en.12.010167.000245.

Uerpmann, H.-P. 1993. Proposal for a separate nomenclature of domestic animals. In *Skeletons in her cupboard: Festschrift for Juliet Clutton-Brock*, eds. Å. T. Ciason, S. Payne, and H.-P. Uerpmann, 239–41. Oxford: Oxbow. ISBN:9780946897643.

Ulicsni, V., I. Svanberg, and Z. Molnár. 2016. Folk knowledge of invertebrates in Central Europe—Folk taxonomy, nomenclature, medicinal and other uses, folklore, and nature conservation. *Journal of Ethnobiology and Ethnomedicine* 12:1–40. DOI:10.1186/s13002-016-0118-7.

Uryson, E. V. 2003. [*Research problems of the linguistic world picture. Analogy in semantics.*] Moscow: YaSK. ISBN:5944571233 (in Russian).
Uzepchuk, S. V. 1956. [Was Linnaeus the creator of "binary nomenclature"?] *Botanical Journal* 41:1056–71. ISSN:0006-8136 (in Russian).
Váczy, C. 1971. Les origines et les principes du développement de la nomenclature binaire en botanique. *Taxon* 20:573–90. DOI:10.2307/1218259.
Valdecasas, A. G., M. L. Pelaez, and Q. D. Wheeler. 2014. What's in a (biological) name? The wrath of Lord Rutherford. *Cladistics* 30:215–23. DOI:10.1111/cla.12035.
Valentine, D. H., and A. Löve. 1958. Taxonomic and biosystematic categories. *Brittonia* 10:153–66. DOI:10.2307/2804945.
van der Hammen, L. 1981. Type concept, classification and evolution. *Acta Biotheoretica* 30:3–48. DOI:10.1007%2FBF00116071.
Van der Hoeven, J. 1864. *Philosophia zoologica*. Lugduni Batavorum: E. J. Brill. ISBN:9781248721575 (2012 reprint).
Van Raamsdonk, L. W. D. 1986. The grey area between black and white: The choice between botanical code and cultivated code. *Acta Horticulturae* 182:153–8. DOI:10.17660/ActaHortic.1986.182.18.
Van Regenmortel, M. H. V. 1997. Viral species. In *Species: The units of biodiversity*, eds. M. F. Claridge, H. A. Dawah, and M. R. Wilson, 17–24. London: Chapman & Hall. ISBN:9780412631207.
Van Regenmortel, M. H. V. 2001. Perspectives on binomial names of virus species. *Archives of Virology* 146:1637–40. DOI:10.1007/s007050170086.
van Steenis, C. G. G. J. 1964. A plea to let stability take precedence over priority where desirable, reasonable, and possible for generic names. *Taxon* 13:154–7. DOI:10.2307/1216132.
Van Valen, L. M. 1973. Are categories in different phyla comparable? *Taxon* 22:333–73. DOI:10.2307/1219322.
Vasilyeva, L. N., and S. L. Stephenson. 2012. The hierarchy and combinatorial space of characters in evolutionary systematics. *Botanica Pacifica* 1:21–30. DOI:10.17581/bp.2012.01103.
Vasudeva Rao, M. K. 2017. Type concept and its importance in plant nomenclature. *Journal of Economic and Taxonomic Botany* 41:91–4. ISSN:0970-3306.
Vialov, O. S. 1972. The classification of the fossil traces of life. In *Proceedings of the 24th International Geological Congress Canada*, Vol. 2, 639–44. Montréal: Geological Association of Canada. ASIN:B000O6XTNK.
Vlachos, E. 2019. Introducing a new tool to navigate, understand and use International Codes of Nomenclature. *PeerJ*:1–17. DOI:10.7717/peerj.8127.
Vlachos, E. 2020. A response to Dubois. *Zoosystema* 42:515–17. DOI:10.5252/zoosystema2020v42a25.
Vrugtman, F. 1986. The history of cultivar names registration in North America. *Acta Horticulturae* 182:225–8. DOI:10.17660/ActaHortic.1986.182.27.
Wakeham-Dawson, A., M. Solene, and P. Tubbs. 2002. Type specimens: Dead or alive? *The Bulletin of Zoological Nomenclature* 59:282–4. www.biodiversitylibrary.org/item/107013#page/308/mode/1up.
Walbaum, J. J. (ed.). 1792. *Petri Artedi sueci genera piscium* […] *Ichthyologiae, Pars 3*. Grypeswaldiae: Ferdin. Rose. https://archive.org/details/petriartedisueci03arte.
Walsingham, Thomas de Grey, and J. H. Durrant 1896. *Rules for regulating nomenclature: With a view to secure a strict application of the Law of Priority in Entomological work* […]. London: Longmans, Green. ISBN:9781275598218 (2012 reprint).
Ward, L. F. 1895. The nomenclature question. *Bulletin of the Torrey Botanical Club* 22:308–29. www.biodiversitylibrary.org/item/8000#page/391/mode/1up.

References

Waterhouse, G. R. 1862. Observations upon the nomenclature adopted in the recently published "Catalogue of British Coleoptera" […]. *Transactions of the Entomological Society of London*, 3rd series 1:328–38. www.biodiversitylibrary.org/item/100203#page/370/mode/1up.

Watson, S. 1892. On nomenclature. *Botanical Gazette* 17:169–70. www.jstor.org/stable/2994165.

Weatherby, C. A. 1949. Botanical nomenclature since 1867. *American Journal of Botany* 36:5–7. DOI:10.2307/2438113.

Weber, H. E., J. Moravec, and J.-P. Theurillat. 2000. International Code of Phytosociological Nomenclature, 3rd edition. *Journal of Vegetation Science* 1:739–68. DOI:10.2307/3236580.

Weir, B. S. 2014. How to get good fungal and bacterial identifications from GenBank sequences. *NZ Rhizobia*. www.rhizobia.co.nz/ids-using-genbank.

Welter-Schultes, F., O, Eikel, V. Feuerstein, et al. 2009. Comment on the proposed Amendment of Articles of the International Code of Zoological Nomenclature to expand and refine methods of publication. *The Bulletin of Zoological Nomenclature* 66:215–19. www.biodiversitylibrary.org/page/12757185#page/20/mode/1up.

Weresub, L. K., and K. A. Pirozynski. 1979. Pleomorphism of fungi as treated in the history of mycology and nomenclature. In *The whole fungus: The sexual–asexual synthesis*, ed. B. Kendrick, 17–30. Ottawa: National Museums of Canada. ISBN:9780660001463.

Wessel, A. 2007. D.E.Z. – A history. 150 years of scientific publishing in entomology. *Deutsche Entomologische Zeitschrift* 54:157–67. DOI:10.1002/mmnd.200700016.

Westwood, J. O. 1836. On the modern nomenclature of natural history. *The Magazine of Natural History* 9:561–6. www.biodiversitylibrary.org/item/19501#page/581/mode/1up.

Westwood, J. O. 1837. On generic nomenclature. *The Magazine of Natural History N.S.* 1:169–76. www.biodiversitylibrary.org/item/87352#page/185/mode/1up.

Wheeler, Q., T. Bourgoin, J. Coddington, et al. 2012. Nomenclatural benchmarking: The roles of digital typification and telemicroscopy. *ZooKeys* 209:193–202. DOI:10.3897/zookeys.209.3486.

Whewell, W. 1847. *The philosophy of the inductive sciences: Founded upon their history*. London: John W. Parker. ISBN:9781230389066 (2013 reprint).

Whitehead, P. J. P. 1972. The contradiction between nomenclature and taxonomy. *Systematic Zoology* 21:215–24. DOI:10.1093/sysbio/21.2.215.

Wilkins, J. 1668. *An essay towards a real character and a philosophical language*. London: Royal Society. ISBN:9781240811144 (2011 reprint).

Will, K. W., B. D. Mishler, and Q. D. Wheeler. 2005. The perils of DNA barcoding and the need for integrative taxonomy. *Systematic Biology* 54:844–51. DOI:10.1080/10635150500354878.

Willdenow, C. 1805. *The principles of botany and of vegetable physiology*. Edinburgh: Blackwood & Cadell. www.biodiversitylibrary.org/item/186882#page/9/mode/1up.

Williams, C. B. 1939. On "type" specimens. *Annals of the Entomological Society of America* 33:621–4. DOI:10.1093/aesa/33.4.621.

Williams, D. M. 2021. Names, taxa and databases: Some aspects of diatom nomenclature. *Diatom Research* 36. DOI:10.1080/0269249X.2021.1873194.

Williams, R. L. 2001. *Botanophilia in eighteenth-century France: The spirit of the Enlightenment*. Boston (MA): Kluwer Academic Publishers. ISBN:9789048156788.

Wimmera, A., and Y. Feinstein. 2010. The rise of the nation-state across the world, 1816 to 2001. *American Sociological Review* 75:764–90. DOI:10.1177/0003122410382639.

Winker, K., J. M. Reed, P. Escalante, et al. 2010. The importance, effects, and ethics of bird collecting. *The Auk* 127:690–5. DOI:1 0.1525/auk.2010.09199.

Winston, J. 1999. *Describing species: Practical taxonomic procedure for biologists.* New York: Columbia University Press. ISBN:9780231068253.

Winston, J. E. 2018. Twenty-first century biological nomenclature—The enduring power of names. *Integrative and Comparative Biology* 58:1122–31. DOI:10.1093/icb/icy060. PMID: 30113637.

Winther, R. G. 2009. Character analysis in cladistics: Abstraction, reification, and the search for objectivity. *Acta Biotheoretica* 57:129–62. DOI: 10.1007/s10441-009-9074-0.

Witteveen, J. 2014. Naming and contingency: The type method of biological taxonomy. *Biology and Philosophy* 30:569–86. DOI:10.1007/s10539-014-9459-6.

Witteveen, J. 2016. Suppressing synonymy with a homonym: The emergence of the nomenclatural type concept in nineteenth century natural history. *Journal of the History of Biology* 49:135–89. DOI:10.1007/s10739-015-9410-y.

Witteveen, J. 2020. Linnaeus, the essentialism story, and the question of types. *Taxon* 69:1141–9. DOI:10.1002/tax.12346.

Witteveen, J. 2021. Taxon names and varieties of reference. *History and Philosophy of the Life Sciences* 43:78. DOI:10.1007/s40656-021-00432-4.

Witteveen, J., and S. Müller-Wille. 2020. Of elephants and errors: Naming and identity in Linnaean taxonomy. *History and Philosophy of the Life Sciences* 42:43. DOI:10.1007/s40656-020-00340-z.

Wolf, N. M. 1776. *Genera plantarum vocabulis characteristicis definita.* Danzig. https://bit.ly/3e0dLvE

Wotton, E. 1552. *De differentiis animalium libri decem* [...]. Paris: Vascosanum. www.biodiversitylibrary.org/item/88019#page/7/mode/1up.

Wüster, W., S. A. Thomson, M. O'Shea, et al. 2021. Confronting taxonomic vandalism in biology: Conscientious community self-organization can preserve nomenclatural stability. *Biological Journal of the Linnean Society* 20:1–26. DOI:10.1093/biolinnean/blab009.

Wyrwoll, T. W. 2003. Still desiderata: Scientific names for domestic animals and their feral derivatives. In *The new panorama of animal evolution*, eds. A. Legakis, S. Sfenthourakis, R. Polymeni, et al., 683–97. Sofia: Pensoft Publishers. ISBN:9789546421647.

Yochelson, E. L. 1966. Nomenclature in the machine age. *Systematic Zoology* 15:88–91. DOI:10.2307/sysbio/15.1.88b.

Yoon, C. K. 2010. *Naming nature: The clash between instinct and science.* New York: W. W. Norton. ISBN:978-0393338713.

Young, J. M. 2009. Legitimacy is an essential concept of the International Code of Nomenclature of Prokaryotes [...]. *International Journal of Systematic and Evolutionary Microbiology* 59:1252–7. DOI:10.1099/ijs.0.011601-0.

Ytow, N., D. R. Morse, and D. M. L. Roberts. 2001. Nomencurator: A nomenclatural history model to handle multiple taxonomic views. *Biological Journal of the Linnean Society* 73:81–98. DOI:l0.1006/bjls.2001.0527.

Zadeh, L. A. 1992. Knowledge representation in fuzzy logic. In *An introduction to fuzzy logic applications in intelligent systems*, eds. R. R. Ygaer, and L. A. Zadeh, 1–25. New York: Springer Science + Business Media. ISBN:9781461536406.

Zalužiansky, A. 1592. *Methodi herbariae libri tres* [...]. Prague: Georgij Daczennni. ASIN:B01JJABJD8 (1940 reprint).

Zhang, Z.-Q. 2012. A new era in zoological nomenclature and taxonomy: ICZN accepts e-publication and launches ZooBank. *Zootaxa* 3450:8. www.mapress.com/zootaxa/2012/f/zt03450p008.pdf.

Zoller, H. 1967. Conrad Gessner als Botaniker. In *Conrad Gessner, 1516–1565. Univers algelehrter Naturforscher*, ed. H. Fischer, 57–63. Zurich: Orell Füssii. ISBN: 9783280001806.

Index

AA Code (American Association Code) 16, 164, 179–80
absolute precedence 70
accepted name 29, 169, 204–5
Aconiti 118
acting *N*-object 25, 54
Adanson, M. x, 59, 98, 134, 137–9, 143, 150, 153, 166, 169
Adanson Rules 137–9
Adansonean Reform 138–9
additional rank/category 61
adelphotype 90
Adopted Lists of Protected Names 72, 196
Aëreorum 140
Aescht, E. 37
Allen, J. A. 148, 177–8
allotype 90
Amadon, D. 201
Amanita 200
ambiregnal protists 184
American Association for the Advancement of Science 168, 179
American Code of Botanical Nomenclature 170
American Ornithologists' Union Code *see* AOU Code
Amomum 121
American Ornithologists' Union, The 173, 177
Amsterdam Code 188
Amsterdam Declaration 183–4
Amyot, C.-G.-B. 62, 161
An essay towards a real character 199
An introduction to botany 154
An outline of the first principles 154
ancestration 36
anatomical nomenclature 99, 117, 165
Anfangsgründe der Naturgeschichte 142
Angelica 120
AOU Code 167–8, 170, 172–7, 180
application of name 133, 136, 204–5
approved name 29
Aquaticorum 140
Aranei 140
Aranei Svecici 140
Arapabaca 117
arbitrariness of names 149–50
Aristotle 105, 107
Armeniaca 121

Artedi, P. 141, 159, 172
Artemisia 108
Arundo 118
Ascherson, P. F. 167
Asclepias 121
Aspergillus 200
assignment/allocation of name 35, 48, 117, 120, 124
Asterostemma/Astrostemma 55
attributive denotator 37
Aurantia 123
Aurelius Augustinus 198
authorization 73–4, 84, 94, 159
automatic typification 77, 79, 87, 139, 148, 157
autonymy 28
auxiliary rank/category 145
available name 29

BA Code 159–60, 171–3, 179, 192
Bachman, A. *see* Rivinus
Bacon, F. 44, 198–9
bacteria 193
Bacteriological Code 67, 193
bacterionymology 183
Balani 123
Barbatus 141
Barcoding 86
Barnhart, J. H. 169
basic rank/category 145
basionym 55, 77, 79, 88–9
Bauhin, C. 111–13, 117, 121–3, 128
Bentham, G. 164–5
Bentham Rule *see* Kew Rule
Bergeret, J.-P. 200–1
Berlin Rules/Code 147, 167, 187–8
Bessey, C. E. 168–9
Bicapsularium 121
Bicingulatus 91
binary nomenclature 42, 129, 176
binomial nomenclature 13, 17, 73, 118, 133, 137, 139–42, 159, 167, 175, 180–1
BioCode 2, 51, 80, 137, 161, 175, 178–80, 183–4, 192, 195–6
biogeography 2, 8
biological diversity (biodiversity, BD) 1–2, 8, 12, 101
biological nomenclature 63, 150, 183
Bionomina 91, 183
bionym 40

The Index includes only those personal names that are mentioned in the main text, and not in the references.

255

biosystematics 11, 18, 197, 202
Birmingham Code 172
Blanchard, R. 175–6
Blanchard Rules 175–6
Blockchain 202
Blumenbach, J. F. 133, 159
Bock, H. 109, 117
BOLD 49, 92
Bologna Code 181
Bologna Code *see* Douvillé Rules
Bonaparte, C. L. 160–1
Bos 28, 55
Botanical Code *see* International Code of Botanical Nomenclature
Botanonymology 183
Briquet, J. I. 187
British Association for Advancement of Science 158
Britton, N. L. 168, 170
Brittonian Code 170
Brunella 112
Brunfels, O. 108–9
Brunfelsia 117
Buccina 123
Buffon, G.-L. 142
Bulletin of Zoological Nomenclature 183, 190–1

Cambridge Code 188
Camomille 109
Camp, W. H. 202
Candolle Jr. Laws/Lois 152, 162
Candolle Jr. *see* de Candolle, A.
Candolle Sr. Règles 142, 152, 158–9, 161, 163–4
Candolle Sr. *see* de Candolle, A.-P.
Carnap, R. 37
Carus, C. G. 176
Caryophyllum 117
category (in hierarchy) 16
Centaurii 118
Cervus 141
Cesalpino, A. 118
Chaper, M. A. 175, 179–81
Chaper Rules/Code 175, 176, 181
Charonia/Charronia 55
chresonym 85
chrononym 40
Cistus 111
Citreum 123
cladistics 11, 13–14, 39, 45, 62, 92, 187, 202–3, 208
cladonomy 203
cladotype 92
class 37, 106, 124, 128
classicality of names 149–50
classification Darwinism 11, 174, 202

classification type 87
classificatory activity 26
classificatory concept 43
classificatory function 39, 57, 201, 208
classificatory task 11, 33–5, 67–8, 79
classifier 39
Clematitis 113
Clements, F. E. 58, 62, 150
Clerck, C. 140
Cobitis 141
Cochlea 123
Code of the American Ornithologists' Union *see* AOU Code
Code of the British Association *see* BA Code
Code of the Deutsche Zoologische Gesellschaft 177
Codex nomenclaturae botanica 166
Codex Westwoodianus 156
cognitive principles 50, 52–7
cognitive situation 8–10, 12, 32, 35, 38–9, 42, 48, 52, 98, 100, 208
cognitive triangle 8
collecting activity 76, 91, 186
collection type 87
collective name 61, 107
collegial authority 159
combination of names 41–2, 163
combined T-designator 41, 44
Committee on Phylogenetic Nomenclature 204
communication function 59
complete name 61, 79, 116, 121
composite T-designator 27, 41–2, 52, 54, 58–9, 61–2, 74, 92, 116, 139
compounded name 28, 139
conceptual history 13, 99, 133, 207
conditionally rigid T-designator 37–9
conservation 72, 91, 186
conserved name 29, 72
Convolvulus 121, 123
coordinate name 28–9
coordinate T-designator 26
Cope, E. 179
Copenhagen Decisions on Zoological Nomenclature 191
Cordia 117
Cordus, V. 110
Cornu cervi 122
correct name 29
Cotula 109
Critica botanica 125, 127
cross-rank homonymy 28
Cruciata 121
cryptozoology 185
cultigens 195

Index

cultivar 194–5
cultivar-groups 78, 194–5
Cultivated Plant Code 58, 194
culton 195
cultonomy 194
Cyperus 112–13

Dall Code 179
Darwin, C. 165
date precedence 70
Dayrat, B. 137
de Candolle, A. (Candolle Jr.) 161–7, 174, 207
de Candolle, A.-P. (Candolle Sr.) 87, 137, 139, 149, 152–5, 157, 159, 160, 162, 169
de Gray, T. 178
de Jussieu, A. L. 113, 137, 143, 148–9, 153, 189
De la Nomenclature des êtres Organisés 175–6
De Materia medica 105
De plantis libri 105, 118
de Queiroz, K. 203
definite genus 164
definite name 128
definiteness of taxa/names 80
denotation 15, 25–6, 29, 33–4, 77
Dentis Leonis 118
deontology 12
deposition (type material) 34
Descartes 135, 198–9
descriptive concept/nomenclature 43
descriptive function 18, 45, 102
descriptor 39
designation 52
designative concept/nomenclature 30, 43
designative function 44
designator 3, 17, 22–3, 37, 39
Deutsche Zoologische Gesellsachaft Code 176–7
diagnosis-based nomenclature 184
diagnosis 9, 26, 36–7, 39, 59, 62, 142, 171
dialect 12, 42, 51, 98–100, 207
Dicotyles 102
Didelphys 142
differential diagnosis 9
Dioscorides 70, 105, 107–8, 110
direct typification 89
disagreeable name 163
diversification trend 51, 98–9, 144, 161, 176, 183–4
DNA 86, 92
Dobzhansky–Mayr species concept 102
DOI 202
Douvillé, J. 179, 181
Douvillé Rules 146, 165, 181
Doxoscopiae physicae 118
Draft BioCode see BioCode

Dresden Code 171
dubious taxonym 72, 82
Dubois, A. 22, 48, 79, 93, 192
duplostensional nomenclature 192

Edinburgh Code 189
eidos 44, 103–4
Élémens de botanique 122
Elleborus 109
emendation of name 149
emended name 29
empire (in nomenclature) 57
empirical concept/nomenclature 42, 43, 97
empty taxonym 35, 82
Encyclopédie 142–3
Engler, A. 73, 89, 147–8, 167
Entomologia systematica 132
Entomological Code 178
epistemology 16
epistemology-based concept 42
epithet 17–18, 27–8, 42, 54–5, 59–60
epitype 90, 92, 185
Ereshefsky, M. 135
Erxleben, J. 132
essence 13, 16, 18, 23, 44, 59, 69, 82, 100, 103–6, 115–16, 126, 128, 133, 197
essentialist concept/nomenclature 14, 37, 44, 48, 57, 69, 100, 115–16, 121, 128, 133, 135, 197
established custom 64
established name 29, 158
ethnobiological nomenclature 101
ethonym 40
E-type 91
Executive Committee of ICNV 194
Executive Committee of Nomenclature 182, 188
exicate 73, 89, 148, 160, 163, 166–7, 179
expediency 32, 93
explicitness of reference 54
extensional definition 10, 37, 84, 147
external circumscription 84

Fabricius, J. C. 132, 134, 140, 149, 171
Familles des plantes 138, 166
family-group 28, 56, 86–7, 153, 175, 177, 189, 192, 196
Filipendula 121
final designation 52
final species 59
fixed precedence 70
fixed rank 40, 43, 68, 78, 80, 83, 101, 108, 200
folk classification 101–3
folk name 101–2
folk nomenclature 26, 43–4, 53, 57–9, 101–3, 104, 109, 115
folk systematics 10, 30, 52–3, 101–2, 123

folk taxon 53, 101–2
Fumariae 118
Fundamenta botanica 125, 127–8
fungi 56, 120, 127, 155, 188–9, 193, 205
fuzzy logic 39

Galilei, G. 198
GenBank 92
Genera plantarum (Bentham & Hooker) 165
Genera plantarum (Jussieu) 143, 189
Genera plantarum (Linnaeus) 126, 134, 199
Genera plantarum (Wolf) 200
General Committee on Botanical Nomenclature 182, 188
generic species 102, 123
genetype 66, 92
Genista 123
Gentiana 112
genuine name 44, 59, 69, 85, 104, 113, 115–16, 128–34, 140
genuineness of taxonyms 69, 80, 149
genus 17–18, 27–28, 40, 54–5, 59, 81, 85–6, 94, 101, 104–10, 112–13, 116, 120–1, 124, 128, 163, 199
genus-group 55–6, 79, 86–7, 170, 175, 177, 188, 191
genus–species binomen 27, 41, 44, 55, 60–3, 129, 139, 141, 142, 160, 164, 165, 167, 169, 171, 177, 191, 201, 205
genus–species scheme 16–17, 44, 59, 77, 105–7, 115–16, 122, 124, 127, 208
genus–species tautonymy 28, 55, 78, 169, 172, 191, 196, 205
Gesetze der entomologischen Nomenclatur 171
Gesner, C. 110–11, 113
Global Biodiversity Information Facility 186
Global Names Architecture 186
Global Taxonomy Initiative 1–2
Glofillaria 123
Gmelin, J. F. 141
GNIDIA 116, 129
Gratiola 118
Gray, S. F. 155
Great Schism 39, 69, 137, 149, 160–1, 173, 178, 184, 195
Greek 62, 102, 104–5, 107–9, 112–13, 121, 150
Greuter, W. 51, 56, 73, 79, 97, 184, 189, 195
group-specific marker 41, 79–80, 201
Grundriss der kräuterkunde 142
Gypsophila 117

Handbuch der Naturgeschichte 133
hapantotype 90, 185
HARTOGIA 130
Harvard Rules 169

Hawksworth, D. 195
Hedera 112
Helleborus 109
hemihomonym 28, 55, 196
herbal 104–10, 113, 115, 117, 120
Herbal epoch 10, 106–11, 113
herbalistics 62, 103, 106, 108, 110
Herbarum vivae eicones 108
Herrera, A. L. 201
Herschel, J. 2, 137, 198
heterodefinitional synonymy 29
hetetotypic synonymy 29
Histoire naturelle du Sénégal 139
Historia animalium 110
Historia Rei Herbariae 108
Historiae animalium Angliae 110
Historiae stirpium 110
historical DNA 86, 92
hologenophore 92
holotype 25, 38, 40, 83, 89–90, 92
homodefinitional synonymy 29
homonym/homonymy 27–29, 50–1, 53–6, 61, 70, 72, 74, 78, 82–3, 121, 153, 158, 160–1, 163, 167, 171–2, 177–8, 180–1, 186, 193, 196, 204–5
homotypic synonymy 28, 86
hoplonym 82
horizontal homonymy 28, 55
Hyacinta 118
hybrid formula 61
Hydrothrix/Hydrotriche 55
Hyocyanus 111
hypodigm 89

IAPT *see* International Association for Plant Taxonomy
ICB *see* International Committee on Bionomenclature
ICBN *see* International Code of Botanical Nomenclature
ichnonomenclature 23–4, 51, 185
ichnotaxon 185
ICNB *see* International Code of Nomenclature of Bacteria
ICZN *see* International Code of Zoological Nomenclature
illegitimate name 29, 68, 82, 91
IMA Fungus 183
improper name 153
inadmissible name 205
inappropriate name 104, 149, 160, 171
inclusive hierarchy 26, 124
Index Kewensis 165
indigenous names/nomenclature 70
indirect typification 88, 158

Index

infraspecies/infrasubspecies 61, 77, 196
Insecta 202
integration trend 98, 100, 136, 183–4
intensional definition 10, 37, 85, 87, 147
intensional nomenclature 10
intercalary name 42, 192
intercalary rank 77, 146
intermediate genus 60, 106
internal circumscription 84
International Association for Plant Taxonomy (IAPT) 188
International Botanical Congress 169, 187–9
International Bulletin of Bacteriological Nomenclature and Taxonomy (ICNB) 183
International Code of Botanical Nomenclature (ICBN), the same as International Code of Nomenclature for Algae, Fungi, and Plants 189, 193
International Code of Nomenclature for Algae, Fungi, and Plants 189, 193
International Code of Nomenclature for Cultivated Plants 194
International Code of Nomenclature of Bacteria (Bacteriological Code, ICNB) 193
International Code of Nomenclature of Bacteria and Viruses 193
International Code of Nomenclature of Prokaryotes 193
International Code of Phylogenetic Nomenclature 204
International Code of Virus Classification and Nomenclature 194
International Code of Zoological Nomenclature (ICZN) 191, 193
International Commission for the Nomenclature of Cultivated Plants 194
International Commission on the Taxonomy of Fungi 182
International Commission on Zoological Nomenclature 14, 178, 182, 190, 192
International Committee on Bionomenclature (ICB) 195
International Committee on Nomenclature of Viruses 182, 194
International Committee on Systematic Bacteriology (= International Committee on Systematics of Prokaryotes) 182, 193
International Committee on Taxonomy of Viruses 194
International Congress on Phylogenetic Nomenclature 204
International Cultivar Registration Authority 194
International Interim Committee 182, 188
International Journal of Systematic Bacteriology 183
International Microbiological Congress 193

International Microbiological Society 193
International Rules for the Scientific Names of Organisms 195
International Rules of Zoological Nomenclature 181, 190
International Society for Horticultural Science 194
International Species Information System 202
International Trust for Zoological Nomenclature 190
International Union of Biological Sciences (IUBS) 182
International Zoological Congress 176, 190
Internet 3, 51, 73, 92, 186, 189, 195, 202, 203–4
interpolated category/rank 77
interpolated name 61, 77
interpretation in nomenclature 1, 13
Introductio generalis in Rem herbariam 120
Iris 104–5
Irregularibus 140
isogenophores 92
isonymy 27
isotype 89–90, 92
IUBS 182–3, 191–2, 194–5
ICVCN 194

Jacea 118
Jordan, A. 165
Jordan, D. S. 177
Jung, J. 115, 117–18
junior homonym 29, 55, 61, 167
junior synonym 29, 70–2, 74, 85, 175–6
juridical principles 63–76

Kew Rule 71, 81, 147, 164–6, 188
Kiesenwetter Code 171
kingdom 57, 101–2, 136, 162, 165, 181, 189
Kinman System 206
Kluge, N. 22, 26, 41, 58, 79–80, 201
Krameria 141
Kuntze Code 166–7, 188
Kuntze, K. 84, 164, 166–8

La phytographie 165
Ladder of Nature 9, 16, 19
Lamarck, J.-B. 136, 143
Lang, K. 123–4, 174, 176
Latin 58–9, 62–3, 73, 82, 104–5, 107–13, 117, 129, 133, 136, 139–41, 149–50, 152, 163, 166, 170, 180–1, 189, 198
Latinization 62–4, 83, 108, 117, 124, 139, 171
Latreille, P. 87, 145, 148, 156
Lavoisier, A. L. 125
law of nomenclature 21, 156, 166
law of priority 31, 47, 154, 157, 159, 166, 172, 178, 191

l'Écluse, Charles de 110, 113
lectotype 53, 89–90
legitimate name 29, 34, 59, 62, 65, 67, 70, 82, 85, 116, 129, 169
Leningrad Code 189
Leptura 132
Leske, N. G. 142
Lewis, A. 147, 172
lex plurimorium 70, 72
lex prioritatis 70–1
Life Science Identifiers (LSID) 186
limited priority 70
Limonia 123
Lindley, J. 67, 94, 154–5
linear hierarchy 31
linguistic principles 30, 48, 57–63, 100, 117
linguistic system 2, 32, 48–9, 53, 98, 187, 199, 207–8
linguistic world picture 11–12, 44, 53, 102
Link, J. 136
Linnaean Canons 35, 52, 62, 68, 80–2, 85, 94, 125–6, 128–9, 133, 139, 151, 155, 164, 173, 208
Linnaean nomenclature x, 133, 164, 166, 202
Linnaean reform 3, 78, 117, 124, 125–34, 135, 150
Linnaeus, C. (also Linné) x, 3, 13, 16–17, 53, 58, 60, 62, 67, 82, 85, 94, 98, 113, 115–18, 120, 122–54, 158–9, 161–2, 164, 166–9, 174, 176, 181–2, 188
Linz ZooCode 35, 77, 192
"list" circumscription 85
Litophyta 120
Locke, J. 44, 135
Lupulus 121, 123
Lychnis 117

Magazine of Natural History 156–7
Magnol, P. 16, 40, 120–1
Malécot, V. 137
Manchester Code *see* BA Code
Melbourne Code 189
Meles 131
Melitotus 121
metagenomic 86
metaphorical *T*-designator 40
method of type 86, 148, 156
Methodi herbariae libri 118
Methodus nova et facilix Testacea 123, 181
Methodus plantarum nova 120
Methodus testacea 124
Minelli, A. 161
mixed nomenclature 58
mnemonic function 59, 202
monomymic method 161

monosemy 27, 36, 47, 54, 56, 69–70, 77, 79, 85, 121, 128, 132, 153, 162, 179, 189, 193, 196
Montreal Code 189
Morison, R. 113, 120–1, 123
morphonym 40, 57, 121
morphotaxon 70
Moscow Code 176
Murray, J. 132, 140, 154
Musca 201
Musci 120
MycoCode 183–4, 189
Mycotaxon 183
Myrtis 111

Nalimov, V. 39
name-bearer 37, 87–8, 129
Natural Arrangement of British Plants 155
natural language 27, 30, 41, 43, 58, 100, 198
natural system 9, 101, 126, 135–6
natural systematics 11, 135, 143, 145, 152, 154, 157
Naturalis Historiae 105
Nautilus 124
Neomys 56
neonym 55, 77, 89
neotype 89–90
Nerita 124
nested hierarchy 31
Neu Kreuterbuch von Underscheidt 109
New Biological Nomenclature 63
Nhandiroba 117
Nicolaus of Damascus 105
N-defined 25–6, 81–2, 85, 147
N-definition 25–7, 29, 31, 34, 36–8, 52, 73, 76–8, 80–1, 83–6, 186
N-object 25–7, 29, 32–9, 42, 52, 76, 80, 83
N-status 24–5, 27, 29, 33–4, 40, 53, 74, 77, 83, 101–2
N-type 27, 37–8, 83, 86–8, 192
nomen 26
nomen dubium 82
nomen nudum 82
nomen vanum 82
Nomenclator aquatilium 110
nomenclatorial rank group 29
nomenclatural act 25, 27, 33, 35, 57, 73–5, 94
nomenclatural activity 3, 30, 33–5
nomenclatural consistency 25, 35, 73, 85, 192
nomenclatural definition 23, 25
nomenclatural illegitimacy 25
nomenclatural inconsistency 67
nomenclatural legitimacy 25
nomenclatural novelty 74
nomenclatural object 24
nomenclatural process 33–4

Index

nomenclatural rank group 29
nomenclatural status *see N-status*
nomenclatural task 3, 11, 23, 27, 30–5, 47, 49, 50, 57, 64, 66–7, 72, 74, 88, 98, 100, 187, 192
nomenclatural taxon 26
nomenclatural type *see N*-type
nomenclatural verity 80
nomenclaturally insignificant/significant 33
nomenclature concept ix, 22–4, 42–5
nomenclature documents 25
Nomenclature in Ichthyology 177
nomenclature norm 30, 32, 148
nomenclature pluralism 32, 207
nomenclature principle ix, 3, 12–15, 30, 47–50, 65, 67, 76, 124, 136, 146, 162, 190, 192
nomenclature regulator 23, 25–7, 29–33, 43, 48–52, 57, 60, 64–6, 73, 76, 80, 93, 99, 136, 151, 208
nomenclature rule 30, 93, 98, 117, 125, 127, 129, 133, 137, 151, 167, 170, 177, 180–1, 184, 188, 190, 193
nomenclature subject 33
nomenclature system 2–3, 12–15, 18, 21–36, 39–47, 50–1, 53, 55, 58, 61, 63–4, 66–7, 71, 76–81, 84–6, 90, 93, 100, 118, 124–6, 132–6, 139, 142–7, 151–2, 157, 160, 163, 168, 171, 179, 182–7, 192–5, 197–203, 206–8
nomenclature theory 7, 21–2
Nomenclaturregeln für die Beamten 167
nomenspecies 90
nominal complex 27, 54, 58–9, 84
nominal taxon 26
nominalist concept/nomenclature 11, 23, 42, 44, 57, 63, 69–70, 98, 128, 133, 135, 137–40, 150–1, 198, 204
nominative nomenclature 2, 31–2, 36, 67, 69, 71–2, 93, 106, 108, 110, 118, 124, 149, 159, 167
nominative task 33–5
nominative verity 80, 82
nominotype 26, 37–8, 40, 79, 85, 87–92, 146–7
nonequilibrium system 98
non-Linnaean nomenclature 67, 134, 152, 174
non-rigid *T*-designator 36
Nova plantarum Americanarum 117
nucleospecies 87

objective synonymy 29
Ochrus 118
Oenanthe 120
Official Lists and Indexes of Names and Works in Zoology 190
Ogilby, W. 151, 158
Oldroyd, H. 87
omophony 28, 55

On the Natural History and Classification of Birds 156
once a synonym always a synonym (rule) 167, 169, 175, 191
one-letter rule 177
onomatophore 52, 87
onomatopoeic *T*-designator 40
ontology 15, 17, 37, 203
ontology-based concept 42
onymology 2–3, 7, 22, 209
onymophoront 87
oolithotaxon 80, 186
open nomenclature 52, 78
Opuscula botanico-physica 118
Ordo Plantarum 120
original authorship 74
original designation 53, 74
Orontium 109
orthonomenclature 23, 186
orthotaxon 56, 185
ostensional/ostensive definition 10
ostensional nomenclature 86

Padua Code 161
Paftinaca 120
Palaeontological Data-Handling Code 185
Pallas, P. S. 141
Panicum 118
Parahomonym 28, 55
Paralectotype 90
paranomenclature 23, 51, 185
parataxon 185
parataxonomy 185
paratype 40, 89–90, 92
Paris Code (botany) 162, 176
Paris Code (zoology) 162, 164, 177, 181, 190
Paris Laws (botany) 16, 162–8, 173
Partonomy 55, 117
Patellae 123
Patronym 28, 40, 94
Pedicularis 121
Permanent Nomenclature Committees 189
Petit-Thouars, M. A. 43, 199–200
Philadelphia Code 170, 187
Philosophia botanica 125, 127–8, 130, 132, 134, 140
Philosophia entomologica 132, 140
Philosophia zoologica 136
Philosophiae botanicae 136
philosophical language 14, 43, 58, 199–201
Philosophie zoologique 136
phototype 66, 91, 185
phraseme 41–3, 58–60, 108, 118, 120
PhyloCode ix, 16, 51, 68, 71, 77, 84–5, 92, 137, 184, 196, 204–6, 208

phylogenetic concept/nomenclature 11, 14, 41, 45, 72–3, 80, 135, 184, 202–6
phylogenetic numericlature 201–2
phylogenetic pattern 9, 11, 19, 45, 203
phylogenetics 3, 12–13
phylonomenclature *see* phylogenetic nomenclature
phylonym 27, 34, 41, 71, 74, 203–4
phyloreference 27, 203
phytocenology 2
Phytonomatotechnie universelle 200
Phytopinax see Φυτόπιναξ
Pica/Picus 28, 55
Pinax see Πιναξ
Pinnipedia 205
Pittonia 28, 117
Plantago 122, 131–2, 200
Plantarum umbelliferarum 120–1
Plasmodiidae 201
Plasmodium 201
Platonic 13–14, 16, 44, 103–5
Pleonasm 28, 55
Plinia 117
Pliny the Elder 105
Plumier, C. 117
Polhem, C. 199
Polygonum 117
population thinking 89
Porus 120
position precedence 71
possible worlds 36–7
post-Linnaean nomenclature 18, 135–7, 139, 142–4, 144, 148, 151–2, 161, 203
pragmatic concept/nomenclature/principles 42, 44, 48, 93
precedence ix, 64, 70–1, 169
preexisting name 204–5
preliminary designation 52
Preliminary Discourse 137
pre-systematics 10, 100
primary folk name 101
primary homonym 28
primary regulator 30
primary tasks 34
primary type 30, 34, 84, 88–90, 92, 98, 101, 103, 108, 120, 124, 145, 178
primary typification 89
Primula 116, 123
principle of accentuation (emphasis) 53, 59
principle of acting certainty 54
principle of adequacy 49, 52
principle of admissible polysemy 54, 56
principle of ancestration 92
principle of arbitrariness 150
principle of authorization 72–6, 84, 94, 159
principle of automaticity 66
principle of binarity 17, 44, 60, 79, 106

principle of binomiality 17, 44, 61, 71, 140, 161, 168, 179, 205
Principles of Botany, the 142
principle of brevity 60
principle of circumscription 85, 147
principle of classicality 71, 149
principle of codification 150
principle of constancy of *T*-designators 54
principle of convenience 44, 93
principle of conventionality 64
principle of depositing 76, 91
principle of designation 52
principle of designative certainty 50, 53–4, 74, 83–93, 103
principle of designative uncertainty 53
principle of diagnosing 15, 84, 147, 154
principle of direct action 66
principle of exhaustiveness 51
principle of first reviser 74, 181
principle of genuineness 149
principle of group non-specificity 80
principle of group specificity 80
principle of homogeneity 78
principle of homonymy suppression 50
principle of independence of Codes 55
principle of indirect action 66
principle of Latinization 108
principle of legitimacy 65
principle of linguistic parsimony 60
principle of literary justice 94
principle of locality 51, 65
principle of mandatory status 65–6
principle of monosemy 27, 36, 54, 56, 77, 85, 128, 179
principle of naturalness 64
principle of neonymy 55
principle of nomenclatural foundation 72
principle of nomenclatural verity 80
principle of nominative verity 82
principle of non-ranking taxonomic hierarchy 77
principle of onomatophores 52
principle of onto-epistemic correspondence 8, 35, 105
principle of overall instability 52
principle of overall stability 51
principle of overall universality 50–1
principle of partial lability 69
principle of partial stability 69
principle of polysemy suppression 54
principle of pragmatism 50, 93
principle of precedence 64
principle of preservation 189, 191
principle of priority 14–15, 32, 50, 57, 65, 70, 72, 75, 78, 121, 129, 133, 138, 146, 158–9, 161, 164–70, 175–7, 181, 186–8, 191, 193, 196
principle of proactivity 67

Index

principle of publication 83, 171
principle of quadrinomiality 61
principle of rank coordination 70–1, 77
principle of rank equivalence 79
principle of rank–category correspondence 9
principle of rank-dependence 15, 78, 80
principle of rank-independence 80
principle of rank specificity 78
principle of ranking taxonomic hierarchy 76–7, 78
principle of rationality 50, 53
principle of recommendatory status 65–6
principle of rejection 72
principle of retroactivity 67
principle of semantic motivation 56
principle of semantic neutrality 57
principle of simplicity 50, 93
principle of subject equality 75
principle of subject inequality 75–6
principle of supremacy of Code 65
principle of symbolness 58, 63
principle of systemity 50
principle of taxonomic certainty 15, 50, 77, 82–4
principle of taxonomic discreteness 77
principle of taxonomic freedom 11, 67
principle of taxonomic non-discreteness 77
principle of taxonomic non-freedom 68
principle of taxonomic verity 81
principle of the independence 51
principle of theoretism 50
principle of trinomiality 174, 197
principle of typification 15, 78, 86, 89, 91–3, 147–8, 175, 185–6, 188, 191, 196
principle of unambiguity 66
principle of unequivocality of names 54
principle of uninomiality 61
principle of universality 188
principle of univocality 54
principle of usage 15, 64–5, 70–2, 146–7, 167, 169, 191, 194
principle of verbalness 58
principle of verity 54, 126
principle of wordness 58–9
principle of zygoidy 54
Priodon 149
priorability of name 54
priority 14–15, 21, 27, 31–2, 47, 50–1, 54, 57, 62, 65, 67, 70–2, 75, 78–9, 93–4, 112, 121, 127, 129, 133–4, 138, 140, 144, 146, 148–50, 153–5, 157–9, 161–78, 180–1, 186, 187–93, 195–6
proactive action 66–7
procedural task 34
Prodromus historiae generalis 121
prokaryotes 28, 182–3, 189, 193, 205
proper name 37–8, 103, 120, 157
protected name 72, 196
protologue 72–3, 205

proto-systematics 10, 97, 100, 103–13
Protozoa 184, 201
Provisional Committee on Nomenclature of Viruses 194
proximal genus 59, 113, 122, 124
pseudoranked nomenclature 79
Psittacus 141
publication 27, 35, 63, 70–6, 82–3, 88, 130, 146, 163, 166–8, 170–1, 179–80, 183 186, 188–9, 192, 196
published work 72–3

qaudrinomial nomenclature 13, 18, 61, 174, 178, 202
qualifying clause 75, 204–5

rank (in hierarchy), 9–10, 14, 16, 19, 26, 40–1, 43, 57, 76–9, 101, 121–2, 144–6, 161, 165, 195–6, 200
rank circumscription 74–5, 85
rank-dependent concept/nomenclature 18–19, 57, 78–9, 81, 83–4, 145
rank-group 26, 28–9, 57, 70, 77–80, 86, 152, 162, 171, 175, 180, 189, 191, 196
rank-independent concept/nomenclature 78
rank-specific marker 41, 79
ranked hierarchy 10, 14, 19, 41, 43, 57, 76–7, 79, 86–7, 101, 121, 144–6, 161, 165, 195–6
ranking 9, 11, 17, 26, 34, 38, 50, 67–8, 76–8, 83, 120
rankless hierarchy 14, 16, 19
ranknym 26–7, 38, 40–1, 43, 57, 76, 83
Ranunculus 110
Rariorum Plantarum Historia 110
rational concept/nomenclature 13, 200
rationalization of nomenclature 98
rational-logical concept/nomenclature 14, 40–1, 43, 45, 80, 100, 137, 198–202
recommendation 30–2, 66, 71, 89, 94, 127, 159–60, 164, 167–70, 172, 176, 180, 193
reference number 201–2
reference of name 204
reference phylogeny 84–5, 205
reference point 26, 92
reference specimen 25, 76, 92
referential denotator 37
Reformed Code of Botanical Nomenclature 184, 189
Regeln für die zoologische Nomenclatur 176–7
registered name 29
registration 29, 35, 48, 73, 194, 196, 202
Registration Center 196
Règles de la nomenclature 137, 152, 162, 175–6, 190
Règles international de la nomenclaure zoologique 190
Règles internationales de la nomenclature botanique 187

Regnum Vegetabile (journal) 183
regulative nomenclature 2, 22, 29–33, 43, 47–8, 69, 106–7, 115, 120, 125, 129, 133, 136, 150, 155, 182, 208
regulatory principles 49–52
rejected name 72, 74
rejection 71–4, 94, 136, 139, 149, 155, 157, 166, 173, 177, 180, 192
replacement 29, 48, 74, 88, 94, 149, 155, 160, 166, 177, 205
replacement name 29, 88
research program 8, 10, 115, 117, 133, 135
Retiariorum 140
retroactive action 60, 67
Revisio generum plantarum 166
rigid *T*-designator 35, 43, 54
Rivinus 115, 120–2, 127–8, 132
RNA 92
Robinson, B. L. 169
Robur 118
Rochester Rules/Code 168–70, 175
rule of availability 82
rule of interpretation 66
rule of legitimacy 82
rule of rank specificity 78
rule of rank–category homonymy 27, 83
Rules for American Pomology 170
Rules of Entomological Nomenclature 178
Rules of Nomenclature of Cultivated Plants, *see* Cultivated Plant Code
Rules of Nomenclature of Viruses 194
Rutilus 102

Sachs, J. 97, 99, 108–9, 122
Sagitta 121
Saint-Lager, J. B. 58, 62, 71, 82, 97, 149–50
Saint-Lager Reform 71, 82, 150
Salix 105
Saltatorum 140
scholastic systematics 10, 13–14, 44, 59, 62, 77, 100, 106, 115–18, 135–6, 145, 149
scientific name 94, 139, 143, 156, 192
scientific systematics 10, 24, 44, 52–3, 59, 64, 94, 98, 100, 106–7, 115
Seattle Code 189
secondary folk name 102
secondary homonym 28, 55
secondary regulator 30–1, 66, 88
secondary tasks 34
secondary type 90
secondary typification 89
semantic triangle 24, 35, 37–8, 87
senior homonym 27–9
senior synonym 29
Shenzhen Code 189
Simia 133

Smith, J. E. 141
Solanum 112, 121
special Code *see* subject-area Code
species 17, 37–8, 59
Species plantarum 84, 116, 126, 129–30, 141, 147, 152, 166–9
species-group 25–8, 38, 54–6, 61, 66, 72, 76, 79, 83, 86–92, 148–9, 154, 170, 175, 179, 183–8, 191, 205
specific epithet 17–18, 28, 42, 55, 59–61, 71, 79, 81–2, 104–5, 110, 116–18, 120–1, 123–4, 136, 139–40, 147, 153, 160, 177, 179–80, 191
specifier 38, 84
spelling of name 28–9, 129
Sporozoa 201
Sprengel, K. 108, 152
St. Louis Code 189
Stearn, W. T. 194
Sterns, E. E. 168
Stirpium Adversaria Nova 110
Stockholm Code 188
strain 86, 90
Strickland Code *see* BA Code
Strickland, H. 38, 156–9
Strickland (Stricklandian) Rules/Code 158–9, 170, 172–3
Strombi 123
Strutio 132–3
subfamily 41, 81, 201
subject-area Code 39, 56, 71, 183–4, 195–6, 204–5
subjective synonymy 8, 23, 29, 150
subsequent authorship 74
subsequent designation 53
subsidiary category 41
subtribe 162, 179
suneg 201
suppressed name 29
suprageneric taxon 18
Svenska spindlar 140
Swainson Rules 156–7
Swainson, W. J. 94, 156, 160
symbolic concept 43
symbolic *T*-designator 35, 43, 54
synonym/synonymy 27–9, 32, 34, 37, 45, 51, 53–4, 56, 69–72, 74, 82, 85–6, 102, 105, 107–9, 112–13, 118, 121, 123–4, 130, 134, 144, 147, 152, 154, 158–60, 165, 167–9, 172, 175–6, 185–6, 191, 202, 204
Synopsis methodica animalium 120
syntype 25, 38, 53, 90
System of Nature 9–11, 18, 44, 104, 107, 115, 122, 125, 127, 130, 136, 164, 199–200
Systema Eleutheratorum 140
Systema Naturae 127, 129–32, 140–1, 159, 166, 172, 174, 176, 181

Systema regni animalis 133
systematic nomenclature 2, 159, 198
systematic taxon 25

taxognosis 25
taxon (in classification) 9, 25
Taxon (journal) 188
taxonomic character 9, 24–6
taxonomic designator 3, 22–3, 24, 48, 201–2, 209
taxonomic diversity (TD) 1–2, 8, 11, 14, 91, 93, 186
taxonomic hierarchy 9, 14–19, 64–6, 76–8, 81, 122–4, 127, 145, 162, 192, 202
taxonomic *N*-objects 80
taxonomic principles 50, 76–93
taxonomic reality (TR) 8, 12–13, 15–17, 22, 24, 37–8, 42, 45, 48–9, 51, 57, 68, 81, 97–100, 187, 207–9
taxonomic synonymy 27–8
taxonomic system (TS) 1, 9, 11, 24, 43, 79–80, 83, 103, 193, 198, 201
taxonomic theory (TT) 7–19, 26, 29, 39, 48–50, 57, 81, 203
taxonomic vandalism 66, 75
taxonomic verity 80–2
taxonominal hierarchy 79
taxonomy 1–2, 7, 13, 55, 86–7, 182–3, 195, 199
taxonymy 2–3, 7, 22, 209
T-designator 24–7, 30–9, 40–5, 52–6, 58–9, 61–3, 69, 73–5, 78, 82, 103, 186
Temminck, C. J. 146, 151
Tentamen methodi ostracologicae 181
Testacea 123–4, 181
Theophrastus 70, 104–5, 107–8, 110
Théorie élémentaire de la botanique 152
theory-dependent nomenclature 42, 208
theory-neutral nomenclature 11, 42, 57, 83, 126, 197
Tokyo Code 189
toponym 40, 57, 172
topotype 89–90
Tournefort, J. P. 13, 53, 64, 104, 112–3, 115–7, 120, 122–4, 126–8, 134–5, 142, 144–5, 164, 174–6, 181
traditional nomenclature 13–14, 17–18, 27, 31, 33–4, 47, 57–60, 78, 135, 155, 182, 184, 195, 198, 200–1, 206, 208
Tragus 109, 117
Tree of Porphyry 105, 110–11, 113, 121, 199
trinomial nomenclature 15, 174, 191
trivial name 82, 117–18, 128–33, 138–40, 142–43, 152, 154
true genus 17, 81, 164–5
Turbines 123
Turbo 124
turbo-taxonomy 86
type culture 76, 86, 90, 94

type fixation/designation 1–2, 18, 24, 27, 30, 41, 43, 47–8, 51–4, 58–9, 61, 68, 74, 78–80, 89, 101–3, 118, 121, 129–30, 138, 141, 144–6, 162, 179, 183, 186, 198–9, 201–2
type locality 90
type series 25, 38, 89–90, 184–5
type-based nomenclature 86, 100, 147, 184
typeless species 91
typification 87–9, 91–3, 148–9
typonym 26–7, 38, 40, 53

Umbelliferarum 120–1
unavailable name 29
Unifolium 121
Ursus 131–2
usage 15, 32, 54, 62, 64–5, 70–2, 93, 103, 111, 119, 121, 138, 144, 146–7, 158, 161, 167, 169, 171, 173, 186, 191, 194–5

valid name 55
valid publication 179, 183, 192
Van der Hoeven, J. 136
verbal concept 43
verbal *T*-designator 35, 43, 54
vernacular name 26, 113, 130
vertical homonymy 55
Verticalibus 140
Vienna Code 187, 189
Vienna Rules 164–5, 170, 187–8
Vindiciae nominum 140
Viruses 57–8, 182–3, 193–4
Vocabulary of bionomenclature 2
von Kiesenwetter, E. H. 171
voucher specimens 76, 86, 90, 92

Walbaum, J. J. 141
Westwood, J. O. 156–7
Wilkins, J. 43, 135, 199–201
Willdenow, C. L. 142
Willdenow Rules 142, 151
Wolf, N. M. 200
wordness 58–60
Wotton, E. 111

Yetus 139

Zalužiansky, A. 118–19, 139
ZooBank 202
Zoological Code *see* International Code of Zoological Nomenclature
zoonymology 183
zygography 27

Πιναξ 112
Φυτόπιναξ 112

Lightning Source UK Ltd.
Milton Keynes UK
UKHW022150190522
403164UK00004B/41